Urban Design Guidelines for Climate Change

Urban Design Guidelines for Climate Change

Editors

Hideki Takebayashi
Jihui Yuan

MDPI • Basel • Beijing • Wuhan • Barcelona • Belgrade • Manchester • Tokyo • Cluj • Tianjin

Editors

Hideki Takebayashi
Department of Architecture
Kobe University
Kobe
Japan

Jihui Yuan
Department of Architecture
and Civil Engineering
Toyohashi University
of Technology
Toyohashi
Japan

Editorial Office
MDPI
St. Alban-Anlage 66
4052 Basel, Switzerland

This is a reprint of articles from the Special Issue published online in the open access journal *Atmosphere* (ISSN 2073-4433) (available at: www.mdpi.com/journal/atmosphere/special_issues/Urban_Design_Climate).

For citation purposes, cite each article independently as indicated on the article page online and as indicated below:

LastName, A.A.; LastName, B.B.; LastName, C.C. Article Title. *Journal Name* **Year**, *Volume Number*, Page Range.

ISBN 978-3-0365-6341-1 (Hbk)
ISBN 978-3-0365-6340-4 (PDF)

Contents

atmosphere

Article

Analysis of a New Index for the Thermal Performance of Horizontal Opaque Building Components in Summer

Hashem Akbari [1], Chiara Lodi [2], Alberto Muscio [2,*] and Paolo Tartarini [2]

1 Gina Cody School of Engineering and Computer Science, Concordia University, Montreal, QC H3G 1M8, Canada; hakbari@encs.concordia.ca

2 Department of Engineering "Enzo Ferrari", University of Modena & Reggio Emilia, 41124 Modena, Italy; chiaralodi4@gmail.com (C.L.); paolo.tartarini@unimore.it (P.T.)

* Correspondence: alberto.muscio@unimore.it

Abstract: The summer behavior of an opaque building component subjected to the solar cycle depends on the combination of its thermal insulation, inertia, and solar reflectance. To rate the component dynamic behavior while an air conditioning system ensures a steady indoor temperature, a 'solar transmittance index' (STI) has been proposed. This is a component-based index calculated from a 'solar transmittance factor' (STF). STI takes into account the radiative properties at the outer surface and the thermophysical properties and layer structure of the materials beneath. It correlates the peak heat flux and temperature at the inner surface, relevant to cooling energy and thermal comfort, to the peak solar irradiance. Similar to the well-known 'solar reflectance index', STI is determined comparing the STF with two reference values, corresponding to a performance relatively low and very high, respectively. Thanks to its simplicity, the approach may allow defining easy to apply requirements to prevent building overheating, improve indoor comfort, reduce cooling energy demand, and mitigate some fallouts of the urban heat island effect. In this work, focused on roofs above occupied attics, peak heat flux and ceiling temperature are calculated by numerical simulation and compared with STF values for a wide range of roof types.

Keywords: building cooling; periodic thermal transmittance; roof; solar reflectance; SRI; thermal inertia; thermal insulation; thermal transmittance

Citation: Akbari, H.; Lodi, C.; Muscio, A.; Tartarini, P. Analysis of a New Index for the Thermal Performance of Horizontal Opaque Building Components in Summer. *Atmosphere* **2021**, *12*, 862. https://doi.org/10.3390/atmos12070862

Academic Editors: Hideki Takebayashi and Jihui Yuan

Received: 18 June 2021
Accepted: 29 June 2021
Published: 2 July 2021

1. Introduction

An approach to calculate the performance of a building relies upon dynamic simulation models, applied on an hourly basis or shorter and taking into account a wide set of parameters. Main relevant parameters are the climate at the installation site, with the relevant meteorological variables and their seasonal and daily cycles, and the dynamic performance of the building envelope. The dynamic performance is influenced by the insulation and the inertia of the building components and their surface properties relevant to solar gains, as well as the characteristics of the transparent components and their shading or sunscreen. Other aspects to be considered are the response of the different components of the heating, cooling, and air conditioning (HVAC) system as controlled by building automation devices and the complex interaction between climate, envelope, HVAC systems, and usage profile. This can provide an objective prediction of the energy needs for HVAC, both in reference conditions and in actual or extrapolated scenarios. Such prediction can be used to assign a class of energy performance to the building, or to evaluate the cost-effectiveness of energy retrofit actions.

An opposite approach is that of calculating the building energy needs with a simplified procedure, e.g., based on an almost stationary heat balance evaluated on a daily or a monthly basis, neglecting the fluctuation of the meteorological parameters and the dynamic interaction of the envelope with the HVAC system. This may, however, lead to a significant discrepancy between the calculated and actual building performance. Therefore, to limit the risk that a required performance target is not really achieved, additional and highly demanding requirements are often imposed for single components of the building envelope and the HVAC system. This is the line of the regulatory system in force in Italy since 2007, where the class of performance is generally based on the results of the simplified calculation method described in the UNI/TS 11300 set of technical rules, i.e., the national implementation of ISO 13790 [1], for building heating and cooling [2]. For new or renovated buildings, stringent minimum requirements have been imposed for the U-value and the periodic thermal transmittance of opaque components, the U-value and the solar factor of transparent components, the solar reflectance of roof surfaces, etc. [3]; even more stringent requirements must be fulfilled to obtain economic incentives for building retrofitting [4]. According to current rules, the calculation of the thermal behavior and the energy performance of the whole building may even be unnecessary in case a partial retrofit is implemented, provided that the minimum requirements are satisfied for the retrofitted components of the envelope and/or the HVAC system.

Of course, intermediate lines are possible between the above two outlined approaches.

The detailed and accurate dynamic calculation of the building thermal behavior, in realistic climatic and usage conditions, requires a substantial effort from the designers, often not supported by adequate preparation and design tools, or not economically justified by the extent of the energy retrofit action. If it is, however, made compulsory by regulation, the probability of a superficial and unreliable implementation is high. On the other hand, a large part of the construction professionals possess preparations and tools to calculate and verify the performance requirements on single building components, which can also be verified with relative ease by supervisory bodies. Moreover, focusing on the requirements at the component level can be effective for the winter heating of buildings with ordinary characteristics, while indoor and outdoor temperatures do not vary significantly along the day, and a proper closed-loop control of the heating system can balance the variability of the relatively weak solar gains. Instead, some concern is justified for summer heating, when fluctuating heat loads induced by the solar cycle become dominant, and therefore, one cannot ignore their dynamic interaction with the building envelope. Setting separate and independent requirements for thermal insulation and thermal inertia, however, may lead to an over-insulated envelope that prevents dissipation of solar gains, to uselessly massive walls and roof that increase cost and seismic risk, or to the use of less durable materials. Furthermore, in the presence of a building component such as a cool roof, characterized by a high solar reflectance at the outer surface, fulfilling additional and separate requirements for both thermal insulation and inertia may not provide additional benefits in terms of the overall thermal response of the component in summer.

Component-based requirements can help obtain a decent behavior of the building in the cold season, when heating is dominant. In fact, in this period, most of the requirements can be inherent to the thermal insulation of opaque and transparent building components, as well as to the prevention of moisture condensation by means of indoor ventilation and the correction of thermal bridges. In this regard, requirements on opaque components can be simply set in terms of minimum limits to the R-value ($m^2 \cdot K \cdot^{-1}$), or maximum limits to its inverse, the U-value or thermal transmittance ($W \cdot m^{-2} \cdot K^{-1}$), calculated according to standard methods such as ISO 6946 [5]. In the hot season, however, the thermal behavior of an opaque building component depends also on its mass and heat capacity. More specifically, the dynamic response of the component is affected by parameters such as modulus of the periodic thermal transmittance, thermal admittance, decrement factor, or time shift, all of them defined in standard methods such as ISO 13786 [6]. Moreover, the reflectance of solar radiation directly at the external surface can be more effective than both thermal insulation and inertia in many cases. Hence, the requirements and/or incentives have been set for solar reflectance, a surface property usually measured according to ASTM C1549 [7] or ASTM E903 [8].

Another surface property, thermal emittance, usually measured according to EN 15976 [9] or ASTM C1371 [10], is seldom considered even if it affects the calculation of solar gains. Indeed, in ordinary buildings with a minimum level of thermal insulation, imposed to limit heat loss in winter, a proper dynamic performance of the opaque components, combined with indoor ventilation and an adequate sizing and shading of transparent components, can reasonably provide a decent summer performance in mild climates. This would limit the energy needed by the air conditioning system to achieve a comfortable indoor condition. More generally, the synergic fulfilling of a few properly identified performance requirements at the component level may lead to a satisfactory overall performance in both winter and summer. This would allow for promoting a faster and generalized enhancement of building lots. For the summer performance of opaque building components, however, requirements at the component level involve parameters that are weakly related or unrelated. It has already been underlined that they may result as largely superfluous—if not penalizing—in terms of cost, mass load, seismic risk, durability, fire risk, etc. On the other hand, the combined effect of both mass and surface properties on the indoor temperature at the building level can be accurately estimated only by numerical [11–14] or experimental techniques [15]. However, these relatively complex assessment approaches can seldom be used for an expeditious selection of the optimal building solution as they are out of reach for many designers and construction firms. In the end, a single performance parameter that reflects the dynamic response of an opaque building component, for which simple requirements can be set, would be highly desirable.

An interesting approach to classify roof and wall components that takes into account both surface and mass properties is that of the 'thermal performance index' (TPI). In this index, a rating value of 100 (the lower the better) is assigned for a given excess of peak temperature at the ceiling of an unconditioned space or for a given peak of heat flow rate per unit of ceiling surface of a conditioned space [16]. A modification of the approach was recently proposed [17] with the 'new' thermal performance index (*TPI), in which the rating value of 100 (the higher the better) is assigned for the decrease of ceiling temperature of an optimal roof solution with respect to the ceiling temperature achieved in the worst case of a galvanized iron roof. Another component-based indicator is proposed in [18], where the R-factor and outer solar reflectance of a roof are combined in an equivalent thermal resistance, increased with respect to the actual one thanks to an enhanced solar reflectance. The need of taking into account not only solar reflectance but also heat capacity is shown in [19]. A comprehensive model of the roof thermal behavior, focused on the double skin roof type, is reported in [20].

With increasing complexity, a set of Environmental Resource Indicators (ERI) is proposed in [21] as building performance indicators, representing the exploitable environmental resources (external air through external convection, natural ventilation, and sky radiation cooling), but their calculation is not immediate. In ref. [22], a methodology and some indexes are proposed to evaluate the building indoor environment and its anthropogenic heat transferred to the environment, yet again based on a simulation approach. Generally speaking, the requirements to prevent building overheating are difficult to identify and verify, be they are conceived as a mere set of limits at the component level or in terms of synthetic comfort indicators such as the 'hot thermal performance index' (TPIh) and the 'hot discomfort degree hour' index (DDHh) analyzed in [13], or the 'thermal deviation index' (TDI) proposed in [23].

In a previous work [24], a 'solar transmittance index' (STI) was proposed to rate the overall dynamic thermal behavior of an opaque building component when its external surface is subjected to the cycle of solar radiation and the indoor temperature is kept constant. STI includes, in a single performance parameter, both the radiative properties at the external surface and the thermophysical properties of the materials under the surface. It is developed by a procedure similar to that of *TPI [17] but with an approach independent of the installation site and aimed to obtain the maximum ease of calculation. More specifically, the solar transmittance index (STI) is based on a multiplicative factor, the solar transmittance factor (STF). This correlates the peak heat flux at the inner surface, relevant to the energy need for air conditioning, to the peak solar irradiance. The radiative properties at the outer surface and the thermophysical properties and layer structure of the materials beneath are taken into account. The peak heat flux is in turn correlated to the inner surface temperature, relevant to thermal comfort. Similar to the well-known 'solar reflectance index' (SRI), as defined in ASTM E1980 [25,26], it has been proposed that STI is determined comparing the abovementioned correlation factor STF with two reference values, corresponding to a performance relatively low and very high, respectively, in order to clearly differentiate the performance of different solutions. With their simplicity, STF and STI may allow defining easy to apply requirements to prevent building overheating, improve indoor thermal comfort, reduce cooling energy demand, and mitigate some fallouts of the urban heat island effect.

In this work, focused on roofs above inhabited attics for which thermal insulation is prescribed, the heat flux and peak temperature are calculated by numerical simulation at the ceiling surface and are compared with the STF values. The comparison is made for a relatively wide range of roof types and layer structures in different environmental conditions in order to verify the existence of a significant correlation. The focus is on the roof as it is the main source of heat gains in the interior heat balance, often higher than the direct solar gains through windows [27].

2. Materials and Methods

2.1. Mass and Surface Performance Parameters

The development of the solar transmittance index is detailed in [24]. The main concepts are summarized below for sake of completeness.

The thermal insulation provided by a building component can be evaluated in terms of U-value ($W \cdot m^{-2} \cdot K^{-1}$), defined for steady-state conditions as the ratio of the heat flux density q ($W \cdot m^{-2}$) to the temperature difference across the component, i.e., between the external temperature T_e and the internal temperature T_i with a positive entering heat flux, as follows:

$$U \equiv \frac{1}{R} = \frac{q}{T_e - T_i} \quad (1)$$

The inverse of the U-value, the R-value, or the thermal resistance ($m^2 \cdot K \cdot W^{-1}$) of the whole component is also defined. Both U and R are calculated according to ISO [5], ASHRAE [28], or other equivalent standards from the sum of the conductive resistances of the component layers with their thickness L (m) and thermal conductivity k ($W \cdot m^{-1} \cdot K^{-1}$), and the heat transfer coefficients, h_e and h_i ($W \cdot m^{-2} \cdot K^{-1}$), at the external and internal surfaces:

$$U \equiv \frac{1}{R} = \frac{1}{\frac{1}{h_e} + \sum \frac{L}{k} + \frac{1}{h_i}} \quad (2)$$

Either h_e or h_i are given by the combination of a convection coefficient, h_{ce} or h_{ci}, and a radiation coefficient, h_{re} or h_{ri}:

$$\begin{aligned} h_e &= h_{ce} + h_{re} = h_{ce} + \varepsilon_e \cdot \left(\frac{T_{sky}^4 - T_{se}^4}{T_{sky} - T_{se}} \right) \\ h_i &= h_{ci} + h_{ri} = h_{ci} + \varepsilon_i \cdot \sigma_0 \cdot \left(\frac{T_{si}^4 - T_i^4}{T_{si} - T_i} \right) \end{aligned} \quad (3)$$

where T_{sky}, T_{se}, and T_{si} (K) are the sky temperature and the external and internal surface temperatures, respectively. The internal convection coefficient h_{ci} is generally induced by the free convection phenomena and it can be an almost-constant value, changing only with the direction of the heat flux. The external convection coefficient h_{ce} also depends on wind velocity. Empirical values or formulas are available for their estimate, e.g., those in [5,28].

The radiation heat transfer coefficients are correlated to a maximum value that depends on T_{se} and T_{si} (K) through the Stefan–Boltzmann constant $\sigma_0 = 5.67 \times 10^{-8}$ W·m^{-2}·K^{-4} and the thermal emittance of the external and internal surfaces, ε_e and ε_i.

The external temperature T_e in Equation (1) is the weighted average of the air temperature T_{air} and the 'effective' sky temperature T_{sky} (which is in turn evaluated from the near-ground air temperature, the partial pressure of water vapor, and the view-factor between the considered surface and the sky by empirical formulas):

$$T_e = \frac{h_{ce} \cdot T_{air} + h_{re} \cdot T_{sky}}{h_{ce} + h_{re}} \tag{4}$$

When the outer surface of the considered building component is subjected to a solar irradiance I_{sol} (W·m^{-2}), T_e must be increased by the non-reflected fraction of the irradiance, divided by h_e. The so-called sol-air temperature is obtained, whose definition includes the solar reflectance ρ_{sol} ($0 < \rho_{sol} < 1$), i.e., the ratio of reflected and incident solar radiation:

$$T_{sol-air} = T_e + \frac{(1 - \rho_{sol})}{h_e} \cdot I_{sol} \tag{5}$$

The combined effects of solar reflectance and external thermal emittance can also be expressed through the 'solar reflectance index' (SRI), a parameter calculated as specified in [25]:

$$SRI = 100 \cdot \frac{T_{sb} - T_{se}}{T_{sb} - T_{sw}} \tag{6}$$

In this case, T_{se} (K) is the temperature that the analyzed surface would steadily reach when irradiated by a reference solar flux $I_{sol,max}$ = 1000 W·m^{-2} at atmospheric air temperature T_{air} = 310 K, sky temperature T_{sky} = 300 K, and with convection heat transfer coefficient h_{ce} for which the values 5, 12, and 30 W·m^{-2}·K^{-1} are specified for low (v_{wind} < 2 m·s^{-1}), intermediate (2 m·s^{-1} < v_{wind} < 6 m·s^{-1}), and high (6 m·s^{-1} < v_{wind} < 10 m·s^{-1}) wind speeds, respectively. T_{sb} (K) and T_{sw} (K) are the temperatures that would be reached by two reference surfaces, a black one ($\rho_{sol,b}$ = 0.05) and a white one ($\rho_{sol,w}$ = 0.80), respectively, both having high thermal emittance (ε_e = 0.90). SRI represents the decrement of surface temperature that, in the reference conditions, the analyzed surface would allow with respect to the reference black one, divided by the decrement allowed by the reference white surface and given in percentage terms. The surface temperature T_{se} (as well as T_{sb} and T_{sw}) is determined by iteratively solving the following surface energy balance:

$$(1 - \rho_{sol}) \cdot I_{sol} = h_{ce} \cdot (T_{se} - T_{air}) + \varepsilon_e \cdot \sigma_0 \cdot \left(T_{se}^4 - T_{sky}^4\right) \tag{7}$$

SRI is considered by voluntary rating systems such as LEED [29] when dealing with the summer performance of opaque building components, so it matches the need of a single performance parameter for different radiative surface properties and allows comparing the performance of different solar reflective or 'cool' solutions for roofs, walls, and pavements. Its main limitation is that it is based on the hypothesis of adiabatic external surface, and it is not affected by either the insulation or the inertia of the materials beneath. On the other end, the SRI works well even with non-adiabatic surfaces since the heat flux conducted inside can be lower by one or two orders of magnitude than the incident solar irradiance.

$T_{sol-air}$ has a periodic time-evolution pattern that follows that of I_{sol}. Their cycles peak at solar noon on a horizontal or almost horizontal surface and they are nil in the night. The external temperatures of the air and the sky may have a periodic time evolution pattern as well, following the cycle of solar irradiance with a short delay. When a strong cycle of the sol-air temperature occurs, the thermal problem becomes unsteady, and the thermal inertia of the envelope components gains significance. In this case, direct solutions to the heat transfer equation are generally unavailable and numerical methods would be needed, but the relatively simple approach to dynamic analysis provided by ISO 13786 [6] is also available and is widely used. It is based on that any periodic function such as the time evolution pattern of the sol-air temperature or solar irradiance can be decomposed into a Fourier series made of the sum of a (possibly infinite) set of simple oscillating functions, namely sines or cosines (or, equivalently, complex exponentials), with base period t_0 = 24h ≡ 86′400 s:

$$\begin{aligned} T_{sol-air}(t) &= \overline{\theta}_{sol-air} + \tfrac{(1-\rho_{sol})}{h_e} \cdot \sum_{n=1}^{\infty} \left|\widehat{\theta}_{sol-air,n}\right| \cdot \left(n \cdot 2\pi \cdot \tfrac{t}{t_0} + \psi_n\right) \cong \\ &\cong \overline{T}_e + \tfrac{(1-\rho_{sol})}{h_e} \cdot \overline{I}_{sol} + \tfrac{(1-\rho_{sol})}{h_e} \cdot \sum_{n=1}^{\infty} \left|\widehat{I}_{sol,n}\right| \cdot \left(n \cdot 2\pi \cdot \tfrac{t}{t_0} + \psi_n\right) \end{aligned} \quad (8)$$

where $\overline{\theta}_{sol-air}$ (°C) is the daily average value of the sol-air temperature and it depends on the daily average values of external temperature $\overline{T}_e$ (°C) and solar irradiance $\overline{I}_{sol}$ (W·m^{-2}). The nth harmonic of the sol-air temperature has an oscillation amplitude $\left|\widehat{\theta}_{sol-air,n}\right|$ (°C) depending on the amplitude of the nth harmonic of the solar irradiance $\left|\widehat{I}_{sol,n}\right|$ (W·m^{-2}), and phase ψ_n (rad). An approximation of Equation (8) consists of neglecting the cycle of the air temperature as its oscillation amplitude is relatively weak over urban and suburban areas (which include most of the building lots), especially in a humid and polluted atmosphere.

The oscillation of sol-air temperature penetrates the building component, and it eventually contributes to the heat flux entering the internal ambient temperature, which has an average density of $\overline{q}_i$ (W·m^{-2}), an oscillation amplitude of the nth harmonic $\left|\widehat{q}_{i,n}\right|$ (W·m^{-2}), and a phase of φ_n (rad):

$$q_i(t) = \overline{q}_i + \sum_{n=1}^{\infty} \left|\widehat{q}_{i,n}\right| \cdot \left(n \cdot 2\pi \cdot \frac{t}{t_0} + \varphi_n\right) \quad (9)$$

The length of penetration of the harmonics decreases with their order n, therefore the analysis can be limited to the first order term (n = 1), which yields the most significant effects in the indoor space (see [24] for details). The deviation arising by the use of non-sinusoidal boundary conditions with respect to the sinusoidal one was found to be small and generally cautionary [30].

At the end, it can be verified that the average entering heat flux density is correlated through the U-value to the difference between the average sol-air temperature and the indoor temperature, the latter of which is assumed to be constant thanks to an ideal air conditioning system:

$$\overline{q}_i = U \cdot \left(\overline{\theta}_{sol-air} - T_i\right) \equiv U \cdot \left(\overline{T}_e + \frac{1-\rho_{sol}}{h_e} \cdot \overline{I}_{sol} - T_i\right) \cong U \cdot \left(\overline{T}_e - T_i\right) + U \cdot \frac{1-\rho_{sol}}{h_e} \cdot \overline{I}_{sol} \quad (10)$$

The first term of the final sum is controlled by the U-value alone and it may be comparatively low if $\overline{T}_e \cong T_i$, as it is often the case in urban areas, while the UHI effect shows up and is enhanced by a high humidity. Moreover, with constant indoor temperature and the analysis limited to the first order components, the oscillation amplitude of the entering heat flux density is correlated to the amplitude of the cycles of sol-air temperature and solar irradiance through the modulus of the (complex) periodic thermal transmittance Y_{ie} (W·m^{-2}·K^{-1}):

$$\left|\widehat{q}_i\right| \leq Y_{ie} \cdot \left|\widehat{\theta}_{sol-air}\right| \cong Y_{ie} \cdot \frac{1-\rho_{sol}}{h_e} \cdot \left|\widehat{I}_{sol}\right| \quad (11)$$

Y_{ie} is calculated according to [6] from the layer structure of the considered component and the thermophysical properties of the layer materials (thermal conductivity, density and specific heat). It is a representation of the heat flow cycle produced at the inner surface by a temperature oscillation at the outer side while the indoor temperature is maintained constant. The higher the thermal inertia is, the lower Y_{ie} is, hence the upper limits have been set in Italy for Y_{ie} [3] alongside those for U.

2.2. The Solar Transmittance Factor (STF) and the Solar Transmittance Index (STI)

A roof with high thermal mass but a dark surface may behave similar to a roof with a relatively low mass and thermal inertia but with a highly reflective surface. A highly reflective surface would almost nullify the effects of the solar cycle; nevertheless, it would be difficult to preserve due to soiling or biofouling, hence the lower solar reflectance of the aged surface must be considered [31,32]. Moreover, highly reflective surfaces are white, but different, less reflective colors may be compulsory for historical or traditional buildings or even entire urban areas. In these cases, thermal insulation and inertia increase their influence. A high thermal inertia, usually achieved by adopting a high mass, has an effective and substantially invariable thermal response, but it may conflict with seismic requirements, maximum allowed roof and wall thickness, costs, and even durability and fire risk if materials such as wooden fibers are used. In general, a proper set of performance parameters is not easy to select. Moreover, an arbitrary set of limiting values could be market distorting. Indeed, this work aims to respond to the need of easily selecting the most effective mix of surface and mass properties.

The daily average cooling power to be supplied by the AC system to offset the transmitted heat is correlated to the quantity calculated in Equation (10). Some supplementary power, however, may be required to ensure a constant operative temperature due to the daily variation of the sol-air temperature, which peaks at noon, affecting the ceiling radiant temperature. Approximating the solar irradiance cycle to a perfectly sine oscillation, in which the average value of the irradiance is made equal to its amplitude of oscillation, the peak cooling power to be provided by an AC system as calculated by Equations (10) and (11) is:

$$q_{i,peak} = \bar{q}_i + \left|\widehat{q}_i\right| \leq U \cdot \frac{1-\rho_{sol}}{h_e} \cdot \bar{I}_{sol} + Y_{ie} \cdot \frac{1-\rho_{sol}}{h_e} \cdot \left|\widehat{I}_{sol}\right| \cong (U + Y_{ie}) \cdot \frac{1-\rho_{sol}}{h_e} \cdot I_{sol,max} \quad (12)$$

A comprehensive index has thus surfaced, which was called the 'solar transmittance factor' (STF) [24]. It includes both the radiative properties at the external surface and the thermophysical properties of the materials under the surface and is calculated as:

$$STF = (U + Y_{ie}) \cdot \frac{1-\rho_{sol}}{h_e} \quad (13)$$

A sketch of the thermal process expressed by Equation (12) is depicted in Figure 1, which shows how the cycle of the external surface temperature as induced by the solar cycle and controlled by the solar reflectance propagates through a building component. It in turn induces a cycle of the internal surface temperature, whose average and peak values are above the indoor temperature and thus yield an entering heat flux. The increase of the average internal surface temperature with respect to the inside air temperature is controlled by the steady-state thermal transmittance U, whereas the oscillation amplitude of the internal surface temperature is controlled by the modulus of the periodic thermal transmittance Y_{ie}.

The solar transmittance factor (STF) is plotted in Figure 2 versus the whole spectrum of values of the solar reflectance ρ_{sol}. Since in Italy and other European countries a minimum insulation level is generally required for new or renovated buildings, two different U-values have been considered: 0.3 $W \cdot m^{-2} \cdot K^{-1}$ (corresponding to a well-insulated roof or wall) and 0.8 $W \cdot m^{-2} \cdot K^{-1}$ (a value often used to distinguish between insulated and non-insulated components). Moreover, Y_{ie} is always lower than U, so values of the Y_{ie}/U ratio have been considered from 1.00 (corresponding to a relatively massive component) down to 0.25 (corresponding to a lightweight component). The same environmental conditions prescribed for calculation of the solar reflectance index [25,26] were used, considering a typical case of intermediate wind conditions and estimating temperature-dependent quantities such as h_e by a recursive approach.

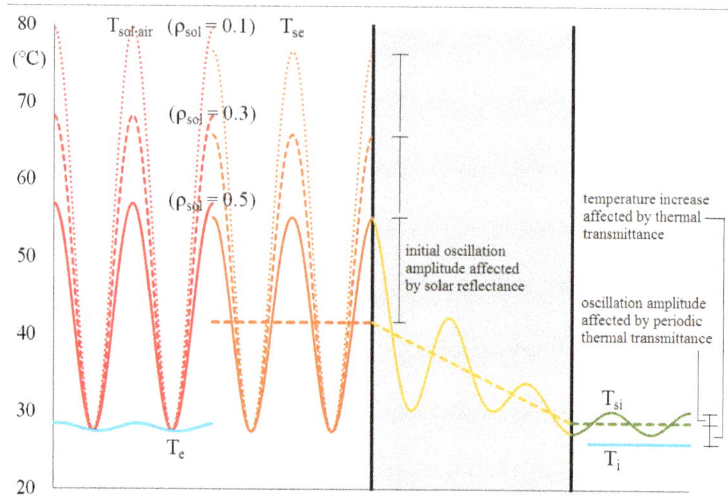

Figure 1. Cycle of the external surface temperature T_{se} induced by the cycle of the sol-air temperature $T_{sol\text{-}air}$ and controlled by the solar reflectance ρ_{sol}: the cycle propagates through a building component and induces a cycle of the internal surface temperature T_{si} that has a peak value well above the internal ambient temperature T_i.

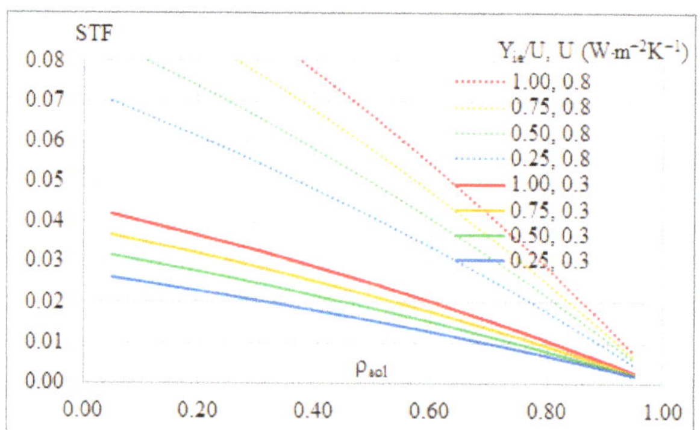

Figure 2. Solar transmittance factor (STF) versus solar reflectance ρ_{sol} for intermediate wind conditions and h_{ce} = 12 W·m^{-2}·K^{-1}, two different values of U (of an almost uninsulated component and a well-insulated component), and several values of Y_{ie}/U (from a relatively massive component down to lightweight component with almost null mass).

It is interesting to observe in Figure 2 the comparatively lower weight of the Y_{ie}/U ratio, i.e., of the thermal inertia, with respect to solar reflectance and thermal insulation. This agrees with the results reported in [18]. Above all, it is worthwhile to note that low values of the solar transmittance factor (e.g., STF < 0.01) can be achieved with very different combinations of U, Y_{ie}, and ρ_{sol}.

The solar transmittance factor (STF) as defined by Equation (12) correlates the peak of the heat flux density that enters the inhabited space to the peak of the solar irradiance cycle, which is approximated to a sine cycle, under the additional approximation of similar internal and average external temperatures. From the peak of the entering heat flux density, one can estimate the peak of radiant temperature of the ceiling as follows:

$$T_{si,peak} - T_i = \frac{q_{i,peak}}{h_i} \approx \frac{STF}{h_i} \cdot \frac{I_{sol,max}}{2} \tag{14}$$

The pre-calculated values of h_i, such as those suggested in [5] or [28], and analogous standards can be used, or a more precise value can be estimated by a recursive approach from Equation (3). The peak of the radiant ceiling temperature affects the operative temperature of the inhabited space (i.e., the weighted mean of air and mean radiant temperature), relevant to thermal comfort as it is the temperature perceived by the occupants. High values of such temperature can be a main source of thermal discomfort, also affecting cooling energy demand by requiring a lower temperature of the indoor air.

The STF would tend to yield very low values for high performance solutions, so it does not seem an effective choice for a performance indicator as it would not allow differentiating alternative building solutions. A similar issue was considered in [16] when the 'new' thermal performance index *TPI was defined as the decrement of the peak of inner surface temperature with respect to a reference worst case, divided by the analogous decrement allowed by a reference optimal case, given in percentage terms. The same concept, clearly inspired by that for SRI calculation [25], can be applied in a more practical way, independent of the local weather conditions, by means of the solar transmittance index (STI) proposed in [24]:

$$STI = 100 \cdot \frac{STF_{worst} - STF_{tested}}{STF_{worst} - STF_{optimal}} \tag{15}$$

Since T_i, h_i, and $I_{sol,max}$ are constant values in Equation (14), one can reason in terms of either inner surface temperature increase or entering heat flux density so the comparison of building solutions can be directly based on STF rather than $T_{si,peak}$. In other words, STI represents the percent fraction of the peak heat flux density transmitted inside in the worst reference case that is cancelled by means of the considered building solution, evaluated in excess to the optimal reference case.

An issue still under investigation is the choice of the worst and optimal reference values of the solar transmittance factor (STF). A truly optimal case could be $STF_{optimal} = 0$. With this setting in Equation (15), values of STI in excess of 100% are avoided. A 100% performance, however, is practically impossible to approach and a different choice may thus be preferable, for example, that of a solution with a high insulation level, a high thermal inertia, and a high solar reflectance. In ref. [24], the choice of an externally insulated concrete slab (density 2400 $kg \cdot m^{-3}$) was proposed, with 20 cm thickness, 10 cm of added foam insulation (thermal conductivity 0.04 $W \cdot m^{-1} \cdot K^{-1}$), and a solar reflective bright white surface ($\rho_{sol} = 0.80$), which achieves an STF value (i.e., $STF_{optimal}$) as low as 0.007. Regarding the reference worst case, a very low performing one such as an uninsulated metal slab with dark surface would cause an STI of high-performance building solutions always close to the best value of 100%, obtaining a situation analogous to the direct use of STF. Therefore, a case with a relatively good performance would be preferable to increase differentiation. In [24], the choice of an externally insulated concrete slab was again proposed, with 5 or 10 cm of foam insulation but with the solar reflectance of a black surface ($\rho_{sol} = 0.05$), which achieves an STF (i.e., STF_{worst}) equal to 0.051 or 0.028, respectively. Values of STF and STI for a wide range of roof types and layer structures were also calculated and are summarized in Table 1.

One can observe that only solutions with a minimum insulation level and without dark coating show a positive STI. The combination of a non-null optimal case ($STF_{optimal} = 0.007$) and the worst case with lower insulation (5 cm, $STF_{worst} = 0.051$) seems to provide higher positive values of STI, yet with a clear differentiation. Solar reflective (i.e., cool) solutions combined with heavy insulation provide high values of STI even if their thermal inertia is low. A solar reflective surface with low emissivity such as a bare metal has performance close to a dark surface. However, the problem of selecting proper reference cases is still open and will be addressed in future research based on statistical analysis of building lots.

Table 1. STI values for different roofing solutions and reference cases (low wind conditions).

Layers are listed from the innermost one. Where not specified, $\varepsilon_e = 0.90$. Empty cells are for negative STI value.			**$STF_{optimal}$**			
			0		0.007 (n.09)	
			STF_{worst}			
			0.051 (n.02)	0.028 (n.03)	0.051 (n.02)	0.028 (n.03)
Type	**Component**	**STF_{tested}**	**STI (%)**	**STI (%)**	**STI (%)**	**STI (%)**
01	Concrete 20 cm, dark ($\rho_{sol} = 0.10$)	0.277				
02	Concrete 20 cm + foam ins. 5 cm, dark	0.051	0		0	
03	Concrete 20 cm + foam ins. 10 cm, dark	0.028	44	0	52	0
04	Concrete 20 cm, light colored ($\rho_{sol} = 0.45$)	0.174				
05	Concrete 20 cm + foam ins. 5 cm, light c.	0.033	35		41	
06	Concrete 20 cm + foam ins. 10 cm, light c.	0.018	64	35	74	47
07	Concrete 20 cm, cool white ($\rho_{sol} = 0.80$)	0.066				
08	Concrete 20 cm + foam ins. 5 cm, cool w.	0.013	75	55	87	73
09	Concrete 20 cm + foam ins. 10 cm, cool w.	0.007	86	75	100	100
10	Metal, bare ($\rho_{sol} = 0.60$, low $\varepsilon_e = 0.20$)	0.381				
11	Metal, dark ($\rho_{sol} = 0.10$, $\varepsilon_e = 0.90$)	0.589				
12	Foam ins. 5 cm + Metal, dark	0.095				
13	Foam ins. 10 cm + Metal, dark	0.051				
14	Metal, light colored ($\rho_{sol} = 0.45$)	0.372				
15	Foam ins. 5 cm + Metal, light c.	0.061				
16	Foam ins. 10 cm + Metal, light c.	0.033	34		40	
17	Metal, cool white ($\rho_{sol} = 0.80$)	0.140				
18	Foam ins. 5 cm + Metal, cool w.	0.024	53	15	62	20
19	Foam ins. 10 cm + Metal, cool w.	0.013	74	54	87	72

3. Correlation of STF with the Component Performance

In this work, the heat flux and the peak temperature were calculated by numerical simulation at the ceiling surface and were compared with the STF values for a relatively wide range of roof types and layer structures in different environmental conditions. TRNSYS 17 was used for the numerical simulation. A 3 m height single-zone building with a 3 m × 3 m roof area was considered in the model, in which a constant inside air temperature of 26 °C was fixed. All surfaces except the roof were set adiabatic, as the focus was on the thermal behavior of the roof alone. The considered locations were Milan (northern Italy, representative of a 'humid sub-tropical climate' (Cfa) according to the Köppen–Geiger climate classification [33]) and Palermo (southern Italy, representative of a 'hot-summer Mediterranean climate' (Csa).

The analysis extended from a light wooden roof, for which different types and levels of insulation were considered, to a concrete roof with different levels and positions of the insulation layer (see Table 2). Moreover, for each roofing solution, five different values were considered for the solar reflectance $\rho_{sol} = 1 - \alpha_{sol}$: 0.1 (representative of a dark black surface), 0.3 (colored surface with relatively dark color), 0.5 (light colored surface),

0.65 (aged white 'cool' surface), and 0.8 (bright white 'cool' surface). For all cases, high-emissivity surfaces were considered, with a thermal emittance 0.9.

Table 2. Roofing solutions considered in the numerical simulation.

Roof Type	Roof Layers	Conductivity ($W \cdot m^{-1} \cdot K^{-1}$)	Density ($kg \cdot m^{-3}$)	Specific Heat ($kJ \cdot kg^{-1} \cdot K^{-1}$)	Thickness (mm)
Light wooden roof + rock wool (low density) insulation LWR + RW	Waterproofing membrane	0.26	1300	1	1
	Wooden panel	0.12	450	2.7	10
	Rock wool panel	0.034	95	1.03	0; 40; 100; 180
	Wooden panel	0.12	450	2.7	10
	U-value ($W \cdot m^{-2} \cdot K^{-1}$)	2.980; 0.661; 0.305; 0.178			
	Y_{ie} ($W \cdot m^{-2} \cdot K^{-1}$)	2.969; 0.655; 0.291; 0.135			
Light wooden roof + wood fiber (low density) insulation LWR + WF	Waterproofing membrane	0.26	1300	1	1
	Wooden panel	0.12	450	2.7	10
	Wood fiber panel	0.038	120	2.4	0; 40; 100; 180
	Wooden panel	0.12	450	2.7	10
	U-value ($W \cdot m^{-2} \cdot K^{-1}$)	2.980; 0.720; 0.337; 0.197			
	Y_{ie} ($W \cdot m^{-2} \cdot K^{-1}$)	2.969; 0.704; 0.265; 0.072			
Concrete roof + polyurethane foam insulation, external CR + PUe	Waterproofing membrane	0.26	1300	1	1
	Polyurethane panel	0.022	36	1.45	0; 50; 100
	Reinforced concrete	1.49	2200	0.88	50
	Concrete	1.61	2200	1	60
	Plaster	0.8	1600	1	1.5
	U-value ($W \cdot m^{-2} \cdot K^{-1}$)	4.318; 0.395; 0.208			
	Y_{ie} ($W \cdot m^{-2} \cdot K^{-1}$)	2.855; 0.153; 0.077			
Concrete roof + polyurethane foam insulation, internal CR + PUi	Waterproofing membrane	0.26	1300	1	1
	Reinforced concrete	1.49	2200	0.88	50
	Concrete	1.61	2200	1	60
	Polyurethane panel	0.022	36	1.45	0; 50; 100
	Plaster	0.8	1600	1	1.5
	U-value ($W \cdot m^{-2} \cdot K^{-1}$)	4.318; 0.395; 0.208			
	Y_{ie} ($W \cdot m^{-2} \cdot K^{-1}$)	2.855; 0.214; 0.108			

The maximum increase of the ceiling temperature T_{si} with respect to the indoor temperature T_i is shown in Figure 3, as calculated for Milan and Palermo in the hottest month, for all considered roofing solutions. The total cooling energy demand along the summer season, from June to September, is also shown in Figure 4. A clear trend is evident in most cases, demonstrating a good correlation between the analyzed parameters and STF. Some scattering occurs for high performing solutions installed in Milan, where the outdoor temperature is often lower than the indoor temperature and reverse heat flows may take place during the day. On the other hand, a sharp correlation is evident throughout the whole range of the considered solutions for a hotter summer season such as that in Palermo, where STF seems the factor driving the thermal response of the roof component.

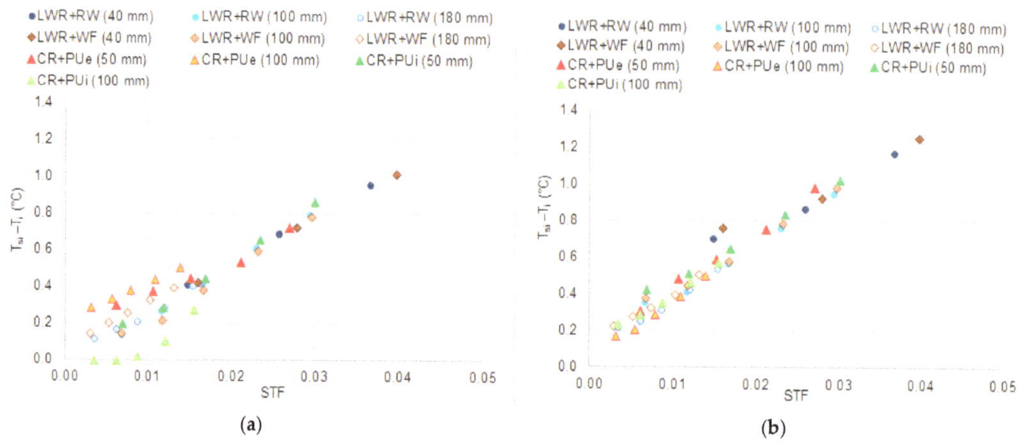

Figure 3. Increase of the ceiling temperature T_{si} with respect to the indoor temperature T_i. for different roofing solutions and insulation thickness. Five different values of STF are considered for each solution, calculated for five different values of the solar reflectance ρ_{sol} (see Table 3): 0.8 (bright white 'cool' surface), 0.65 (aged white 'cool' surface), 0.5 (light colored surface), 0.3 (colored surface with relatively dark color), and 0.1 (dark black surface). (**a**) Milan, maximum temperature difference in July between ceiling and indoor temperature. (**b**) Palermo, maximum temperature difference in August between ceiling and indoor temperature.

Table 3. Solar transmittance factor (STF) for the considered roofing solutions and five different values of the solar reflectance (with $h_{ce} = 12\ W{\cdot}m^{-2}{\cdot}K^{-1}$, $\varepsilon_e = 0.9$).

Roof Type and Insulation	U ($W{\cdot}m^{-2}{\cdot}K^{-1}$)	Y_{ie} ($W{\cdot}m^{-2}{\cdot}K^{-1}$)	ρ_{sol}				
			0.80	**0.65**	**0.50**	**0.30**	**0.10**
			Solar Transmittance Factor (STF)				
LWR	2.980	2.969	0.067	0.117	0.166	0.230	0.294
LWR + RW 040 mm	0.661	0.655	0.015	0.026	0.037	0.051	0.065
LWR + RW 100 mm	0.305	0.291	0.007	0.012	0.017	0.023	0.029
LWR + RW 180 mm	0.178	0.135	0.004	0.006	0.009	0.012	0.015
LWR + WF 040 mm	0.720	0.704	0.016	0.028	0.040	0.055	0.070
LWR + WF 100 mm	0.337	0.265	0.007	0.012	0.017	0.023	0.030
LWR + WF 180 mm	0.197	0.072	0.003	0.005	0.008	0.010	0.013
CR	3.896	2.855	0.076	0.133	0.188	0.261	0.333
CR + PUe 050 mm	0.395	0.153	0.006	0.011	0.015	0.021	0.027
CR + PUe 100 mm	0.208	0.077	0.003	0.006	0.008	0.011	0.014
CR + PUi 050 mm	0.395	0.214	0.007	0.012	0.017	0.024	0.030
CR + PUi 100 mm	0.208	0.108	0.004	0.006	0.009	0.012	0.016

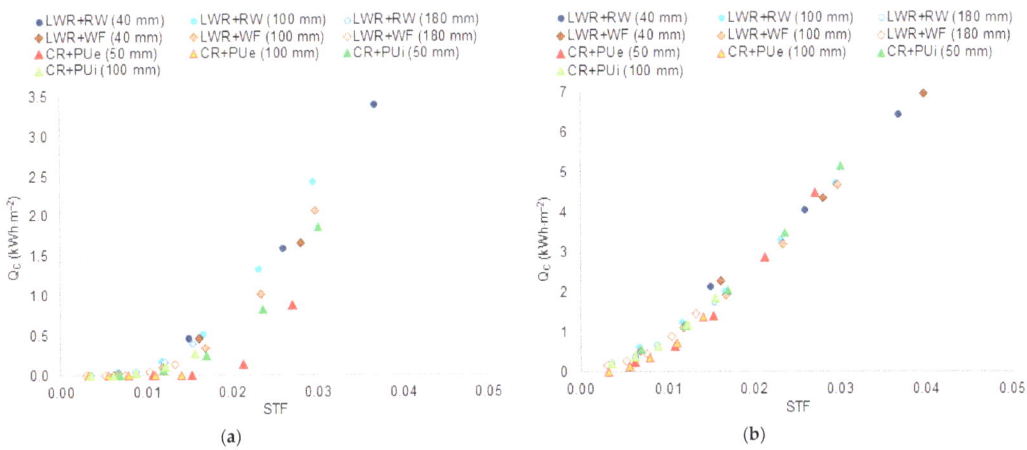

Figure 4. Total cooling energy demand in the summer season (June–September) for different roofing solutions and insulation thicknesses. Five different values of STF are considered for each solution, calculated for five different values of the solar reflectance ρ_{sol} (see Table 3): 0.8 (bright white 'cool' surface), 0.65 (aged white 'cool' surface), 0.5 (light colored surface), 0.3 (colored surface with relatively dark color), and 0.1 (dark black surface). (**a**) Milan, seasonal cooling load. (**b**) Palermo, seasonal cooling load.

4. Concluding Remarks

A 'solar transmittance index' (STI) has been proposed for the energy rating of opaque building components. It is aimed to take into account, in a single performance parameter specifically targeted at the cooling season, the solar reflectance and thermal emittance of a component's external surface together with the thermophysical properties of the component materials relevant to thermal insulation and inertia. This may allow for a quick comparison of different solutions, thus helping the definition of component-based requirements and policies for building energy retrofit. The solutions can be easily implemented by designers and constructors or verified by regulatory institutions. The STI amplifies and enhances the significance of another purposely developed parameter on which it is built, the 'solar transmittance factor' (STF). This is calculated by a very simple formula from the U-value, the modulus of the periodic thermal transmittance, and the radiative properties of the considered component.

The relationship was investigated between STF and the ceiling temperature, which is relevant to the operative temperature indoors, as well as the relationship between STF and cooling energy demand over the cooling season, in Italian locations with different climates. The analysis extended from a light wooden roof, for which different types and levels of insulation were considered, to a concrete roof with different levels and positions of the insulation layer. A significant correlation was found over a relatively wide range of roof types and STF values, especially for a summer Mediterranean climate such as that in Palermo, southern Italy.

We plan to widen the range of considered roof types and extend the investigation to outer vertical walls. The analysis of new roof types would allow to further investigate the relationship among the relevant variables. Moreover, statistical analysis will be carried on existing roofs and their thermal characteristics, in order to identify proper reference values of STF to be used as worst case and optimal case in the calculation of STI.

Author Contributions: Conceptualization, A.M. and H.A.; methodology, A.M. and C.L.; software, C.L.; writing—original draft preparation, A.M., C.L. and P.T.; writing—review and editing H.A. and P.T. All authors have read and agreed to the published version of the manuscript.

Funding: This research received no external funding.

Conflicts of Interest: The authors declare no conflict of interest.

References

1. International Organization for Standardization. *ISO 13790:2008—Energy Performance of Buildings—Calculation of Energy Use for Space Heating and Cooling*; Withdrawn; International Organization for Standardization: Geneva, Switzerland, 2008.
2. Libbra, A.; Muscio, A.; Siligardi, C. Energy performance of opaque building elements in summer: Analysis of a simplified calculation method in force in Italy. *Energy Build.* **2013**, *64*, 384–394. [CrossRef]
3. Italian Government. Inter-Ministerial Decree 26 June 2015 "Minimum Requirements". Available online: www.mise.gov.it (accessed on 20 April 2021).
4. Italian Government. Ministerial Decree 6 August 2020 "Energy Efficiency". Available online: www.mise.gov.it (accessed on 20 April 2021).
5. *ISO 6946:2017—Building Components and Building Elements—Thermal Resistance and Thermal Transmittance—Calculation Methods*; International Organization for Standardization: Geneva, Switzerland, 2017.
6. *EN ISO 13786:2017—Thermal Performance of Building Components—Dynamic Thermal Characteristics—Calculation Methods*; International Organization for Standardization: Geneva, Switzerland, 2017.
7. *ASTM C1549-16 Standard Test Method for Determination of Solar Reflectance Near Ambient Temperature Using a Portable Solar Reflectometer*; ASTM International: West Conshohocken, PA, USA, 2016.
8. *ASTM E903-20—Standard Test Method for Solar Absorptance, Reflectance, and Transmittance of Materials Using Integrating Spheres*; ASTM International: West Conshohocken, PA, USA, 2020.
9. *EN 15976:2011—Flexible Sheets for Waterproofing—Determination of Emissivity*; European Committee for Standardization (CEN): Brussels, Belgium, 2011.
10. *ASTM C1371-15 Standard Test Method for Determination of Emittance of Materials Near Room Temperature Using Portable Emissometers*; ASTM International: West Conshohocken, PA, USA, 2015.
11. Zinzi, M.; Agnoli, S. Cool and green roofs: An energy and comfort comparison between passive cooling and mitigation urban heat island techniques for residential buildings in the Mediterranean region. *Energy Build.* **2012**, *55*, 66–76. [CrossRef]
12. Barrios, G.; Huelsz, G.; Rechtman, R.; Rojas, J. Wall/roof thermal performance differences between air-conditioned and non air-conditioned rooms. *Energy Build.* **2011**, *43*, 219–223. [CrossRef]
13. Barrios, G.; Huelsz, G.; Rojas, J.; Ochoa, J.M.; Marincic, I. Envelope wall/roof thermal performance parameters for non air-conditioned buildings. *Energy Build.* **2012**, *50*, 120–127. [CrossRef]
14. Barrios, G.; Casa, J.M.; Huelsz, G.; Rojas, J. Ener-Habitat: An online numerical tool to evaluate the thermal performance of homogeneous and non-homogeneous envelope walls/roofs. *Sol. Energy* **2016**, *131*, 296–304. [CrossRef]
15. Cheng, V.; Ng, E.; Givoni, B. Effect of envelope colour and thermal mass on indoor temperatures in hot humid climate. *Sol. Energy* **2013**, *78*, 528–534. [CrossRef]
16. Chandra, P. Rating of Wall and Roof Sections—Thermal Considerations. *Build. Environ.* **1980**, *15*, 244–251. [CrossRef]
17. Kabre, C. A new thermal performance index for dwelling roofs in the warm humid tropics. *Build. Environ.* **2010**, *45*, 727–738. [CrossRef]
18. Zingre, K.T.; Kiran Kumar, D.E.V.S.; Wan, M.P. Analysing the effect of substrate properties on building envelope thermal performance in various climates. *Energies* **2020**, *13*, 5119. [CrossRef]
19. Thomas, L.P.; Marino, B.M.; Munoz, N. Steady-state and time-dependent heat fluxes through building envelope walls: A quantitative analysis to determine their relative significance all year round. *J. Build. Eng.* **2020**, *29*, 101122. [CrossRef]
20. Zingre, K.T.; Wan, M.P.; Wong, S.K.; Toh, W.B.T.; Lee, I.Y.L. Modelling of cool roof performance for double-skin roofs in tropical climate. *Energy* **2015**, *82*, 813–826. [CrossRef]
21. Omar, A.I.; David, D.; Vergnault, E.; Virgone, J.; Idriss, A.I. A new set of indicators to evaluate the bioclimatic performance of air conditioned buildings in a hot and humid climate. *J. Build. Eng.* **2020**, *31*, 101350. [CrossRef]
22. Kaboréa, M.; Bozonnet, E.; Salagnac, P.; Abadie, A. Indexes for passive building design in urban context—Indoor and outdoor cooling potentials. *Energy Build.* **2018**, *173*, 315–325. [CrossRef]
23. Pisello, A.L.; Goretti, M.; Cotana, F. A method for assessing buildings' energy efficiency by dynamic simulation and experimental activity. *Appl. Energy* **2012**, *97*, 419–429. [CrossRef]
24. Muscio, A.; Akbari, H. An index for the overall performance of opaque building elements subjected to solar radiation. *Energy Build.* **2017**, *157*, 184–194. [CrossRef]
25. *ASTM E1980-11(2019)—Standard Practice for Calculating Solar Reflectance Index of Horizontal and Low-Sloped Opaque Surfaces*; ASTM International: West Conshohocken, PA, USA, 2019.
26. Muscio, A. The solar reflectance index as a tool to forecast the heat released to the urban environment: Potentiality and assessment issues. *Climate* **2018**, *6*, 12. [CrossRef]
27. Torres-Quezada, J.; Coch, H.; Isalguè, A.; López, J. The roof impact on the heat balance of low height buildings at low latitudes, PLEA 2018—Smart and Healthy within the Two-Degree Limit. In Proceedings of the 34th International Conference on Passive and Low Energy Architecture (PLEA 2018), Hong Kong, China, 10–12 December 2018; Volume 3, pp. 937–938.
28. *ASHRAE Handbook: Fundamentals*; American Society of Heating, Refrigeration and Air-Conditioning Engineers: Atlanta, GA, USA, 2005.

29. *LEED® Green Building Program v4 for Building Design and Construction*. U.S. Green Building Council. 2019. Available online: www.usgbc.org (accessed on 20 April 2021).
30. Gasparella, A.; Pernigotto, G.; Baratieri, M.; Baggio, P. Thermal dynamic transfer properties of the opaque envelope: Analytical and numerical tools for the assessment of the response to summer outdoor conditions. *Energy Build.* **2011**, *43*, 2509–2517. [CrossRef]
31. Paolini, R.; Terraneo, G.; Ferrari, C.; Sleiman, M.; Muscio, A.; Metrangolo, P.; Poli, T.; Destaillats, H.; Zinzi, M.; Levinson, R. Effects of soiling and weathering on the albedo of building envelope materials: Lessons learned from natural exposure in two European cities and tuning of a laboratory simulation practice. *Sol. Energy Mater. Sol. Cells* **2020**, *205*, 110264. [CrossRef]
32. Santunione, G.; Ferrari, C.; Siligardi, C.; Muscio, A.; Sgarbi, E. Accelerated biological ageing of solar reflective and aesthetically relevant building materials. *Adv. Build. Energy Res.* **2019**, *13*, 264–281. [CrossRef]
33. Beck, H.E.; Zimmermann, N.E.; McVicar, T.R.; Vergopolan, N.; Berg, A.; Wood, E.F. Present and future Köppen-Geiger climate classification maps at 1-km resolution. *Sci. Data* **2018**, *5*, 180214. [CrossRef] [PubMed]

atmosphere

MDPI

Article

Effect of Urban Heat Island and Global Warming Countermeasures on Heat Release and Carbon Dioxide Emissions from a Detached House

Daisuke Narumi [1,2,*], Ronnen Levinson [2] and Yoshiyuki Shimoda [3]

[1] Graduate School of Environmental and Information Sciences, Yokohama National University, Yokohama 240-8501, Japan
[2] Heat Island Group, Lawrence Berkeley National Laboratory, Berkeley, CA 94720, USA; rmlevinson@lbl.gov
[3] Division of Sustainable Energy and Environmental Engineering, Osaka University, Suita 565-0871, Japan; shimoda@see.eng.osaka-u.ac.jp
* Correspondence: narumi-daisuke-rs@ynu.ac.jp; Tel.: +81-45-339-3719

Abstract: Urban air temperature rises induced by the urban heat island (UHIE) effect or by global warming (GW) can be beneficial in winter but detrimental in summer. The SCIENCE-Outdoor model was used to simulate changes to sensible heat release and CO_2 emissions from buildings yielded by four UHIE countermeasures and five GW countermeasures. This model can evaluate the thermal condition of building envelope surfaces, both inside and outside. The results showed that water-consuming UHIE countermeasures such as evaporative space cooling and roof water showering provided positive effects (decreasing sensible heat release and CO_2 emissions related to space conditioning) in summer. Additionally, they had no negative (unwanted cooling) effects in winter since they can be turned off in the heating season. Roof greening can provide the greatest space- conditioning CO_2 emissions reductions among four UHIE countermeasures, and it reduces the amount of heat release slightly in the heating season. Since the effect on reducing carbon dioxide (CO_2) emissions by UHIE countermeasures is not very significant, it is desirable to introduce GW countermeasures in order to reduce CO_2 emissions. The significance of this study is that it constructed the new simulation model SCIENCE-Outdoor and applied it to show the influence of countermeasures upon both heat release and CO_2 emissions.

Keywords: urban heat island; sensible heat release; carbon dioxide emissions; building simulation; detached house; countermeasure

Citation: Narumi, D.; Levinson, R.; Shimoda, Y. Effect of Urban Heat Island and Global Warming Countermeasures on Heat Release and Carbon Dioxide Emissions from a Detached House. *Atmosphere* **2021**, *12*, 572. https://doi.org/10.3390/atmos12050572

Academic Editors: Hideki Takebayashi and Jihui Yuan

Received: 24 March 2021
Accepted: 23 April 2021
Published: 28 April 2021

1. Introduction

Recently, increasing urban temperatures due to the urban heat island effect (UHIE) and global warming (GW) have become remarkable around the world. During the 20th century, GW increased the annual average air temperature in Japan by about 1 K [1]. During the same period, high temperatures in the summer have risen about 1–2 K, and low temperatures in winter have risen about 3–6 K in some large cities [1]. These extreme temperature rises in some large cities result from UHIE and GW, but currently, the influence of UHIE in Japan is larger than that of GW. Based on past trends, it is easy to imagine that the temperature rise will continue for some time into the future [2]. UHIE and GW do more than increase temperatures; they also have other diverse impacts on urban dwellers that affect energy and resources [3,4], human health [5,6], and air pollution [7,8]. Therefore, we must take action to mitigate these two phenomena.

Under such circumstances, it is highly desirable that countermeasures reduce both UHIE and GW. However, some measures taken to mitigate UHIE can affect carbon dioxide (CO_2) emissions negatively, and measures taken to mitigate GW can affect heat release negatively. For this reason, it is necessary to evaluate the efficacies of UHIE countermeasures and GW countermeasures, and then examine whether UHIE countermeasures will increase GW and whether GW countermeasures will increase the UHIE. In this study, we used a simulation model to quantify the effect of UHIE countermeasures and GW countermeasures upon heat release and CO_2 emissions, respectively, and investigated features of various countermeasures related to a detached house.

Here, it is necessary to consider the implications of reducing heat release during winter where winter temperatures are low (like East Asian countries). It has been shown that temperature rise during winter in Japan helps to reduce the adverse effects of energy consumption and health impacts [9–11]. To maintain these indirect benefits, it is not desirable to reduce the amount of heat released during winter.

Much research has been conducted on the relationship between UHIE and energy consumption, and it is roughly divided into three groups. The first focuses on the influence of UHIE on energy consumption, and many studies have been conducted to obtain the temperature sensitivity from seasonal energy consumption data [12–15]. The second group investigates the influence of energy consumption as a cause of UHIE, and much research has been done to calculate the anthropogenic waste heat from the detailed energy consumption of the region and to evaluate the temperature effect by the climate model, especially in Japan [16–19]. The last group investigates the influence of UHIE countermeasures on regional temperature by reducing energy consumption. Numerous studies have been conducted thanks to the development of user-friendly models typified by MM5 (Fifth-Generation Penn State/NCAR Mesoscale Model) [20] and WRF (Weather Research and Forecasting Model) [21], as well as the improvement of computer performance in recent years [22–25]. While the impacts on CO_2 emissions from reducing energy consumption are well analyzed in GW studies [26–28], most researchers do not pay attention to reducing heat release by implementing GW countermeasures.

Most UHIE and GW studies have only examined the effects of UHIE and GW countermeasures individually. Some advanced studies have examined both effects of UHIE and GW when evaluating roof greening [29] or air source heat pump water heaters [30] or photovoltaic panels [31], but few have evaluated heat release to the atmosphere and energy consumption simultaneously with a single simulation model. Ihara et al. used the CM-BEM model to quantify the impacts of various measures on annual energy consumption and ambient temperature for typical Japanese office blocks [32]; we are unaware of any such study targeting residential buildings. To successfully implement urban and regional planning measures to address these issues, it is important to examine not only the positive effects of countermeasures but also their potential adverse effects and to use the same model to compare heat release and CO_2 emissions on an equal footing.

From the above background, in this study, through an examination using the SCIENCE-Outdoor simulation model, we quantify the effect of UHIE countermeasures and GW countermeasures on heat release and CO_2 emissions. This study's purpose is to evaluate various technologies for their potential to mitigate UHIE and GW, primarily in houses, and then to propose their proper implementation.

2. Methods

2.1. Simulation Model

In this study, the SCIENCE-Outdoor model was used for evaluation. This model is based on the computational fluid dynamics (CFD) SCIENCE model [33] that can evaluate thermal and fluid conditions inside a building. This model was modified to yield the SCIENCE-Vent model by expanding the fluid and radiant analysis outside the building and by adding an indoor climate control behavior model [34,35]. The SCIENCE-Vent model can predict energy consumption of space conditioning (SC) and lighting for a building. It also considers the relationship between the inside and outside environments, as well as occupant indoor thermal environment control behavior (e.g., cross-ventilation by opening windows and space conditioning use).

For this study, we modified the SCIENCE-Vent model to yield the SCIENCE-Outdoor model by adding outside heat release analysis and by omitting fluid analysis. Because the SCIENCE-Outdoor model was developed based on the CFD model, it can evaluate the thermal condition of building surfaces both inside and outside at each detailed mesh. As shown in Figure 1, the SCIENCE-Outdoor model consists of the three submodels: (1) radiant, (2) inside thermal environment, and (3) outside heat release.

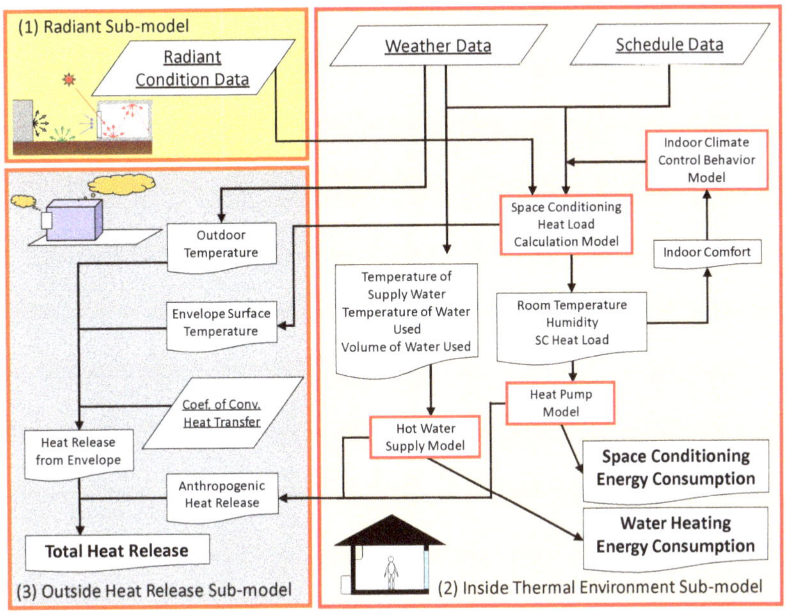

Figure 1. Outline of the SCIENCE-Outdoor model showing the flowchart of data between or inside each submodel: (**1**) radiant, (**2**) inside thermal environment, and (**3**) outside heat release.

The radiant submodel calculates the radiation field both inside and outside buildings by considering the influence of surrounding buildings and trees. To evaluate radiation outside of buildings, it can obtain the sky view factor of each building's outer surface, the shape factor between each outer surface and the ground or surrounding buildings, and the reception ratio for direct solar radiation. To evaluate radiation inside buildings, it can obtain the sky factor of the building's inner surface, the shape configuration of the ground and surrounding buildings, the Gebhart absorption coefficient [36] for solar radiation, and long-wavelength components. These data are used to calculate the envelope surface temperature and radiant heat flux for the inside thermal environment submodel.

The inside thermal environment submodel is used to predict an indoor thermal environment and to calculate energy consumption of the space conditioning and water heating (WH) by using the space conditioning heat load calculation model, the heat pump model [37], the hot water supply model [38], and the indoor climate control behavior model [39]. The space conditioning heat load calculation model assumes complete mixing of indoor air. The thermal conduction calculation through the envelopes uses a one-dimensional finite difference method. In the heat pump model, the coefficient of performance (COP) is corrected in consideration of the outdoor air temperature and the amount of space conditioning heat load [40]. Space conditioning energy consumption is calculated from the corrected COP and the space conditioning heat load. In the hot water supply model, it is necessary to know the supply water temperature, the amount of water used, and the temperature of that water. In the indoor climate control behavior model, the standard effective temperature (SET*) at 1.2 m above the room center floor is used as an evaluation index of the comfort of the indoor thermal environment, and the heating, ventilation, and air conditioning (HVAC) method (e.g., cross-ventilation by opening windows and space conditioning use) in each room is determined.

In the outside heat release submodel, the sensible heat release derived from the building outer envelope and glass surface is calculated based on the data provided by the inside thermal environment submodel and the outdoor air temperature obtained from weather data. The sensible heat release derived from the envelope surface is calculated from the outdoor air temperature, the outer envelope and glass surface temperature, and the convective heat transfer coefficient. Here, the indoor convective heat transfer coefficient was constant at 5 W/m^2, and the outdoor convective heat transfer coefficient was set as a value that depends on the external wind speed according to the Jurges empirical formula [41].

In this study, the heat release is defined as the total amount of convective sensible heat transfer from the building envelope, plus waste sensible heat produced outside the building by heat pumps used to condition the occupied space, plus waste sensible heat produced by the water heater used to make hot water. Since the latent heat does not directly affect the air temperature change near the ground, it was not evaluated in this paper. Figure 2 shows the schematic diagram of the heat release defined in this study.

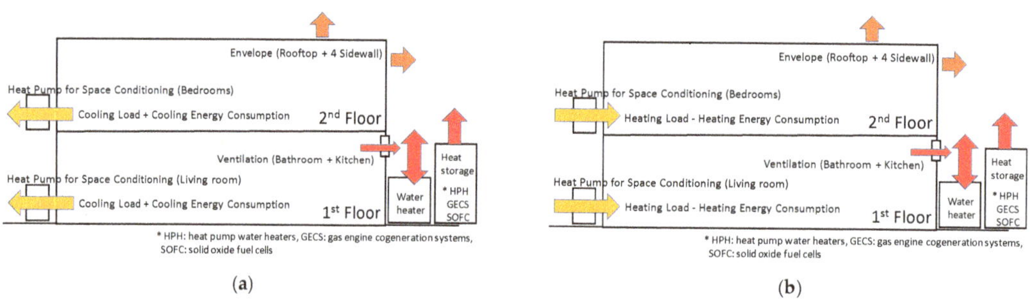

Figure 2. Schematic diagram of the heat release defined in this study for (**a**) the cooling season and (**b**) the heating season.

To determine the effect of countermeasures on UHIE mitigation, we calculated the change of sensible heat release when applying countermeasures. Because the influence of the decrease in heat release on ambient air temperature varies with time and place [42], it is difficult to evaluate its effectiveness on temperature precisely. To determine the effect of countermeasures on GW mitigation, we calculated the change of energy consumption and CO_2 emissions when applying countermeasures. Here, energy consumption included both space conditioning and water heating. The details of each countermeasure targeted in this research are described in Section 2.3. Regarding the CO_2 emission change by introduction of various measures mentioned above, 0.512 kg CO_2/kWh was used for site electricity [43] and 0.05 kg CO_2/MJ was used for site gas [44].

2.2. Building and Weather Condition

This simulation used the Japanese prototype detached house model by the Architectural Institute of Japan (AIJ) [45]. Figure 3 shows the plan view of this house model. The total floor area of the house is about 125 m^2. The house includes a living room (including dining area) and three individual rooms.

Table 1 shows the computational condition and Table 2 shows the wall composition of the buildings to be evaluated. Two kinds of building structures were evaluated: wooden and reinforced concrete (RC). Three levels of insulation performance were considered: (1) no insulation (where thermal transmittance of the wooden outer wall is 2.5 W/m^2·K, with single-pane windows); (2) low insulation, equivalent to the old 1980 Japanese energy-saving code (where the thermal transmittance of the wooden outer wall is 0.9 W/m^2·K and windows are single-paned); and (3) high insulation, equivalent to the current 1999 Japanese energy-saving code (where the thermal transmittance of the wooden outer wall is

0.3 W/m^2·K and windows are double-paned). An air-cooled heat pump system was set to provide the living room (cooling capacity: 3.6 kW), main bedroom (cooling capacity: 3.6 kW), and both child bedrooms (cooling capacity: 2.2 kW). The solar reflectance (albedo) of the outer envelope for the base condition was set to 0.20.

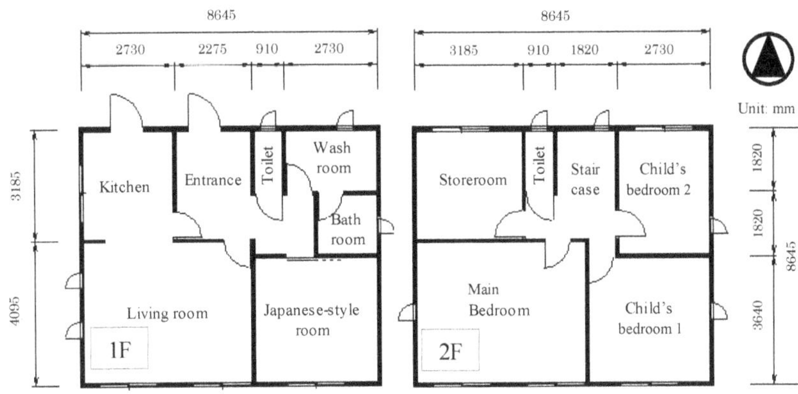

Figure 3. Plan view of the targeted detached house.

Table 1. Outline of the computational condition.

	Property	Value
Outdoor	Climate condition	Expanded AMeDAS weather data [46]): standard year (Osaka city, Japan)
	Ground albedo	0.16
Building	House model	Standard residential house model [45]
	Structure	Wooden or reinforced concrete
	Insulation	3 levels: no = no insulation; low = insulation equivalent to the old 1980 Japanese energy-saving code; high = insulation equivalent to the next-generation 1999 Japanese energy-saving code
	Envelope albedo	0.20 (base condition)
	Space conditioning unit	Air-cooled type heat pumps; cooling capacity in living room, main bedroom: each 3.6 kW; cooling capacity in child bedrooms 1 and 2: each 2.2 kW
Occupant	Household	Two adults (one employed outside the home, the other a homemaker) and two schoolchildren
	Preset temp. and relative humidity	27 °C and 60% RH in the cooling season; 22 °C and uncontrolled humidity in the heating season
	Opening pattern	Determined by the indoor climate control behavior model [39] depending on the weather conditions
	Schedule of occupancy and heat generation	Set by applying the automatic setup scheduling program SCHEDULE [47]

The household was considered to be a family of four: two adults (one employed outside the home, the other a homemaker) and two schoolchildren. The schedule of occupancy and heat generation were set by applying "SCHEDULE" [47], which was an automatic setup scheduling program. An air-cooled heat pump system was used for the cooling and heating, with temperature setpoints of 27 °C in the cooling season and 22 °C in the heating season. The relative humidity setpoint was 60% in the cooling season; relative humidity was not controlled in the heating season.

Table 2. Wall composition of the building to be evaluated.

Part	Insulation Level	Wooden			
		Layer	Thermal Conductivity (W/m·K)	Volumetric Heat Capacity (kJ/m^3·K)	Thickness (m)
Outer Wall	Common	Gypsum board	2.14×10^{-1}	8.54×10^{2}	1.2×10^{-2}
	High	Insulator	3.60×10^{-2}	2.70×10^{1}	4.4×10^{-2}
	Low		5.11×10^{-2}	8.41	3.2×10^{-2}
	High	Air layer	7.00×10^{-2} *	1.20	4.2×10^{-2}
	Low				5.4×10^{-2}
	No				8.6×10^{-2}
	Common	Plywood	1.29×10^{-1}	1.11×10^{3}	9.0×10^{-3}
	Common	Mortar	1.09	2.31×10^{3}	3.0×10^{-2}
Rooftop	Common	Gypsum board	2.14×10^{-1}	8.54×10^{2}	1.2×10^{-2}
	High	Insulator	3.60×10^{-2}	2.70×10^{1}	4.4×10^{-2}
	Low		5.11×10^{-2}	8.41	3.2×10^{-2}
	High	Air layer	7.00×10^{-2} *	1.20	4.2×10^{-2}
	Low				5.4×10^{-2}
	No				8.6×10^{-2}
	Common	Plywood	1.29×10^{-1}	1.11×10^{3}	1.2×10^{-2}
	Common	Slate	9.63×10^{-1}	1.52×10^{3}	1.2×10^{-2}
		Reinforced concrete			
Part	Insulation level	Layer	Thermal conductivity (W/m·K)	Volumetric heat capacity (kJ/m^3·K)	Thickness (m)
Outer Wall	Common	Gypsum board	2.14×10^{-1}	8.54×10^{2}	1.2×10^{-2}
	High	Insulator	3.60×10^{-2}	2.70×10^{1}	4.4×10^{-2}
	Low		5.11×10^{-2}	8.41	3.2×10^{-2}
	High	Air layer	7.00×10^{-2} *	1.20	4.2×10^{-2}
	Low				5.4×10^{-2}
	No				8.6×10^{-2}
	Common	Plywood	1.29×10^{-1}	1.11×10^{3}	9.0×10^{-2}
	Common	Mortar	1.09	2.31×10^{3}	3.0×10^{-2}
Rooftop	Common	Gypsum board	2.14×10^{-1}	8.54×10^{2}	1.2×10^{-2}
	High	Insulator	3.60×10^{-2}	2.70×10^{1}	4.4×10^{-2}
	Low		5.11×10^{-2}	8.41	3.2×10^{-2}
	High	Air layer	7.00×10^{-2} *	1.20	4.2×10^{-2}
	Low				5.4×10^{-2}
	No				8.6×10^{-2}
	Common	Plywood	1.29×10^{-1}	1.11×10^{3}	1.2×10^{-2}
	Common	Slate	9.63×10^{-1}	1.52×10^{3}	1.2×10^{-2}

* This cell shows the value of thermal resistance (m^2·K)/W.

Expanded AMeDAS (Automated Meteorological Data Acquisition System) weather data for HVAC systems [46] were used for the climate condition. The interval time of the calculation can be set arbitrarily; in this study, the interval was set to 15 min. However, since AMeDAS data are provided in 1-h intervals, it was interpolated at 15-min intervals (calculation time-step) using a Lagrange polynomial interpolation. The targeted area was set in Osaka, the second-largest city in Japan, located near the center of the country. Osaka has very severe weather conditions in the summer and is one of the hottest metropolises in Japan.

2.3. Countermeasures

Table 3 outlines the countermeasures evaluated in this study. For this research, the countermeasures for UHIE and GW that are expected to become common in the near future were selected. High-albedo roof, roof greening, roof water showering, and evaporative space cooling were selected as the UHIE countermeasures. Condensing water heater, heat pump water heater, gas engine cogeneration system, solid oxide fuel cell, and photovoltaic power generation were selected as the GW countermeasures.

Table 3. Outline of the evaluated countermeasures.

Countermeasure	Main Target	Computational Condition
High-albedo roof (HAR)	UHIE	Raising the rooftop albedo to 0.60 from 0.20
Roof greening (RG)	UHIE	Improving evaporation efficiency of the rooftop to 0.3 from 0.0 and albedo of the rooftop to 0.25 Adding a greening and soil layer on rooftop surface Setting for the reinforced concrete structure only Setting the condition for withering during winter
Roof water showering (RWS)	UHIE	·Setting evaporation efficiency of rooftop to 0.7 from 0.0 when the rooftop surface temperature exceeds 40 °C in the daytime until 5 p.m. Evaporation efficiency will gradually decrease in the nighttime
Evaporative space cooling (ESC)	UHIE	Improving indoor thermal comfort by spraying dry fog jet Cooling effect is equivalent to 1 K decrease in SET * Jetting will be stopped when behavior model judges AC is required Installing only in the air-conditioned room—9 nozzles each in the living room and the main bedroom, and 4 nozzles each in the child bedrooms The amount of water used per nozzle was 1.34 L per minute (L/min)
Condensing water heater (CWH)	GW	Improving the efficiency to 95% from 78%
Heat pump water heater (HPH)	GW	Setting rated generation output of the hot water at 4.5 kW Improving the efficiency Absorbing heat from the ambient atmosphere Changing the COP due to outside air temperature
Gas engine cogeneration system (GECS)	GW	Setting rated power generation output at 1.0 kW and rated power generation efficiency at 20% Heat exhaust efficiency at 57% Operating in accordance with the heat demand Number of operations per day is unlimited but excessive start/stop is restricted
Solid oxide fuel cell (SOFC)	GW	Setting rated power generation output at 0.7 kW and rated power generation efficiency at 45% Heat exhaust efficiency at 36% (depend on the load) Setting hourly power generation as for fitting electricity load without start/stop
Photovoltaic power generation (PV)	GW	Setting rated power generation efficiency at 13% Considering the influence of decreasing the albedo on the rising temperature of the rooftop surface and increasing heat release

A high-albedo roof can reduce both sensible heat release from a rooftop and cooling energy consumption in the building by using reflective paint or material to reflect solar radiation. Although its effect is largest on summer afternoons, there is some concern about an increase in heating energy consumption during the heating season [48]. For this calculation, the rooftop surface albedo was changed to 0.60 from 0.20 throughout the year.

Roof greening—placing greenery (vegetation) on the roof—can reduce sensible heat release from the rooftop, as well as cooling energy consumption. The greenery can reduce the roof surface temperature via evaporation and transpiration; greening units can also provide insulation [49]. For this calculation, layers of vegetation and soil were added on the rooftop surface (with a thermal conductivity of 1.85 W/m·K). Rooftop surface evaporation efficiency was changed to 0.3 from 0.0, and the rooftop surface albedo was changed to 0.25 from 0.20 [50,51]. The roof greenery was assumed to be withered in winter. The evaporation efficiency while withered was set at 0.05, and only the soil layer was considered. Roof greening was evaluated only on the reinforced concrete structure because the greenery was too heavy for the roof of the wooden house.

Installing a roof water shower (RWS) can reduce the amount of sensible heat release from a rooftop, as well as cooling energy consumption. Showering water can reduce the roof surface temperature via evaporation. It is not necessary to use a motor pump since the waterdrops can be large. Because this system is not used in winter, there is no concern about an increase in heating energy consumption [52]. For this calculation, the RWS was set to represent showering water on the rooftop surface only in the daytime until 5 p.m., when its surface temperature exceeds 40 °C in the cooling season. Evaporation efficiency of the rooftop surface was set at 0.7 while showering, gradually decreasing after stopping the shower [52].

An evaporative space cooling system can reduce sensible heat release from space conditioning, as well as cooling energy consumption, by installing a fine fog jetting outside windows on an outer wall and taking the fog into the room. It can reduce the indoor air temperature and atmospheric sensible heat through evaporation. Because the particles of the mist are exceptionally fine, they evaporate immediately. However, it is necessary to use a motor pump to increase the water pressure and reduce mist size. Because this system is not used during winter, there is no concern about an increase in heating energy consumption [52]. For this calculation, it was installed only in the air-conditioned room—nine nozzles each in the living room and the main bedroom, and four nozzles each in the child bedrooms. Each nozzle delivers water at the rate of 1.34 L/min. In the cooling season, it was assumed that this system would operate as needed to cool occupants. If unable to maintain occupant comfort, the system would be stopped and mechanical cooling would be turned on. Based on the results of our feasibility study [52], the cooling effect was almost equivalent to a 1 K decrease in SET*, and the reduction of cooling time and the heat absorption from the atmosphere due to evaporation were considered.

A condensing water heater can reduce water heating energy consumption and heat release from water heating by using waste-heat recovery at the secondary heat exchanger to improve heat exchange efficiency [38]. This system does not require installation of a hot water storage tank. In recent years, condensing water heaters have become standard in Japan. For this calculation, the heat exchange efficiency was improved to 0.95 from 0.78.

A heat pump water heater can reduce water heating energy consumption and heat release from water heating by absorbing atmospheric heat with a compression heat pump. It can obtain a large reduction effect of atmospheric sensible heat, especially at night, because this system usually works during the night by utilizing low-rate midnight power [38]. In recent years, the share of this system has been increasing rapidly as a substitute for a normal electric-resistance water heater in Japan. There is a concern about heat release from the hot water storage tank to the atmosphere. For this calculation, the rated generation output of the hot water was set at 4.5 kW. The effect of improvement in energy efficiency compared to conventional water heater and heat absorption from the ambient atmosphere was considered. Regarding the COP, the characteristic change due to outdoor air temperature was considered.

A gas engine cogeneration system can reduce consumption of water heating energy, as well as grid electricity consumption, by generating onsite electricity with a gas engine. This system is operated in accordance with the heat demand. The number of operations per day is unlimited, but excessive start/stop must be restricted [38]. There is some concern about heat release from the hot water storage tank and heat release from the engine to burn the fuel on site. For this calculation, the rated power generation output was set at 1.0 kW. The rated power generation efficiency was set at 20%, and the exhaust heat utilizing efficiency was set at 57%.

A solid oxide fuel cell (SOFC) can reduce the amount of water heating energy, as well as grid power consumption, by generating onsite electricity. Hourly power generation will be set to fit the electricity load without start/stop. Heat demand that is not supplemented by this system will be handled by an auxiliary heat source [38]. There is a concern about heat release from the hot water storage tank and generator. For this calculation, as a result of determining the optimum capacity from the thermal demand and the electric power demand, the rated power generation output was set at 0.7 kW. The rated power generation efficiency was set at 45%, and the exhaust heat utilizing efficiency was set at 36%, and they were changed depending on the operation status from moment to moment.

Photovoltaic power generation (PV) can reduce the amount of grid electricity consumption by generating electricity from solar energy. In addition, due to its small heat capacity, the surface temperature of PV will decrease during the night. There is a concern about the surface temperature of PV rising during the daytime, as the solar reflectance is lower than the general building surface. For this calculation, the installed capacity was 3.0 kW. It was assumed that all the generated electric power could lead to a reduction of the grid power supply. The rated power generation (conversion) efficiency was set at 13%. The influence of decreasing the albedo on the rising temperature of the rooftop surface and increasing heat release was taken into consideration [31].

3. Results

3.1. Base Condition (No Countermeasure)

Figure 4a shows the sensible heat release from each path on a representative sunny summer day (August 5) for the wooden structure with the low insulation level. Figure 5a shows the weather condition. For the base condition, the heat from space conditioning is the sum of the cooling load and the consumption energy for space conditioning; it is equivalent to the amount of heat release to the outside through the heat pump outdoor unit, as shown in Figure 2. The heat from the water heating equals the total amount of heat released to the outside through the water heating and inside the house at the time of hot water use. Hereafter, the basis area is equal to the building area.

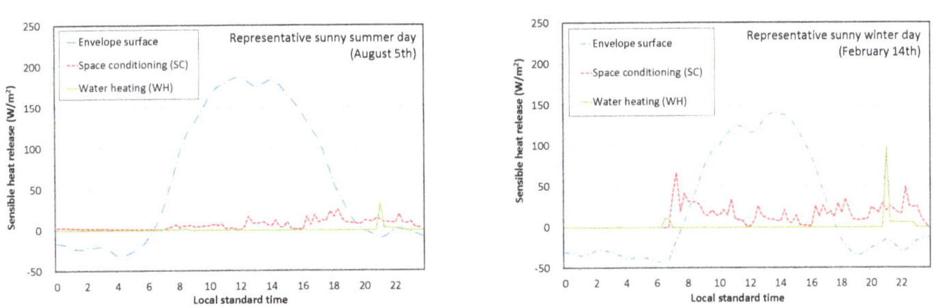

(a) (b)

Figure 4. Sensible heat release from each path for a wooden structure with low insulation on a representative sunny day in (**a**) summer and (**b**) winter.

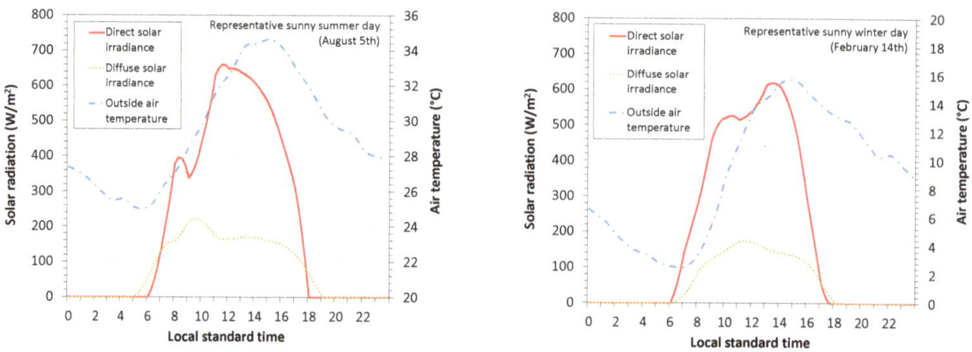

(a) (b)

Figure 5. Weather conditions on a representative sunny day in (**a**) summer and (**b**) winter.

The maximum rate of sensible heat release during the daytime was 180 W/m^2. The daily heat from the envelope surface was by far the largest contribution, accounting for about 87% of the total. The anthropogenic heat was very slight; the space conditioning accounted for 12% and the water heating accounted for only 1%. However, when the heat from the envelope surface decreased in the evening, the proportion of anthropogenic heat became much greater than it was in the daytime. Because on the representative hot summer days the indoor climate control behavior model showed the space conditioning being used for almost the entire day, the heat from the space conditioning system occurred even at midnight. The exhaust heat from the water heating rose temporarily around 9 p.m. because of hot water needed for bathing. Due to the atmospheric radiation cooling, the heat release from the envelope surface showed a negative value at night until dawn.

Figure 4b shows the sensible heat release from each path on a representative sunny winter day (14 February). Figure 5b shows the weather condition. Here, the heat from space conditioning is the sum of the heating load and the space conditioning consumption energy; it is equivalent to the amount of heat release to the inside through the heat pump outdoor unit, as shown in Figure 2.

The maximum sensible heat release rate during winter daytime was 150 W/m^2. The heat release from the envelope surface was relatively low throughout the day, compared to daytime release in summer, and it showed a negative value from evening to dawn. The heat absorption through the space conditioning increased, especially during the period when the heat load was great in the morning and at night. Since the space conditioning is not used during sleeping hours in the heating season, no heat absorption occurs at night. The heat release from the water heating showed a large value compared to the summer, as the hot water demand increased in winter.

3.2. *Variation of the Heat Release by Applying Countermeasures*

3.2.1. Envelope Surface

Figure 6 shows the effectiveness of countermeasures influencing the heat release from the envelope surface on a representative sunny summer day (August 5). Compared to the base condition, the heat release from the envelope surface of reinforced concrete structure decreased for high-albedo roof (HAR), roof greening (RG), and roof water showering (RWS), but increased for photovoltaic power generation (PV). The reduction rate for the whole day became larger, with roof water showering providing the largest reductions, then roof greening, then high-albedo roof, in declining order. When looking at daytime reduction alone, the largest reduction rate was from roof water showering, then high-albedo roof, and then roof greening. In contrast, at nighttime, the reduction rate was largest with roof greening, then roof water showering, then high-albedo roof. When comparing the wooden structure with a low insulation level to the reinforced concrete structure, there was no significant difference in results; the reduction rate for the whole day was largest with roof water showering, and then high-albedo roof.

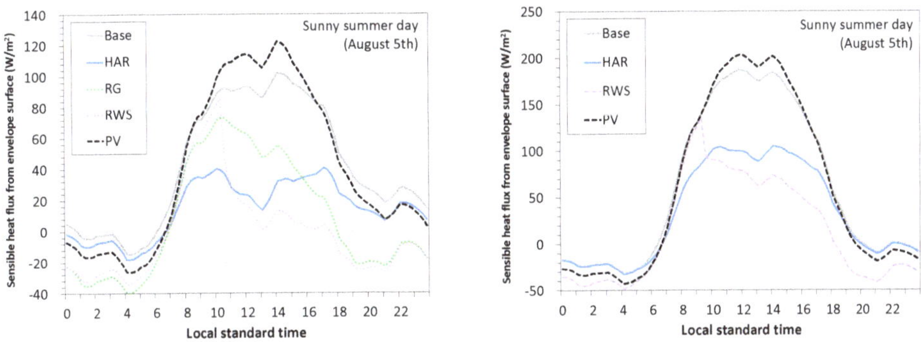

(a)

(b)

Figure 6. Time series of the sensible heat release from an envelope surface on a representative sunny summer day for (**a**) reinforced concrete structure and (**b**) wooden structure with low insulation level.

The roof water showering scenario that achieved the maximum effect had water showering starting at 10 a.m., when the roof surface temperature exceeded 40 °C, and the sensible heat release decreased rapidly. At 6 p.m., the heat release almost reached zero. Although the showering was stopped at 5 p.m., the effect continued throughout the night. As for the high-albedo roof that achieved a notable effect during the daytime, the heat release decreased considerably from the early morning toward evening, and it slightly continued at night due to the high thermal mass of the roofing material. For the roof greening, the effect was approximately in the middle between roof water showering and high-albedo roof in the daytime, and almost the same as that of roof water showering at night. Because of the large solar absorptance (1—solar reflectance—conversion efficiency) of the PV material, the heat release increased during the day and decreased at night.

Figure 7 shows the effectiveness of countermeasures influencing heat release from the envelope surface on a representative sunny winter day (14 February). Since the soil layer of withered roof greening has less thermal mass and higher thermal resistance than a normal concrete roof, the amount of heat release during the day increased compared to the base condition. The other countermeasures show almost the same change compared to the summer result, so detailed consideration is omitted.

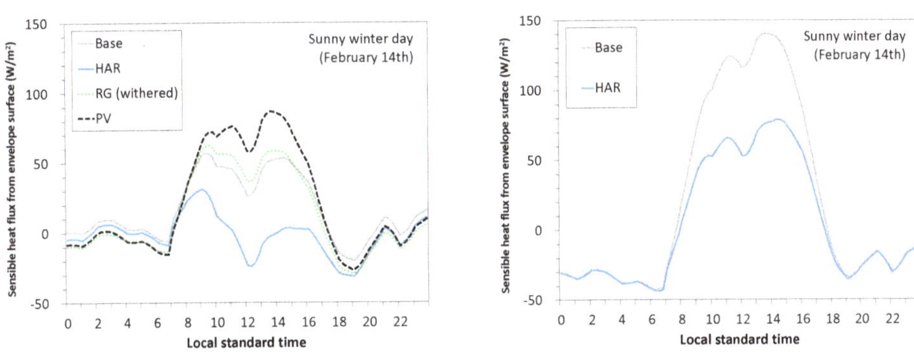

(a)

(b)

Figure 7. Time series of the sensible heat release from an envelope surface on a representative sunny winter day for (**a**) reinforced concrete structure and (**b**) wooden structure with low insulation level.

3.2.2. Space Conditioning System

Figure 8 shows the effectiveness of countermeasures on heat release from a space conditioning system, showing heat release with each countermeasure minus heat release without countermeasures, averaging each hourly value over sunny days in August. Compared with the base condition, the heat release from space conditioning decreased for all countermeasures used with the reinforced concrete structure. The reduction rate for the whole day became largest with roof greening (RG), then roof water showering (RWS), evaporative space cooling (ESC), and high-albedo roof (HAR), in declining order. The reduction rate for the whole day for the wooden structure with low insulation level was the largest with evaporative space cooling, then roof water showering, and then high-albedo roof.

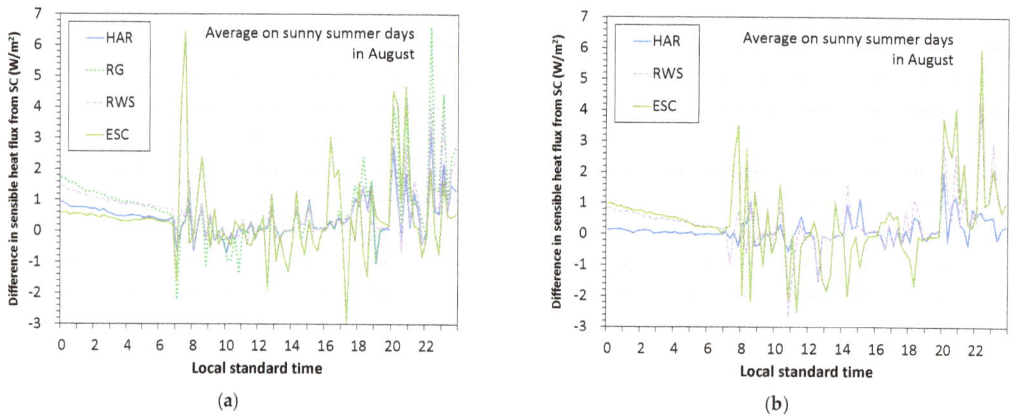

Figure 8. Time series of the difference in sensible heat release from the space conditioning (with countermeasure minus without countermeasures), on an average value of sunny days in August, for (**a**) reinforced concrete and (**b**) wooden structure with low insulation level.

The reduction rate of roof greening was the largest, due to the improved insulation provided by installing the greening and soil layer. The reduction rate reached up to only 11%, so there was not much effect, especially in the daytime. Most countermeasures (except evaporative space cooling) were targeted at the roof, but the occupants were on the first floor when the space cooling load was relatively large compared to the nighttime, so the countermeasures were not effective. Moreover, because the outdoor air temperature was very high during the summer in Osaka, space conditioning was required even if countermeasures were implemented, and the duration of time the space conditioning was used could not be reduced. For that reason, the countermeasure was less effective. Although it was slight, the effect of evaporative space cooling exceeded that of roof water showering in the wooden structure. The wooden structure has a low thermal mass, so room air temperature tended to drop at night, and the effect of reducing cooling time of evaporative space cooling was relatively large. Figure 8 only shows the heat reduction from space conditioning, but because the fine jet fog sprayed can convert the sensible heat into latent heat, it is expected that a large heat reduction effect could be obtained with evaporative space cooling. Concerning the total number of hours of space conditioning operation during August by time of day, roof greening was the most effective. The reduction rate for the whole day and night was 28%; for the nighttime, it was 44%.

Figure 9 shows the effectiveness of countermeasures influencing heat release from the space conditioning system (with countermeasure minus without countermeasures), averaging each hourly value over sunny days in February. In both the reinforced concrete and wooden structures, the amount of heat absorbed from the roof surface decreased as a result of the high-albedo roof, and the amount of absorbed heat through the space conditioning increased slightly. In any case, the impact, including the withered roof greening, was negligible.

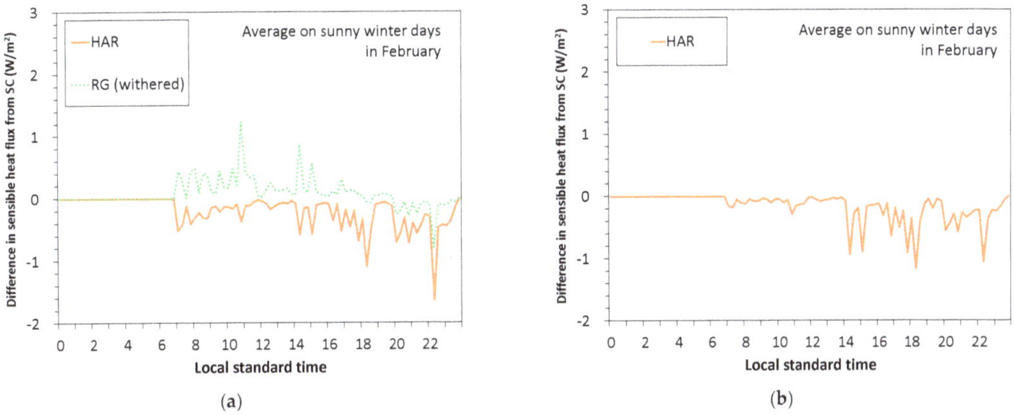

Figure 9. Time series of the difference in sensible heat release from the space conditioning in the case with or without utilizing countermeasures (with-without), on an average value of sunny days in February, for (**a**) reinforced concrete and (**b**) wooden structure with low insulation level.

3.2.3. Water Heater System

Figure 10ashows the result of the effectiveness of countermeasures on heat release from the water heating system (with countermeasure minus without countermeasures), averaging each hourly value over sunny days in August. Compared with the base condition, heat release from the water heating system decreased for heat pump water heater (HPH) and condensing water heater (CWH). The reduction rate for the whole day was larger for the heat pump water heater than for the condensing water heater. Since the heat pump absorbs and accumulates heat from the atmosphere, clearly a large reduction was seen around dawn, but a slight heat release from the hot water tank increased during the day. The condensing water heater always reduced heat loss by improving efficiency, but the effect was minimal.

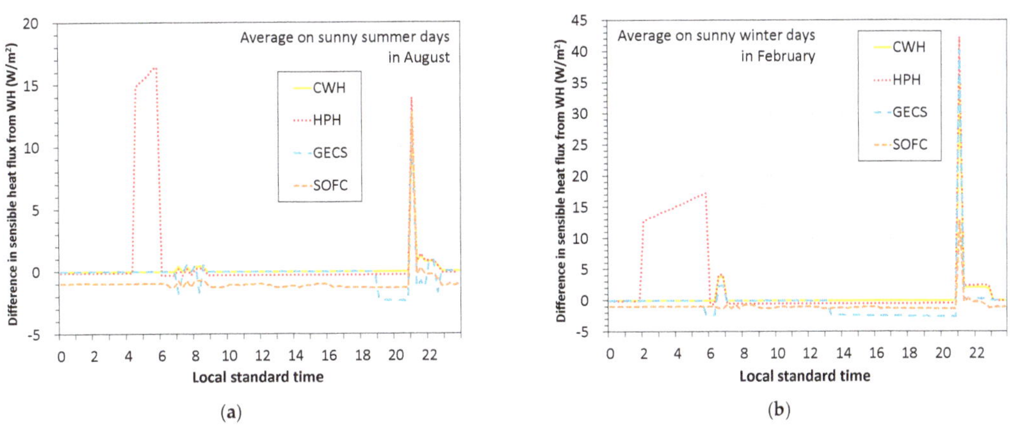

Figure 10. Time series of the difference in sensible heat release from water heating in the case with or without utilizing countermeasures (with-without), on an average value of sunny days, for (**a**) August and (**b**) February.

Heat was always released during the operation of the gas engine cogeneration system (GECS) and solid oxide fuel cell (SOFC) equipment, so the heat release from the system was increased compared to the base condition. The solid oxide fuel cell without start/stop always generated heat release, but the gas engine cogeneration system operated in accordance with the heat demand generated heat release, especially during times with great heat demand. However, the effects for both were slightly small; neither system had a significant impact on the total heat release.

Figure 10b shows the result of the effectiveness of countermeasures on heat release from the water heating system (with countermeasure minus without countermeasures), averaging each hourly value over sunny days in February. As for the heat pump water heater, since the hot water demand was larger during winter than in summer, it influenced heat release more during the winter season. The number of hours the hot water system operates became greater due to an increased hot water demand (the absorbed heat started at around 2 a.m.). As for the gas engine cogeneration system, the effect of prolonged operation time was also seen. The temperature of the water being used rose during winter season, so the amount of heat released through the ventilation during that time also increased overall.

3.3. *Performance Evaluation Concerning the Heat Release and Carbon Dioxide Emissions*

3.3.1. Cooling Season

Figure 11 shows the relationship between the sensible heat release reduction and the CO_2 emissions reduction by applying each countermeasure for the cooling season during daytime and nighttime. The plot color varies according to the evaluated countermeasures. This graph is helpful when considering UHIE and GW adaptation, as it offers the chance to examine both the amount of heat release and the amount of CO_2 emissions, using the same calculation model. The reduction of sensible heat release (vertical axis) sums contributions from the envelope, space conditioning, and water heating per building area, and that sum is time averaged over sunny days in August. For the CO_2 emissions (horizontal axis), the effects of UHIE countermeasures incorporate reduction of space conditioning energy only, while the effects of GW countermeasures incorporate energy reductions achieved by atmospheric heat absorption, power generation, waste heat utilization, and other factors. The value of the horizontal axis shows the reduction of total CO_2 emissions during cooling season (from July to September). For photovoltaic power generation, all of the generated electric power can be available, and it is evaluated on the premise that the same amount of system power supply can be reduced.

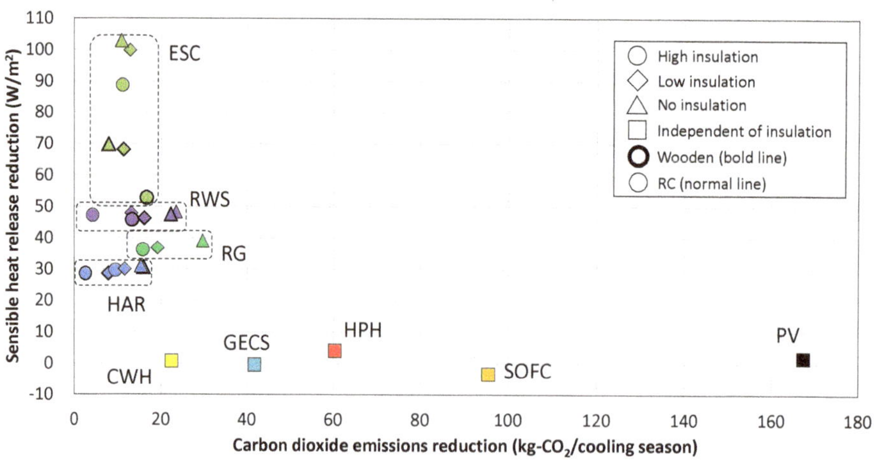

Figure 11. Diurnal (24-h) average reduction in sensible heat release on an average value of sunny days in August versus total summer (July–September) reduction in CO_2 emissions upon applying each countermeasure.

Based on the results, the plots were roughly classified into two technology groups: (1) effective for heat release reduction and (2) effective for CO_2 emissions reduction. No countermeasure served both ends. As mentioned above, evaporative space cooling (ESC) contributed most to the reduction of heat release, then roof water showering (RWS), roof greening (RG), and high-albedo roof (HAR). However, no GW countermeasure contributed considerably to heat release reduction. Although evaporative space cooling contributed the most to reduce the amount of time the space conditioning was used, the building structure and the insulation level also influenced the result significantly, because the amount of heat release reduction was dependent on the space conditioning time and load.

Among the GW countermeasures, photovoltaic power generation (PV) contributed most to reduce CO_2 emissions, followed by solid oxide fuel cell (SOFC), heat pump water heater (HPH), gas engine cogeneration system (GECS), and condensing water heater (CWH), in order of reductions. Among the UHIE countermeasures, roof greening with no insulation level reduced CO_2 emissions the most, although the reduction was almost the same compared with the gas engine cogeneration system and condensing water heater. Regarding the UHIE countermeasures, heat insulation and building structure somewhat influenced the CO_2 emissions reductions, but they had no considerable influence on reducing heat release. Meanwhile, although some GW countermeasures (e.g., heat pump water heater) reduced the heat release, the amount of change was very small, and it was less than that of high-albedo roof, which was the least effective of the UHIE countermeasures. Some GW countermeasures (e.g., solid oxide fuel cell) increased the heat release, but the amount of change was very small as well.

During the day, the overall trend is roughly the same as it was in Figure 11, but the effect of evaporative space cooling, in particular, was considerably smaller. Evaporative space cooling is used as the substitute equipment for cooling, but because the thermal environmental condition is severe during the daytime, space conditioning is required because it is unpleasant to use only evaporative space cooling, so the effect is reduced. High-albedo roof was a more effective countermeasure than roof greening. Although it had little influence on GW countermeasures, photovoltaic power generation increased the amount of heat release, and heat pump water heater had almost no effect. At night, the effect of the evaporative space cooling was much larger. Unlike in the daytime, this was because the space conditioning is used many fewer hours when evaporative space cooling is used. The high-albedo roof was less effective than the evaporative space cooling. Regarding the GW countermeasures, photovoltaic power generation changed from increasing heat release to decreasing heat release, and heat pump water heater had a considerable effect.

3.3.2. Heating Season

Figure 12 shows the relationship between the sensible heat release reduction and the CO_2 emissions reduction by applying each countermeasure for the heating season during daytime and nighttime. The reduction of sensible heat release (vertical axis) is averaged over sunny days in February. The CO_2 emissions (horizontal axis) show sum reductions over the heating season (from December to February). The meaning of the graph and its effectiveness are the same as those for the cooling season graph.

Based on the results, the plots were roughly classified into two groups: (1) effective for heat release reduction and (2) effective for CO_2 emissions reduction, just as they were for the cooling season. However, only the heat pump water heater countermeasure reduced both heat release reduction and CO_2 emissions. High-albedo roof contributed most to the reduction of heat release, followed by heat pump water heater, roof greening (withered), and condensing water heater. However, high-albedo roof increased CO_2 emissions due to an increase in heating load. Since roof water showering and evaporative space cooling stopped operation in the heating season, they had no influence on heat release or CO_2 emissions during heating season. Since the hot water demand was greater in winter, the water heating systems had greater influence in winter than in summer. Some GW countermeasures (e.g., photovoltaic power generation) increased the heat release. Since the overall trend is the same as it was in the cooling season, the result of separating during daytime and nighttime in winter season is omitted.

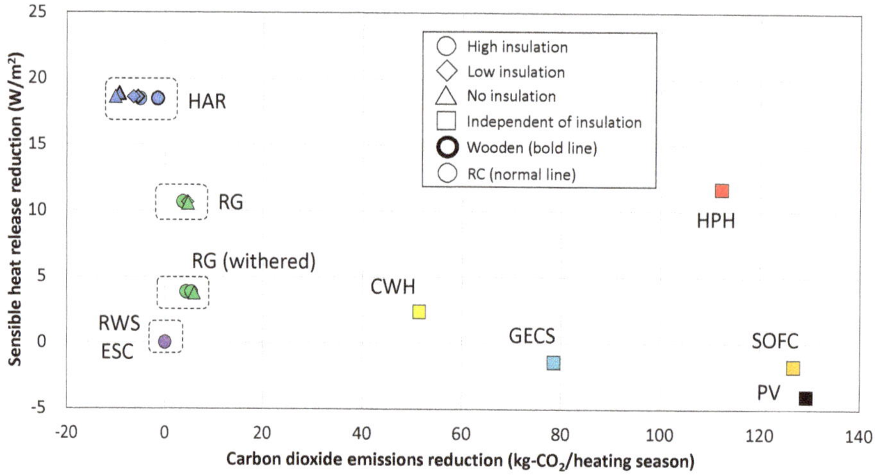

Figure 12. Diurnal (24-h) average reduction in sensible heat release on an average value of sunny days in February versus total winter (December–February) reduction in CO_2 emissions upon applying each countermeasure.

Figure 13 shows the annual reduction of CO_2 emissions from space conditioning by applying the UHIE countermeasure. Roof water showering and evaporative space cooling were not operated during the heating season, so only benefits for the cooling season are shown. Since roof greening has merits throughout the year, the sums of benefits for the cooling and heating seasons are shown. As for high-albedo roof, there is a demerit in the heating season, so the offsets of benefits for the cooling season and demerits for the heating season are shown. As a result, roof greening achieved the largest reduction effect of CO_2 emissions, followed by measures using water, such as roof water showering and evaporative space cooling. The effect of the high-albedo roof was positive but slight.

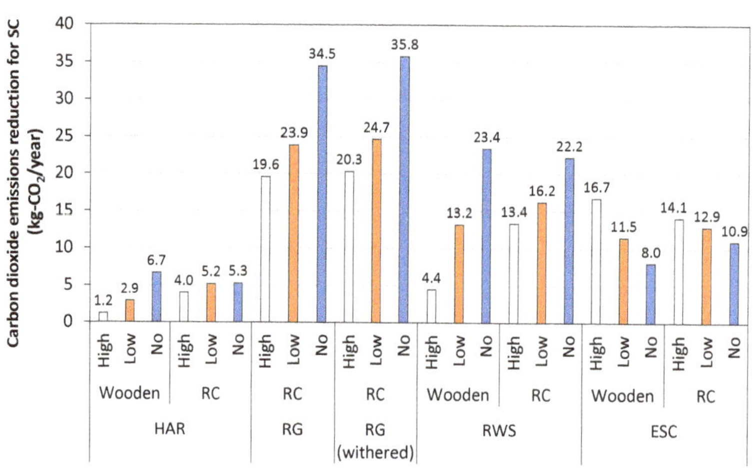

Figure 13. Annual reduction of CO_2 emissions from space conditioning by applying the UHIE countermeasures.

4. Discussion

Based on the results obtained up to Section 3.3, we propose some guidelines for designing a detached house that considers the benefits and impacts of both UHIE and GW. These are especially effective where summer temperature is high (like East or Southeast Asian countries).

For the cooling season, it is desirable to introduce measures to achieve both heat release and CO_2 emissions reductions, but this study did not identify any measures that were very effective in both areas at the same time. Therefore, it is necessary to apply countermeasures for each purpose individually. It was shown that countermeasures using an evaporative cooling effect, such as evaporative space cooling and roof water showering, are desirable to reduce heat release. Countermeasures such as photovoltaic power generation and solid oxide fuel cell are effective in reducing CO_2 emissions. The UHIE countermeasure that can reduce CO_2 emissions the most is roof greening with no insulation; in fact, this can reduce CO_2 emissions more than condensing water heater. Regarding UHIE countermeasures for the envelope surface, it was shown that the heat insulation level of the building greatly affects the CO_2 emissions reduction, but not the heat release reduction. A difference in the heat release reduction was seen between the insulation levels only in the case of evaporative space cooling. A larger heat release reduction effect was obtained with the low insulation level due to the long use time of evaporative space cooling.

For the heating season, only the heat pump water heater countermeasure had effects on both heat release reduction and CO_2 emissions, so the heat pump water heater is positioned as an effective countermeasure to reduce both heat release and CO_2 emissions. However, although heat pump water heaters are effective in reducing CO_2 emissions, it is counterproductive to reduce heat release in the winter season. Moreover, high-albedo roofs should not be adopted as much as measures such as roof greening and roof water showering in Japan because the heat release from high-albedo roofs is greater than the heat releases from roof greening and roof water showering during the heating season. However, from the results shown in Figure 13, since the winter penalty does not exceed the summer merit, it is not a countermeasure that should be avoided. In this regard, countermeasures that use water, such as evaporative space cooling and roof water showering, can stop during winter, so there is an advantage in that they do not affect the amount of heat release and CO_2 emissions in winter at all.

These results show that the water-using UHIE countermeasures (such as evaporative space cooling or roof water showering) can provide positive effects during the cooling season and no negative effects during the heating season. Although roof greening reduces the amount of heat release slightly in the heating season, it has the most potential to reduce CO_2 emissions due to changes in space conditioning. However, since their effects on reducing CO_2 emissions are not very significant, it is desirable to introduce GW countermeasures such as solid oxide fuel cells and photovoltaic power generation, which significantly effect CO_2 emissions reduction.

As described above, we have provided guidelines that take into consideration the benefits and impacts of both UHIE and GW. However, it is necessary to show the constraints on the results obtained in this study.

First of all, the examined area, Osaka, has a particularly high temperature in summer, but also a very large demand for heating in winter. Therefore, UHIE countermeasures chosen for Osaka must avoid negative effects (overcooling) in winter. However, winter penalties are less important in the other regions such as the West Coast of the United States, East Asia, and Australia, where the weather is warm throughout the year. It is necessary to clearly indicate that the guidelines presented in the current study are suitable for weather conditions like Japan where the four seasons are distinct. In addition, there are differences in lifestyles of the people around the world. For example, in Japan, people often take baths every day, so the demand for hot water supply is greater than in other countries. Moreover, in this study, we have simulated assuming a four-person household, but differences in family composition greatly affect the amount of energy consumption and time patterns.

For example, in households of young, single persons, who are often absent during the daytime, the effect of high-reflection material is expected to be particularly small because space conditioning would not be operated when a home is unoccupied.

Regarding the restrictions of the model constructed in this study, as described in Section 2, this model uses the empirical formula of Jurges [41]. This formula is a model customarily used in the field of building environmental engineering as the convective heat transfer coefficient from the indoor and outdoor walls to the surrounding atmosphere. This formula is related to heat transfer on a single horizontal plane, so this model is not sufficient to evaluate heat transfer from complex exterior surfaces of the building. In assessing the absolute amount of flux reduction and its impact on the ambient temperature in the future, it is desirable to make an evaluation considering the complexity of the airflow around the buildings [53]. In regard to this issue, this study assumed a space with no other buildings around. However, in general, there are various buildings in the surroundings, and the above-mentioned complicated airflow and radiant environment with surrounding buildings affect heat release from the building surface and energy consumption in the building. It should be noted that the simulation of this study has been carried out without assuming such conditions. In the future, it is necessary to extend to a calculation model that can assume urban blocks using a convective heat transfer coefficient that takes into account the complex airflow around the building.

Finally, regarding the area and period to be evaluated in this study, Sangiorgio et al. [54] showed that not only the land cover but also various parameters such as population density, spatial geometry, and daily weather conditions influence the heat island intensity. The current study evaluates the heat release and energy consumption in a single standard detached house on a representative sunny day, as well as the effects of various countermeasure technologies. However, since the effects are likely to vary with the above parameters, these guidelines could be made more versatile by simulating other regions with varying urban geometries, densities, and climates.

5. Conclusions

In this study, through an examination using the SCIENCE-Outdoor simulation model, we quantified the effect of urban heat island (UHIE) countermeasures and global warming (GW) countermeasures on heat release and CO_2 emissions, and we investigated various features of countermeasures related to buildings. This study's purpose was to evaluate various technologies for their potential to mitigate UHIE and GW and then to propose their proper implementation. The results of this research are described below.

(1) We constructed the SCIENCE-Outdoor model for evaluating UHIE countermeasures and GW countermeasures. This model can evaluate the thermal condition of building envelope surfaces, both inside and outside, and consists of the three submodels: (a) radiant, (b) inside thermal environment, and (c) outside heat release.

(2) The maximum heat release rate for a wooden detached house with a low insulation level, on a representative sunny summer day for the base condition (no countermeasures), was 180 W/m^2. The breakdown of the cumulative daily heat was almost all from the envelope surface, accounting for about 87% of the total. The anthropogenic heat was very slight: space conditioning accounted for 12% and the water heating system accounted for only 1%.

(3) Concerning the effectiveness of countermeasures influencing the heat from envelope surface, the reduction rate of heat release for day and nighttime was largest with roof water showering, then roof greening, and then high-albedo roof in the cooling season.

(4) Concerning the effectiveness of countermeasures influencing the heat from space conditioning, the reduction rate for day and nighttime was largest with roof greening, then roof water showering, then evaporative space cooling, and then high-albedo roof in the cooling season. In the heating season, the amount of absorbed heat through the space conditioning increased slightly under the high-albedo roof countermeasure.

(5) The effectiveness of countermeasures influencing the heat from the water heating system decreased for heat pump water heaters and condensing water heaters but increased for gas engine cogeneration systems and solid oxide fuel cells in the cooling season. During the heating season, hot water demand rose, so water heating had a greater influence than in the cooling season.
(6) As the result of evaluating the relationship between heat release reduction and CO_2 emissions reduction, and by applying each countermeasure for the cooling season, the plots were roughly classified into two technology groups: those effective for heat release reduction and those effective for CO_2 emissions reduction.
(7) As the result of evaluating the same relationship used for the heating season, the plots were roughly classified into the same two groups as those for the cooling season, but only the heat pump water heater countermeasure was found to effect heat release reduction and CO_2 emissions reduction.
(8) The results showed that it is best to introduce water-using countermeasures (evaporative space cooling and roof water showering), which can provide positive effects during summer but no negative effects during winter, to plan for UHIE and GW considerations. However, since water-based countermeasures do not significantly reduce CO_2 emissions, it is desirable to introduce GW countermeasures such as solid oxide fuel cell and photovoltaic power generation that substantially decrease CO_2 emissions.

This study modeled a detached house in Osaka, Japan. Since results will vary with building construction, building operation, building equipment, and climate, it is difficult to make a final judgment on the superiority or inferiority of countermeasures to be adopted in the future based on this study's research results. However, further studies should recognize that it is important to evaluate both heat release and CO_2 emissions with the same model. The significance of this study is that it constructed the simulation model and showed the influence of countermeasures upon both heat release and CO_2 emissions. The results can be used when considering the implementation of urban and regional building planning that is considering UHIE and GW adaptation by examining the amount of both heat release and CO_2 emissions. Potential tasks for future study include establishing a database that supports selection of countermeasures, targeting further countermeasures, evaluating the impact of local climate characteristics, and evaluating the impacts of building use or size.

Author Contributions: Conceptualization, Y.S. and D.N.; methodology, validation, investigation, and data curation, D.N.; writing—original draft preparation, D.N.; writing—review and editing, R.L.; supervision, Y.S. and R.L.; funding acquisition, D.N. and R.L. All authors have read and agreed to the published version of the manuscript.

Funding: This work was supported by JSPS KAKENHI Grant Number JP18KK0123. It was also supported by the Assistant Secretary for Energy Efficiency and Renewable Energy, Building Technologies Office of the U.S. Department of Energy under Contract No. DE-AC02-05CH11231.

Institutional Review Board Statement: Not applicable.

Informed Consent Statement: Not applicable.

Data Availability Statement: Data sharing not applicable.

Conflicts of Interest: The authors declare no conflict of interest.

References

1. Japan Meteorological Agency. *Climate Change Monitoring Report 2016*; Chapter 2 Climate Change; Japan Meteorological Agency: Tokyo, Japan, 2017.
2. Japan Meteorological Agency. *Global Warming Forecast Information*; Global Warming Forecast Information; Japan Meteorological Agency: Tokyo, Japan, 2017; Volume 9.
3. Narumi, D.; Niurao, Y.; Shimoda, Y.; Mizuno, M. Effects of increasing temperature on the regional energy consumption in Osaka Pref. *J. Environ. Eng. Trans. AIJ* **2007**, *72*, 71–78. [CrossRef]

4. Santamouris, M.; Papanikolaou, N.; Livada, I.; Koronakis, I.; Georgakis, C.; Argiriou, A.; Assimakopoulos, D.N. On the impact of urban climate on the energy consuption of building. *Sol. Energy* **2001**, *70*, 201–216. [CrossRef]
5. Fujii, H.; Fukuda, S.; Narumi, D.; Ihara, T.; Watanabe, Y. Fatigue and sleep under large summer temperature differences. *Environ. Res.* **2015**, *138*, 17–21. [CrossRef]
6. Narumi, D.; Ihara, T.; Fukuda, S. Study on the impact for sleep disturbance due to changing urban outdoor temperature. *AIJ J. Technol. Des.* **2016**, *52*, 1045–1048. [CrossRef]
7. Narumi, D.; Kondo, A.; Shimoda, Y. The effect of the increase in urban temperature on the concentration of photochemical oxidants. *Atmos. Environ.* **2009**, *43*, 2348–2359. [CrossRef]
8. Sarrat, C.; Lemonsu, A.; Masson, V.; Guedalia, D. Impact of urban heat island on regional atmospheric pollution. *Atmos. Environ.* **2006**, *40*, 1743–1758. [CrossRef]
9. Kiyokawa, Y.; Narumi, D. Regional and secular characteristics on temperature sensitivity of power supply. *J. Environ. Eng. Trans. AIJ* **2018**, *83*, 1015–1024. [CrossRef]
10. Narumi, D.; Ihara, T.; Fukuda, S.; Shimoda, Y. Comprehensive evaluation of the influence of outdoor temperature change on human health around the urban area. *J. Environ. Eng. Trans. AIJ* **2019**, *84*, 205–214. [CrossRef]
11. Shimoda, Y.; Narumi, D.; Mizuno, M. Environmental impact of urban heat island phenomena—Cause-effect chain and evaluation in Osaka City. *J. Life Cycle Assess. Jpn.* **2005**, *1*, 144–148. [CrossRef]
12. Franco, G.; Sanstad, A.H. Climate change and electricity demand in California. *Clim. Chang.* **2008**, *87*, 139–151. [CrossRef]
13. Hirano, Y.; Fujita, T. Evaluation of the impact of the urban heat island on residential and commercial energy consumption in Tokyo. *Energy* **2012**, *37*, 371–383. [CrossRef]
14. Sailor, D.J. Relating residential and commercial sector electricity loads to climate—Evaluating state level sensitivities and vulnerabilities. *Energy* **2001**, *26*, 645–657. [CrossRef]
15. Santamouris, M.; Cartalis, C.; Synnefa, A.; Kolokotsa, D. On the impact of urban heat island and global warming on the power demand and electricity consumption of buildings—A review. *Energy Build.* **2015**, *98*, 119–124. [CrossRef]
16. Ashie, Y.; Tokairin, T.; Kono, T. High resolution numerical simulation on the urban heat island in a ten-kilometer square area of central Tokyo by using the earth simulator. *J. Environ. Eng. Trans. AIJ* **2007**, *72*, 67–74. [CrossRef]
17. Ichinose, T.; Shimodozono, K.; Hanaki, K. Impact of anthropogenic heat on urban climate in Tokyo. *Atmos. Environ.* **1999**, *33*, 3897–3909. [CrossRef]
18. Kondo, H.; Tokairin, T.; Kikegawa, Y. Calculation of wind in a Tokyo urban area with a mesoscale model including a multi-layer urban canopy model. *J. Wind Eng. Ind. Aerodyn.* **2008**, *96*, 1655–1666. [CrossRef]
19. Narumi, D.; Kondo, A.; Shimoda, Y. Effects of anthropogenic heat release upon the urban climate in a Japanese megacity. *Environ. Res.* **2009**, *109*, 421–431. [CrossRef]
20. Chen, F.; Dudhia, J. Coupling an advanced land surface–hydrology model with the Penn State–NCAR MM5 modeling system. Part I: Model implementation and sensitivity. *Mon. Weather Rev.* **2001**, *129*, 569–585. [CrossRef]
21. Skarmaeock, W.C.; Klemp, J.B.; Dudhia, J.; Gill, D.O.; Barker, D.M.; Wang, W.; Powers, J.G. *A Description of the Advanced Research WRF Version 2*; NCAR Techniacal Note TN-468+STR; Defense Technical Information Center: Fort Belvoir, VA, USA, 2005; p. 88.
22. Akbari, H.; Konopacki, S. Energy effects of heat-island reduction strategies in Toronto, Canada. *Energy* **2004**, *29*, 191–210. [CrossRef]
23. Ohashi, Y.; Ihara, T.; Kikegawa, Y.; Sugiyama, N. Numerical simulations of influence of heat island countermeasures on outdoor human heat stress in the 23 wards of Tokyo, Japan. *Energy Build.* **2016**, *114*, 104–111. [CrossRef]
24. Sailor, D.J.; Dietsch, N. The urban heat island Mitigation Impact Screening Tool (MIST). *Environ. Model. Softw.* **2007**, *22*, 1529–1541. [CrossRef]
25. Santamouris, M. Cooling the cities—A review of reflective and green roof mitigation technologies to fight heat island and improve comfort in urban environments. *Sol. Energy* **2014**, *103*, 682–703. [CrossRef]
26. Shimoda, Y.; Yamaguchi, Y.; Okamura, T.; Taniguchi, A.; Yamaguchi, Y. Prediction of greenhouse gas reduction potential in Japanese residential sector by residential energy end-use model. *Appl. Energy* **2010**, *87*, 1944–1952. [CrossRef]
27. Iwata, T.; Kuwasawa, Y.; Murakami, S.; Ikaga, T. Study on Scenarios towards Low Carbon Society in Residential Sector. *J. Environ. Eng. Trans. AIJ* **2011**, *76*, 839–846. [CrossRef]
28. Wakiyama, T.; Kuromochi, T. Scenario analysis of energy saving and CO_2 emissions reduction potentials to ratchet up Japanese mitigation target in 2030 in the residential sector. *Energy Policy* **2017**, *103*, 1–15. [CrossRef]
29. Hirano, Y.; Ihara, T.; Gomi, K.; Fujita, T. Simulation-Based Evaluation of the Effect of Green Roofs in Office Building Districts on Mitigating the Urban Heat Island Effect and Reducing CO_2 Emissions. *Sustainability* **2019**, *11*, 2055. [CrossRef]
30. Yamaguchi, K.; Yamada, K.; Masumura, H.; Endo, Y.; Ihara, T. Evaluation for heat island mitigation potential of air source heat pump water heaters. In Proceedings of the Renewable Energy 2010 Conference and Exhibition, Berlin, Germany, 22–24 November 2010.
31. Genchi, Y.; Ishisaki, M.; Ohashi, Y.; Kikegawa, Y.; Takahashi, H.; Inaba, A. Impacts of large scale photovoltaic panel installation on the heat island effect in Tokyo. In Proceedings of the 5th International Conference on Urban Climatology, Lodz, Poland, 1–5 September 2003; pp. 1–4.

32. Ihara, T.; Kikegawa, Y.; Asahi, K.; Genchi, Y.; Kondo, H. Changes in year-round air temperature and annual energy consumption in office building areas by urban heat-island countermeasures and energy-saving measures. *Appl. Energy* **2008**, *85*, 12–25. [CrossRef]
33. Onishi, J.; Takeya, N.; Mizuno, M. Study on numerical prediction methods for indoor air-flow and thermal environments Part 1 calculation procedures used in the simulation code "SCIENCE". *Trans. Soc. Heat. Air-Cond. Sanit. Eng. Jpn.* **1995**, *20*, 23–34. [CrossRef]
34. Habara, H.; Narumi, D.; Kobayashi, S.; Shimoda, Y.; Mizuno, M. Development of a method to estimate indoor thermal condition and air conditioning energy consumption in residential house using natural ventilation. *J. Environ. Eng. Trans. AIJ* **2004**, *69*, 107–114. [CrossRef]
35. Narumi, D.; Taketa, A.; Shimoda, Y. Development of a method to estimate indoor climate and air conditioning energy consumption in residential house considering the influence of cross ventilation. In Proceedings of the 29th Air Infiltration and Ventilation Centre Conference, Kyoto, Japan, 14–16 October 2008.
36. Gebhart, B. A new method for calculating radiant exchanges. *ASHRAE Trans.* **1959**, *65*, 321–332.
37. Fujii, T.; Shimoda, Y.; Morikawa, T.; Mizuno, M. Development of city scale residential energy end-use model including heat load calculation: Development and application of city scale residential energy end-use model by considering with various household categories part 1. *J. Environ. Eng. Trans. AIJ* **2005**, *70*, 51–58. [CrossRef]
38. Shimoda, Y.; Okamura, T.; Yamaguchi, Y.; Yamaguchi, Y.; Taniguchi, A.; Morikawa, T. City-level energy and CO2 reduction effect by introducing new residential water heaters. *Energy* **2010**, *35*, 4880–4891. [CrossRef]
39. Habara, H.; Narumi, D.; Shimoda, Y.; Mizuno, M. A study on determinants of air conditioning on/off control in dwellings based on survey. *J. Environ. Eng. Trans. AIJ* **2005**, *70*, 83–90. [CrossRef]
40. Shimoda, Y.; Fujii, T.; Morikawa, T.; Mizuno, M. Residential end-use energy simulation at city scale. *Build. Environ.* **2004**, *39*, 956–967. [CrossRef]
41. Watanabe, K. *Architectural Planning Principles Part 2*; Maruzen: Tokyo, Japan, 1964.
42. Narumi, D.; Shimoda, Y.; Habara, K.; Mizuno, M.; Kondo, A. Development of an environmental thermal load evaluation system for the purpose of mitigating urban thermal environment: Part 1: Effects of differences in environmental thermal load discharge conditions on temperature change near the ground surface. *Trans. Soc. Heat. Air-Cond. Sanit. Eng. Jpn.* **2006**, *31*, 71–78. [CrossRef]
43. Ministry of the Environment & Ministry of Economy, Trade and Industry. *Emission Factor for Each Electric Utility (for Calculating Total GreenHouse Gas Emissions in the Government and Local Government Action Plan)—FY2016 Results*; EPA: Washington, DC, USA, 2018.
44. Ministry of the Environment. *Guidelines for Calculating Total Greenhouse Gas Emissions Version 1*; EPA: Washington, DC, USA, 2017.
45. Udagawa, M. Proposal of a standard calculation condition for a residential house. In Proceedings of the 15th Heat Transfer Symposium of AIJ, Nagoya, Japan, 28–29 November 1985; pp. 23–33.
46. Architectural Institute of Japan Expanded AMeDAS Weather Data 1981–2000. Available online: https://www.metds.co.jp/product/ea/ (accessed on 14 March 2015).
47. Ishida, K. Program for scheduling of indoor heat generation rate due to livelihood. In Proceedings of the SHASE Symposium, Japan, 23 October 1996.
48. Global Cool Cities Alliance and R20 Regions of Climate Action. *Practical Guide to Cool Roofs and Cool Pavements;* CC Alliance: Washington, DC, USA, 2012.
49. Castleton, H.F.; Stovin, V.; Beck, S.B.M.; Davison, J.B. Green roofs; Building energy savings and the potential for retrofit. *Energy Build.* **2010**, *42*, 1582–1591. [CrossRef]
50. Matsushita, I.; Hatano, S.; Narumi, D. Evaluation on the ability of roofing unit covered with greens in mitigating outdoor thermal environment and reducing energy consumption for air-conditioning system. *Pap. Environ. Inf. Sci.* **2005**, *19*, 117–122. [CrossRef]
51. Yamaguchi, T.; Yokoyama, H.; Ishii, K. Mitigating the urban heat island effect by light and thin rooftop greening. *J. Jpn. Inst. Landsc. Archit.* **2005**, 509–512. [CrossRef]
52. Narumi, D.; Shigematsu, K.; Shimoda, Y. Effect of the evaporative cooling techniques by spraying mist water on reducing urban heat flux and saving energy in an apartment house. *J. Heat. Isl. Inst. Int.* **2012**, *7*, 175–181.
53. Hagishima, A.; Tanimoto, J.; Narita, K. Review of the former research on the convective heat transfer coefficient of urban surfaces. *J. Jpn. Soc. Hydol. Water Resour.* **2004**, *17*, 536–554. [CrossRef]
54. Sangiorgio, V.; Fiorito, F.; Santamouris, M. Development of a holistic urban heat island evaluation methodology. *Sci. Rep.* **2020**, *10*. [CrossRef]

atmosphere

Article

Temperature of Paved Streets in Urban Mockups and Its Implication of Reflective Cool Pavements

Yi Zhang [1], Peiyuan Wei [2], Lei Wang [1] and Yinghong Qin [1,2,*]

[1] College of Civil Engineering and Architecture, Guilin University of Technology, Guilin 541004, China; yzhang@glut.edu.cn (Y.Z.); lwang@glut.edu.cn (L.W.)
[2] College of Civil Engineering and Architecture, Guangxi University, 100 University Road, Nanning 530004, China; peiyuanwei@st.gxu.edu.cn
* Correspondence: yqin1@mtu.edu; Tel.: +86-771-323-2464

Abstract: In summer, urban heat islands increase building cooling demands, aggravate air pollution, and cause heat-related illnesses. As a mitigation strategy, reflective cool pavements have been deemed an effective measure to decrease the temperature in urban areas. However, the reflection of paved streets in an urban area will be different from that in an open area. It remains unknown which fraction of paved streets needs to be cooled upmost, and if increasing the albedo of paved streets can effectively reduce their temperature. This study measured the skin temperature of two urban mockups, of which one contained white streets and the other, gray streets. The streets were orientated at different strikes. It was found that in summer the East-West street was hotter than both the cross street and the South-North street. At nighttime, the heat released from building blocks kept the paved street about 0.2 °C hotter than paved areas in open spaces. It was also found that street orientation controlled the skin temperature of an urban street while the sky view factor (or building height and street width) acted in a secondary role only. Increasing the albedo of the paved street in an urban canyon effectively reduced the skin temperature of the street. Reflective pavements should be built preferentially on East-West streets and the cross streets.

Keywords: urban street; temperature; thermal comfort; sky view factor; cool pavements; street orientation

Citation: Zhang, Y.; Wei, P.; Wang, L.; Qin, Y. Temperature of Paved Streets in Urban Mockups and Its Implication of Reflective Cool Pavements. *Atmosphere* **2021**, *12*, 560. https://doi.org/10.3390/atmos12050560

Academic Editors: Hideki Takebayashi and Jihui Yuan

Received: 31 March 2021
Accepted: 21 April 2021
Published: 26 April 2021

1. Introduction

Urbanization alters the thermal environment from its natural conditions by replacing natural evaporable soils and green coverings with paved streets and building blocks. Both the buildings and the streets create a space that is similar to a canyon, which is termed an urban canyon. Solar irradiance to an urban canyon is entirely different from solar irradiance on natural ground in an open space. Photons reflected from paved streets can be captured by the adjacent buildings [1–4]. Heat absorbed by the urban street cannot discharge as it does in rural area because the radiant heat re-radiates between buildings and because the sensible heat is retained in the canyon due to the reduction of air flow [5]. As a consequence, cities experience elevated air temperature, a phenomenon that is termed an urban heat island. The urban heat island effect directly reduces the thermal comfort of the cities' residences [6–10], reduces the urban environmental quality [11–13], increases the urban energy usage [14,15], and causes other heat-related health problems [16–18].

As paved surfaces make up 20–40% of urban texture, the implementation of cool pavements as a strategy for mitigating the urban heat island effect has recently gained momentum [19–22]. Reflective pavement is a type of "cool pavement" that can reflect more sunlight than convectional pavements. Due to this feature, reflective pavements are expected to be built in places that have long isolation time, and that are exposed to strong solar irradiance. As the solar irradiance to an urban street is different from that to a paved surface in open area, whether a reflective pavement in an urban canyon can effectively reduce the skin temperature remains unknown. By studying the shady area of an urban space, Takebayashi et al. [23] found that reflective cool pavements should be considered on the north sides of East-West streets and the centers of North–South streets. It was also found that the thermal comfort of an urban canyon is jointly influenced by the sky view factor, the ratio of building height and street width, and the street orientation [24–27]. While the skin temperature of an urban area has been widely modelled, the skin temperature of an urban street is seldom measured, which is important to precisely decide which fraction of urban texture needs reflective pavements most.

This study measures the skin temperature of two square urban mockups to understand the temperature of paved streets. Gray concrete cubes were placed on a flat paved surface to represent the building blocks in an urban area. The paved surface under the cubic blocks was set as a gray substrate and a white substrate, respectively, for representing a gray paved street and a white one. Adjacent to the square mockups, a gray paved square and a white paved one were prepared to compare the paved skin temperature in a paved street to a paved surface in open area. The skin temperature of the two mockups and the two open squares were measured simultaneously on typical summer days to conclude whether reflective pavements in urban area can effectively decrease the skin temperature of paved streets.

2. Experiments

Typically, an urban canyon consists of building blocks and urban streets. To make the street high-reflective, we painted a 2.2 m $\times$ 2.2 m square on the flat roof white. After the paint hardened, concrete cubes with an edge length of 0.15 m were arranged on the white square such that the space between two adjacent blocks was set as 0.15 m. In this setup, the cubes represented the building blocks and the area uncovered by the cubes stood for high-reflective paved streets. Considering that urban buildings are usually gray, we painted the hardened concrete cubes gray before placing them on the white square. At 1.5 m from the white square, a new square area was painted gray with the same color as the paint on the concrete cubes. Upon the gray square, the same type of concrete cubes were placed in order to mimic the urban area with a gray street. The morphology of the two urban mockups was exactly the same. Parallel with the urban mockup with the white paved street, another white square with the same length was painted to mimic a high-reflective pavement in an open area. Similarly, parallel with the gray urban mockup, a new gray open square with the same length was prepared. The geometry of the four urban squares can be seen in Figure 1.

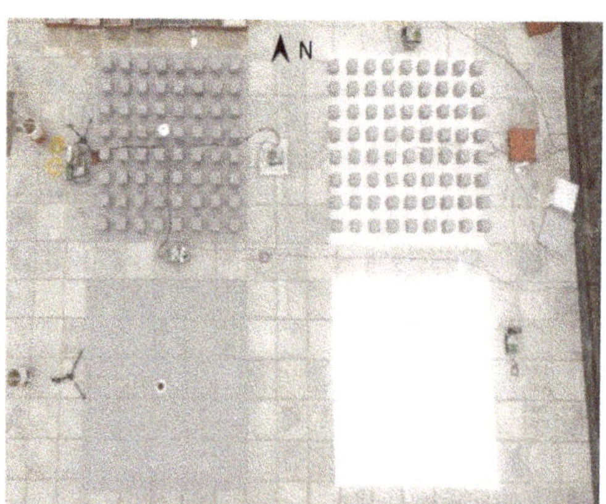

Figure 1. Two urban mockups and two open squares with the same size side-by-side for determining the temperature difference between pavements in urban areas and in open areas.

The solar reflectance spectrum of the white paint and of the gray paint on concrete surfaces was tested using a Lambda 900 with an integral sphere. Each paint was sprayed onto six hardened concrete slices. Weighing the solar reflectance with the spectrum of air-mass on global horizontal solar irradiance, it was found that the albedo of the white paint was 0.55 + 0.03 and of the gray paint was 0.30 + 0.02, see Appendix A. To get the temperature of the paved street in an urban area, thermocouples were anchored to the paved street on the urban mockup to measure the local temperature. Considering the thermal symmetry of the mockup, temperatures of the East-West street, South-North street and cross street were measured. In addition, the skin temperature at the centers of the white open square and of the gray open square were measured simultaneously. To get the true temperature, the thermocouple was mounted on a copper plate, which was then anchored to the designed place. The plates, thermocouple tips, and lines were then painted unicolor to minimize the thermal disturbance caused by the measurement. Thermocouples were deployed as indicated in the Figure 2.

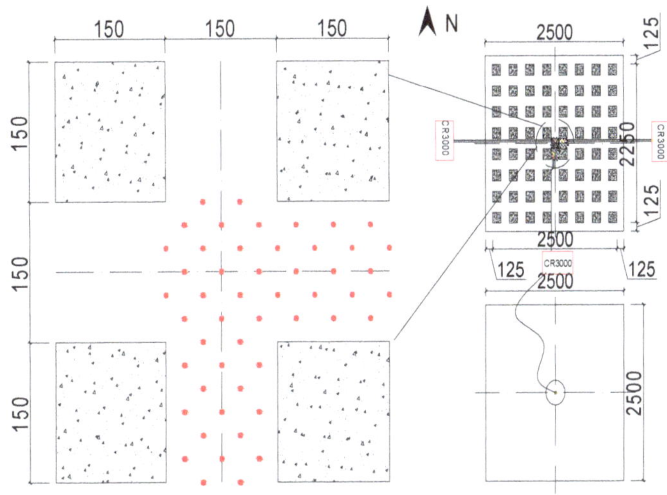

Figure 2. Schematic show of thermocouples that were installed to log the temperature of places in the urban mockup with gray streets. While the thermocouples did not cover the entire street in the representative area, the temperature of the entire street can be extrapolated due to the thermal symmetry of the streets.

The experiment was conducted on the rooftop of a building in Guangxi University, China (latitude, 22.84 and longitude 108.29). The temperature data were logged by three Campbell CR3000s simultaneously at an interval of 60 s. The experiment began from 19 June 2018 to 24 June 2019, a period that was partially sunny.

3. Results

3.1. Skin Temperatures of an Urban Street

The skin temperatures at different places in the gray streets were different by less than 2 °C at solar noon (Figure 3). As indicated in Figure 3, the temperature contour was different place to place, but a closer look found that the highest temperature was approximately 49 °C, while the lowest temperature was close to 47 °C. The maximum difference was about 2 °C. As a result, there are some small difference at temperature contours across the street, possibly because of the measurement error of the thermocouple sensors. The small difference in the skin temperature is because at solar noon, the sun is right above the urban street so the solar irradiance on the street is similar. While the temperature difference is small, one can find that the East-West street is hotter than both the cross street and the South-North street.

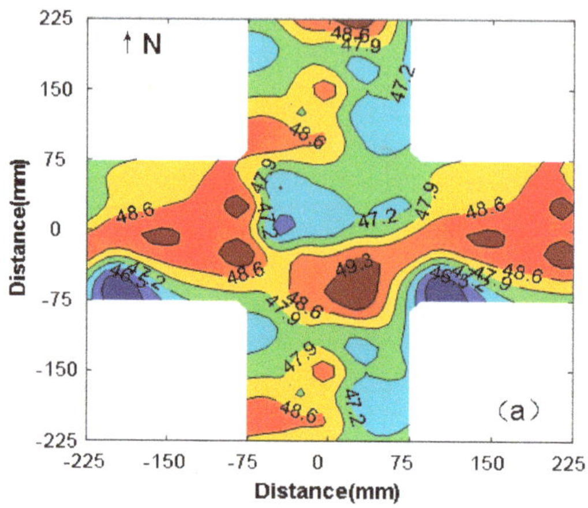

Figure 3. Skin temperature of the typical gray street of the mockup at the solar noon on 21 June.

The skin temperature of the gray street told a different story at 15:00 (Figure 4). At this time, the skin temperature of the urban street peaked because the surface temperature of an un-evaporable surface usually peaks three hours behind the solar noon [28]. At 15:00, the sun stays at the West; as a result, the West side of the urban street is shaded by the blocks and thus is the coolest place across the street. At this time, the cross street and the East-West street is still insolated by the sun and thus is still hot. The difference of the skin temperature in this study was magnified, with a maximum temperature approaching 49 °C and a minimum close to 43 °C. This different temperature and temperature contour infers that during daytime, the most effective way to reduce the skin temperature of an urban street is to increase the shady area.

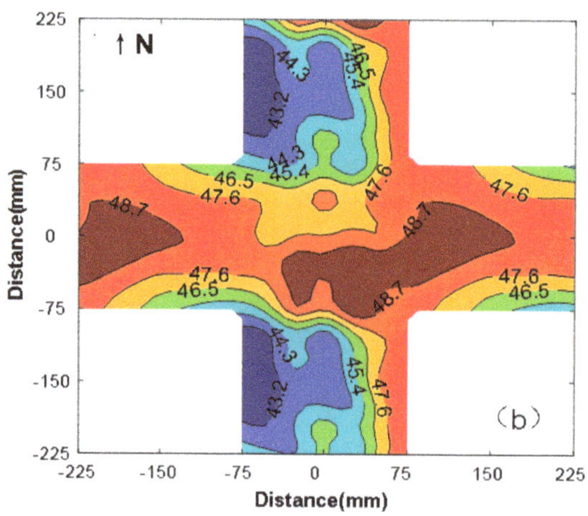

Figure 4. Skin temperature of the typical gray street of the mockup at 15:00 on 21 June.

The skin temperature of the gray street at 24:00 was different from that at other times (Figure 5). The magnitude of the difference of the skin temperature at different places decreased to a range of 0–1.0 °C. At this time, the cross street was the coolest place because the sky view factor of the cross street was greatest compared to that of both the East-West street and South-North street. The influence of the sky view factor on the skin temperature of the paved street could be seen from the skin temperature of the place that was closest to the building wall, which was 0.1–0.3 °C hotter than the skin temperature at the middle of the street. Similarly, the East-West street was about 0.1–0.2 °C cooler than that the South-North one. This difference was due to a marginal measurement error. The difference may also be because turbulence convection at the measurement time was different. This experiment reveals that the sky view factor has a limited effect on the skin temperature of paved streets.

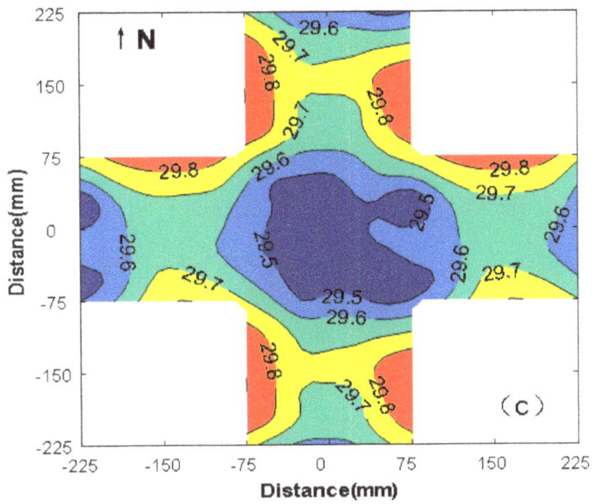

Figure 5. Skin temperature of the typical gray street of the mockup at 24:00 on 21 June.

The mean daily skin temperature across the street on 21 June is shown in Figure 6. It was found that the South-North street was the coolest place on the urban street. This is reasonable, because some parts of the South-North street were shaded, except for at the solar noon. This further substantiates that increasing the shady area is the most effective strategy for reducing the temperature of urban streets. The hottest place was the East-West street. The reason for this was that the East-West street received greater solar irradiance than the South-North street, but had a lower sky view factor than the cross street. The hottest place on the East-West street was the middle of the street for the reason that the places closer to the building blocks were shaded for some time due to the variation of the solar position during the course of the day. The cross street had the greatest sky view factor to dissipate heat at nighttime, but it received the greatest solar irradiance at daytime; the mean daily temperature of the cross street was thus somewhere between that of the South-North and East-West street.

It is a surprise that the East side of a South-North street was cooler than the West side. This thermal asymmetry may be caused by the different solar irradiance during the measured day, at which the solar irradiance at the afternoon is smaller than that at the morning. To better understand the cause for this thermal asymmetry, the mean skin temperature of the urban street during four continual measured days are plotted in Figure 7. It was found that the four-day mean skin temperature of the street, in pattern and in distribution, was highly similar to the mean skin temperature of the street on 21 June. This similar temperature contour further confirms the phenomenon that the East site of the South-North street was cooler than the West site. The difference was 0.3–0.6 °C, but the reason behind this difference needs further study.

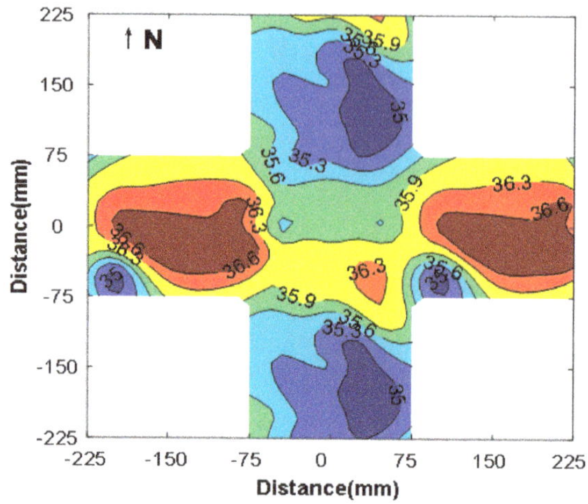

Figure 6. Mean daily skin temperature of typical gray streets of the urban mockup on 21 June 2019.

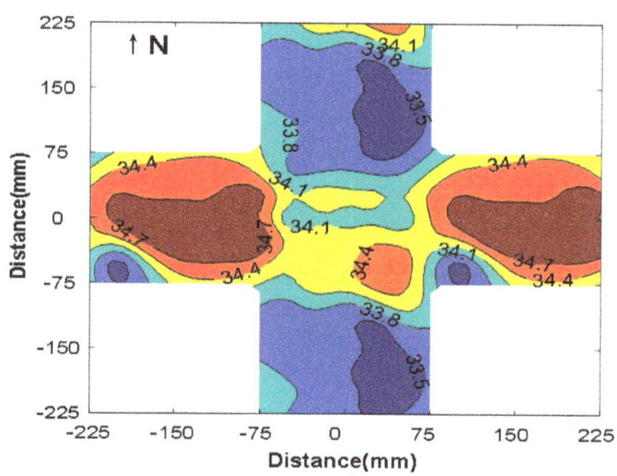

Figure 7. Four-day mean skin temperature of the typical paved street in the urban street mockup from 19–23 June 2019.

3.2. Reducing Skin Temperature by Increasing the Street Albedo

It is well-known that increasing the albedo of a paved area in an open space reduces the skin temperature effectively. The temperature reduction is compromised somewhat when the paved area in an urban area is made high-reflective. This study measures temperatures at the centers of a gray urban mockup, of a white urban mockup, of a gray open square, and of a white open square simultaneously. As shown in Figure 8, the center of the gray urban mockup was 5–10 °C cooler than that of the gray open square. This difference is reasonable, because compared to the gray streets at the center of an urban area, the center of the gray open square receives more sunlight and is exposed to longer isolation. It is interesting that at daytime, the temperature at the center of the white urban mockup was almost equal to the temperature at the center of the white open square. This equality of temperature means that if the urban street was made high-reflective, the solar trapping effect of an urban area would play a secondary role in the skin temperature variation. Another interesting result in Figure 8 is that at nighttime, the paved areas in the open squares were always about 1–2 °C cooler than those in the urban mockup. One reason for this difference is that concrete cubes reduce the sky view of the paved areas in the urban canyon. Another reason is that in the urban mockup, the heat emitted from the concrete cubes was absorbed by the paved area, making the paved area in the urban canyon hotter.

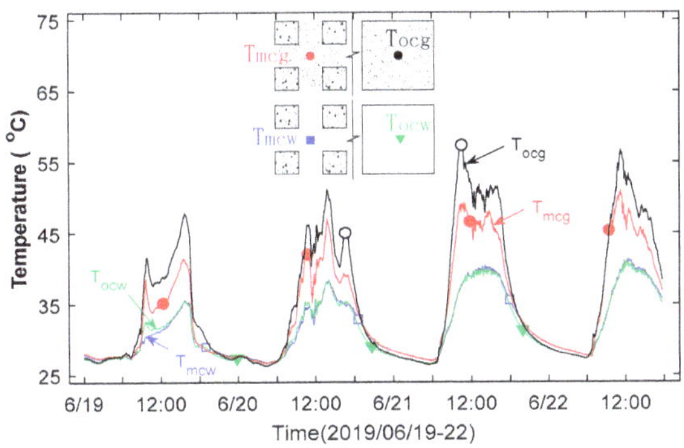

Figure 8. Daily temperature series at the center of the gray unban mockup, of the white urban mockup, of the gray open square and of the white open square. Noted: subscripts o = open square, m = mockup, c = center, g = gray, and w = white

Due to the shading effect of the urban canyon, the reflective paved street in the urban mockup could not be cooled in the same way as those pavements in an open area. To estimate this difference, Figure 9 compares the time serials of $T_{ocg} - T_{ocw}$ and $T_{mcg} - T_{mcw}$ (subscripts o = open square, c = center, g = gray, and w = white). It was found that making the pavement high-reflective reduces the skin temperature up to 20 °C in an open area and up to 10 °C in an urban canyon, substantiating the claim that increasing the pavement in an urban area effectively cools the pavement. At nighttime, the influence of albedo on the skin temperature of paved streets vanishes. To further demonstrate this effectiveness, the ratio $\delta = (T_{mcg} - T_{miw})/(T_{ocg} - T_{ocw})$ was calculated to illustrate the effect of using reflective streets in urban areas. The value of δ was set from 0 to 1.0, with 0 for full discount and 1.0 for no discount. It was found that δ varied somewhere between 0 and 1.0. The average of δ was 0.786; the discount was 0.214. That is, if increasing the albedo of a paved surface in an open area decreases the skin temperature by 10 °C, the same albedo increment for a paved street in an urban area would be 7.86 °C. One can conclude that increasing the albedo of paved streets effectively reduces the skin temperature of the street in an urban morphology such as the mockup.

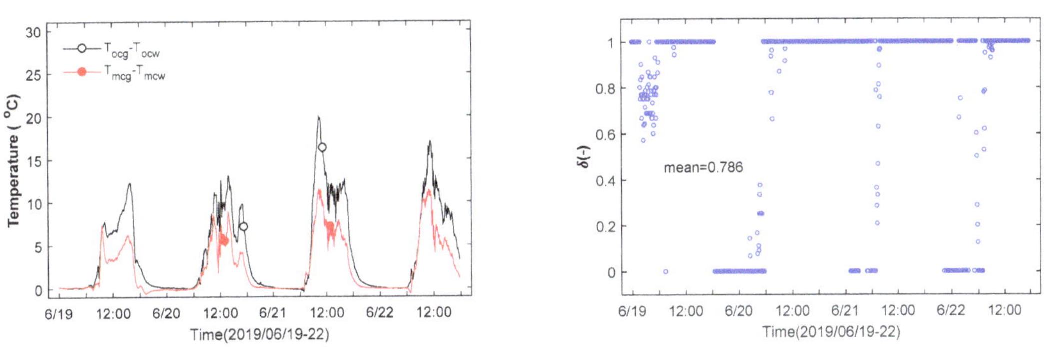

Figure 9. Time serials of $T_{ocg} - T_{ocw}$ and $T_{mcg} - T_{mcw}$ (subscripts o = open square, m = mockup c = center, g = gray, and w = white).

4. Discussion

The results in the above section show similarities and differences to the studies on the H/W (H = building height, W = street width) aspect ratio, street orientation, and sky view factor on urban microclimates (Table 1). Theoretically, the H/W ratio is proportional to the sky view factor of the street. As indicated in some studies [27,29–31], pedestrians feel thermally comfortable in cases of H/W > 1.0, while the region and the local climate act in a secondary role. For instance, Harbich et al. [29] found that a H/W ratio up to 2.0 is enough to improve the thermal comfort in urban area, and that this ratio can be lowered to about 0.5 in cases of shading. This finding is actually coincident with the findings in this study, which indicate that the cross street is not the hottest place, although the solar irradiance on this street is the largest. However, the difference between temperatures in the cross street and the East-West street was very small, except for the time close to the solar noon. This infers that the orientation of the urban street is more important to determining the thermal comfort of an urban canyon than the sky view factor or the H/W ratio. According to Toudert et al. [32] and Kruger et al. [26], street orientation dominates the microclimate in an urban area, while the sky view factor plays a secondary role. As shown by the temperature measurements in this study (Figures 4, 6 and 7), pedestrians on South-North streets would be more thermal comfortable than those on East-West streets. While the skin temperature cannot equate to the thermal comfort, which is determined jointly by air temperature, wind, humidity and insolation, thermal comfort decreases with the increase of the skin temperature of the streets. Therefore, the findings in this study are coincident with published articles [33–36].

The findings in this study also confirm the importance of increasing the shady areas in an urban space to improve the thermal comfort of the urban area. As indicated by the temperature contours in the above section, daily solar radiation on a paved street controls the skin temperature. In published articles [35–39], mean radiant temperature is scalar to determine the thermal comfort in an urban area. Without shading, pedestrians will be exposed to sunlight directly, reducing their thermal comfort [40]. When thermal comfort is considered, the shading factor can dilute the importance of other variables such as urban albedo, green ratio, and orientation [23,41]. According to Yuan et al. [38], the optional shading ratio in an urban area is 40%, while a high ratio could suffocate air convection. A summary on the influence of shady areas on the thermal comfort of an urban area can be found in Table 1.

Keeping in mind that this study reports the skin temperature of urban streets while others focus on thermal comfort, one can find that the urban geometry used in this study is very limited and cannot fully represent the cases of other urban morphologies. The building blocks are concrete cubes that do not have hollow spaces. Further works are expected to explore the impact of different canyon geometries, concrete block sizes, and pavement colors (i.e., albedo of paved streets) on the pavement surface temperature. Further studies are also expected to understand the temperature of paved streets over a long time and in different regions. In this study, the experiment time ranged from June 19–24. Within this time spell, the sun was almost right above the experiment location. As a result, the East-West street was exposed to long and strong sunlight during the daytime. At other dates, the solar position would have been different, so the sunlight falling on the paved street would have been different. The temperature of a paved street in an urban canyon would be different accordingly. In addition, this experiment was conducted in a city near the Tropic of Cancer. At other locations (different latitudes), the sunlight falling on the street would be different. However, as the skin temperature is closely correlated to the solar irradiance on the surface, it is believed that the skin temperature shown in this study is similar to the surface temperature of paved streets in real urban areas.

Table 1. Studies on the H/W aspect ratio, street orientation, and sky view factor on urban microclimates.

Reference and Location	Results
[34] Camaguey, Caba	Thermal comfort of a street is dictated by aspect ratio and street orientation; South-North streets provide better thermal comfort than East-West ones.
[35] De Bilt, Netherlands	Urban form determines duration of direct sun and mean radiant temperature, with the hottest situation on the East-West street.
[33] Stuttgart, Germany	The sky view factor of urban streets in the Northern hemisphere is decisive to the heat stress of the urban canyon. A street canyon with an H/W-ratio of at least 1.5 seems to be the best option to reduce heat stress.
[29] Campinas, Brazil	An H/W ratio up to 2 increases shade and improves thermal comfort during the daytime more than other ratios. Forestry management and green areas increase the shade on façades and pedestrian routes and enhance the microclimate, particularly for H/W less than 0.5.
[42] Rome, Italy	Shaded areas are directly correlated to thermal comfort values, which are depicted by the physiologically equivalent temperature.
[30] Tunis, Tunisia	For all configurations, a South-North street exhibits the best thermal comfort, while the thermal comfort of an East-West street is the worst.
[31] Island of Tinos, Greece	South-North streets exhibit good thermal comfort in cases of H/W < 1.3, but in case of H/W > 2.0 show similar thermal conditions to other streets.
[36] Freiburg, Germany	North–South and East-West streets are the two extrema, with the highest mean radiant temperature for North–South and the lowest values for East-West.
[37] Colombo, Sri Lanka	The temperature differences between sunlit and shaded urban surfaces reached 20 °C, highlighting shade as the main strategy for lowering air and radiant temperatures.
[25] Putrajaya Boulevard, Malaysia	The sky view factor of an urban street is not the decisive factor influencing urban canyon microclimates or heat islands, unless the orientation of the urban street is considered.
[26] Curitiba, Brazil	The sky view factor correlates strongly to the nocturnal heat island but weakly to the daytime temperature, which is dominated by solar irradiance.
[32] Ghardaia, Algeria	South-North streets cause less heat stress in its duration and intensity, whereas East-West streets are thermally uncomfortable for a much longer time.
[38] Osaka, Japan	Increasing the urban green covering around a city up to 40% improves the urban microclimate, but further increments are compromised of the air convection.
[41] Shanghai, China	At daytime the shading factor overwhelms the green ratio and surface albedo; at nighttime, anthropogenic heat and sky view factors control thermal comfort.
[23] Osaka, Japan	Aspect ratio W/H dominates daily net solar-irradiance gains on roads. Reflective cool pavements should be considered on the North sides of East-West streets and at the centers of North–South streets.
[27] Constantine, Algeria	H/W ratio and the sky view factor are decisive factors for the observed air temperature in urban areas.

Note: H = building height, W = street width.

5. Conclusions

This study measured the skin temperature across an urban mockup as an attempt to understand the surface temperature distribution in a real urban canyon. It was found that in summer the East-West street was hotter than the cross street and the South-North street, with a difference of about 3–5 °C at daytime, and 0.1–1.0 °C at nighttime, and with 0.1–0.3 °C daily mean skin temperature. At nighttime, the heat release from buildings can keep the paved street at about 0.2 °C hotter than paved areas in open spaces. It is also found that a paved street in an open square is hotter than a paved street in an urban mockup due to the shading effect. Our findings substantiate the claim that street orientation controls the skin temperature of an urban street while the sky view factor (or building height and street width) acts only in a secondary role. Increasing the albedo of a paved street in an urban canyon can effectively reduce the skin temperature of the street, although there is some discount (about 0.124). Further studies are also expected to understand the temperature of paved streets over a long time and in different regions.

Author Contributions: Conceptualization, Y.Q.; methodology, L.W.; formal analysis, P.W.; writing, Y.Z. All authors have read and agreed to the published version of the manuscript.

Funding: Not applicable.

Institutional Review Board Statement: Not applicable.

Informed Consent Statement: Not applicable.

Data Availability Statement: Not applicable.

Acknowledgments: This paper is jointly supported by the National Natural Science Foundation of China (41762022), by the High-level innovation team and outstanding scholar program in Guangxi colleges (granted to Yinghong Qin), by the Science Foundation of Guangxi (Grant nos. 2018GXNSFAA294070, 2018GXNSFDA138009, 2017 GXNSFBA198052), and by the Guilin City Technology Development Project (20170222).

Conflicts of Interest: The authors declare no conflict of interest.

Appendix A

Figure A1 shows the global horizontal solar irradiation during the experiment, which lasted for four days. Days are attributed to partially sunny days.

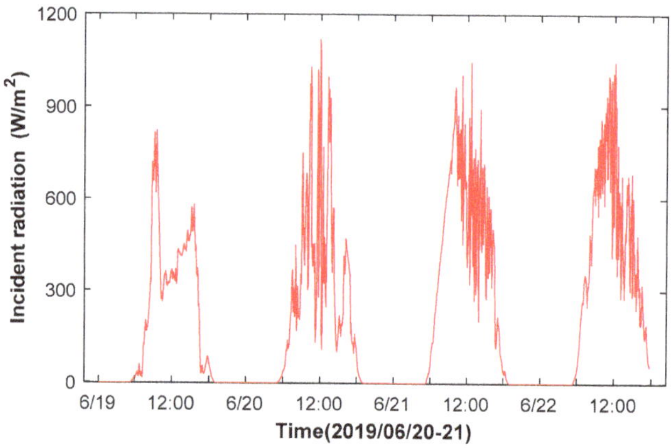

Figure A1. The global horizontal solar irradiance during the experiment.

References

1. Qin, Y. Urban canyon albedo and its implication on the use of reflective cool pavements. *Energy Build.* **2015**, *96*, 86–94. [CrossRef]
2. Aida, M. Urban albedo as a function of the urban structure—A model experiment. *Bound. Layer Meteorol.* **1982**, *23*, 405–413. [CrossRef]
3. Chen, J.; Wang, H.; Zhu, H. Analytical approach for evaluating temperature field of thermal modified asphalt pavement and urban heat island effect. *Appl. Therm. Eng.* **2017**, *113*, 739–748. [CrossRef]
4. Panão, M.N.O.; Gonçalves, H.P.; Ferrão, P.C. A matrix approach coupled with Monte Carlo techniques for solving the net radiative balance of the urban block. *Bound. Layer Meteorol.* **2007**, *122*, 217–241. [CrossRef]
5. Santillán-Soto, N.; Garcia-Cueto, R.; Haro-Rincon, Z.; Ojeda-Benitez, S.; Quintero-Nunez, M.; Velazquez-Limon, N. Radiation balance of urban materials and their thermal impact in semi-desert region: Mexicali, México study case. *Atmosphere* **2015**, *6*, 1578–1589. [CrossRef]
6. Parker, J. The Leeds urban heat island and its implications for energy use and thermal comfort. *Energy Build.* **2021**, *235*, 110636. [CrossRef]
7. Sharma, R.; Pradhan, L.; Kumari, M.; Bhattacharya, P. Assessing urban heat islands and thermal comfort in Noida City using geospatial technology. *Urban Clim.* **2021**, *35*, 100751. [CrossRef]
8. He, B.-J.; Wang, J.; Liu, H.; Ulpiani, G. Localized synergies between heat waves and urban heat islands: Implications on human thermal comfort and urban heat management. *Environ. Res.* **2021**, *193*, 110584. [CrossRef]

9. Qaid, A.; Lamit, H.B.; Ossen, D.R.; Shahminan, R.N.R. Urban heat island and thermal comfort conditions at micro-climate scale in a tropical planned city. *Energy Build.* **2016**, *133*, 577–595. [CrossRef]
10. Fahed, J.; Kinab, E.; Ginestet, S.; Adolphe, L. Impact of urban heat island mitigation measures on microclimate and pedestrian comfort in a dense urban district of Lebanon. *Sustain. Cities Soc.* **2020**, *61*, 102375. [CrossRef]
11. Ballinas, M.; Barradas, V.L. The Urban Tree as a tool to mitigate the urban heat island in Mexico City: A simple phenomenological model. *J. Environ. Qual.* **2016**, *45*, 157–166. [CrossRef]
12. Nichol, J.; Wong, M.S. Modeling urban environmental quality in a tropical city. *Landsc. Urban Plan.* **2005**, *73*, 49–58. [CrossRef]
13. Henao, J.J.; Rendón, A.M.; Salazar, J.F. Trade-off between urban heat island mitigation and air quality in urban valleys. *Urban Clim.* **2020**, *31*, 100542. [CrossRef]
14. Yuan, J.; Farnham, C.; Emura, K. Inter-building effect and its relation with highly reflective envelopes on building energy use: Case study for cities of Japan. *Atmosphere* **2017**, *8*, 211. [CrossRef]
15. Kubilay, A.; Allegrini, J.; Strebel, D.; Zhao, Y.; Derome, D.; Carmeliet, J. Advancement in urban climate modelling at local scale: Urban heat island mitigation and building cooling demand. *Atmosphere* **2020**, *11*, 1313. [CrossRef]
16. Shahmohamadi, P.; Che-Ani, A.I.; Etessam, I.; Maulud, K.N.A.; Tawil, N.M. Healthy environment: The need to mitigate urban heat island effects on human health. *Procedia Eng.* **2011**, *20*, 61–70. [CrossRef]
17. Heaviside, C.; Macintyre, H.; Vardoulakis, S. The urban heat island: Implications for health in a changing environment. *Curr. Environ. Health Rep.* **2017**, *4*, 296–305. [CrossRef]
18. Dong, W.; Liu, Z.; Zhang, L.; Tang, Q.; Liao, H.; Li, X. Assessing heat health risk for sustainability in Beijing's urban heat island. *Sustainability* **2014**, *6*, 7334–7357. [CrossRef]
19. Wang, J.; Meng, Q.; Zhang, L.; Zhang, Y.; He, B.-J.; Zheng, S.; Santamouris, M. Impacts of the water absorption capability on the evaporative cooling effect of pervious paving materials. *Build. Environ.* **2019**, *151*, 187–197. [CrossRef]
20. Bao, T.; Liu, Z.; Zhang, X.; He, Y. A drainable water-retaining paver block for runoff reduction and evaporation cooling. *J. Clean. Prod.* **2019**, *228*, 418–424. [CrossRef]
21. Li, H. *Evaluation of Cool Pavement Strategies for Heat Island Mitigation*; University of California: Davis, CA, USA, 2012; p. 387.
22. Cao, X.; Tang, B.; Zhu, H.; Zhang, A. Cooling principle analyses and performance evaluation of heat-reflective coating for asphalt pavement. *J. Mater. Civ. Eng.* **2011**, *23*, 1067–1075. [CrossRef]
23. Takebayashi, H.; Moriyama, M. Relationships between the properties of an urban street canyon and its radiant environment: Introduction of appropriate urban heat island mitigation technologies. *Solar Energy* **2012**, *86*, 2255–2262. [CrossRef]
24. Touchaei, A.G.; Wang, Y. Characterizing urban heat island in Montreal (Canada)–Effect of urban morphology. *Sustain. Cities Soc.* **2015**, *19*, 395–402. [CrossRef]
25. Qaid, A.; Ossen, D.R. Effect of asymmetrical street aspect ratios on microclimates in hot, humid regions. *Int. J. Biometeorol.* **2015**, *59*, 657–677. [CrossRef]
26. Krüger, E.L.; Minella, F.O.; Rasia, F. Impact of urban geometry on outdoor thermal comfort and air quality from field measurements in Curitiba, Brazil. *Build. Environ.* **2011**, *46*, 621–634. [CrossRef]
27. Bourbia, F.; Boucheriba, F. Impact of street design on urban microclimate for semi arid climate (Constantine). *Renew. Energy* **2010**, *35*, 343–347. [CrossRef]
28. Sun, T.; Wang, Z.-H.; Ni, G.-H. Revisiting the hysteresis effect in surface energy budgets. *Geophys. Res. Lett.* **2013**, *40*, 1741–1747. [CrossRef]
29. Abreu-Harbich, L.V.; Labaki, L.C.; Matzarakis, A. Thermal bioclimate in idealized urban street canyons in Campinas, Brazil. *Theor. Appl. Climatol.* **2014**, *115*, 333–340. [CrossRef]
30. Achour-Younsi, S.; Kharrat, F. Outdoor thermal comfort: Impact of the geometry of an urban street Canyon in a Mediterranean subtropical climate—Case study Tunis, Tunisia. *Procedia Soc. Behav. Sci.* **2016**, *216*, 689–700. [CrossRef]
31. Andreou, E. Thermal comfort in outdoor spaces and urban canyon microclimate. *Renew. Energy* **2013**, *55*, 182–188. [CrossRef]
32. Ali-Toudert, F.; Mayer, H. Numerical study on the effects of aspect ratio and orientation of an urban street canyon on outdoor thermal comfort in hot and dry climate. *Build. Environ.* **2006**, *41*, 94–108. [CrossRef]
33. Ketterer, C.; Matzarakis, A. Human-biometeorological assessment of heat stress reduction by replanning measures in Stuttgart, Germany. *Landsc. Urban Plan.* **2014**, *122*, 78–88. [CrossRef]
34. Rodríguez Algeciras, J.A.; Gómez Consuegra, L.; Matzarakis, A. Spatial-temporal study on the effects of urban street configurations on human thermal comfort in the world heritage city of Camagüey-Cuba. *Build. Environ.* **2016**, *101*, 85–101. [CrossRef]
35. Taleghani, M.; Kleerokoper, L.; Tenpierik, M.; van den Dobbelsteen, A. Outdoor thermal comfort within five different urban forms in The Netherlands. *Build. Environ.* **2015**, *83*, 65–78. [CrossRef]
36. Herrmann, J.; Matzarakis, A. Mean radiant temperature in idealised urban canyons–Examples from Freiburg, Germany. *Int. J. Biometeorol.* **2012**, *56*, 199–203. [CrossRef]
37. Rohinton, E.; Erik, J. Influence of urban morphology and sea breeze on hot humid microclimate: The case of Colombo, Sri Lanka. *Clim. Res.* **2006**, *30*, 189–200.
38. Yuan, J.; Emura, K.; Farnham, C. Is urban albedo or urban green covering more effective for urban microclimate improvement?: A simulation for Osaka. *Sustain. Cities Soc.* **2017**, *32*, 78–86. [CrossRef]
39. Salata, F.; Golasi, I.; de Lieto Vollaro, A.; de Lieto Vollaro, R. How high albedo and traditional buildings' materials and vegetation affect the quality of urban microclimate. A case study. *Energy Build.* **2015**, *99*, 32–49. [CrossRef]

40. Watanabe, S.; Ishii, J. Effect of outdoor thermal environment on pedestrians' behavior selecting a shaded area in a humid subtropical region. *Build. Environ.* **2016**, *95*, 32–41. [CrossRef]
41. Yang, F.; Lau, S.S.Y.; Qian, F. Urban design to lower summertime outdoor temperatures: An empirical study on high-rise housing in Shanghai. *Build. Environ.* **2011**, *46*, 769–785. [CrossRef]
42. Martinelli, L.; Lin, T.-P.; Matzarakis, A. Assessment of the influence of daily shadings pattern on human thermal comfort and attendance in Rome during summer period. *Build. Environ.* **2015**, *92*, 30–38. [CrossRef]

atmosphere

Review

Untangling Urban Sprawl and Climate Change: A Review of the Literature on Physical Planning and Transportation Drivers

Qiu Feng * and Pierre Gauthier

Department of Geography, Planning and Environment, Concordia University, 1455 Boulevard de Maisonneuve O, Montréal, QC H3G 1M8, Canada; pierre.gauthier@concordia.ca
* Correspondence: qiu.feng@mail.concordia.ca

Abstract: Significant efforts have been dedicated to studying the linkages between urban form, fossil energy consumption, and climate change. The theme of urban sprawl helped to federate a significant portion of such efforts. Yet, the research appears fragmented, at stems from different disciplines and mobilizes different methods to probe different aspects of the issue. This paper seeks to better understand the status of knowledge concerning the linkages between sprawl and climate change through a critical review of the literature published between 1979 and 2018. The exercise entailed revisiting how sprawl has been defined, characterized and measured, and how such parameters have informed the research themes and the approaches mobilized to study its impacts on climate change. For, sprawled environments contribute the climate change directly and indirectly, due to the individual or combined effects of its land use, land cover, urban form, and transportation characteristics. The results indicate that sprawl's impacts have been mainly investigated in three principal streams of research and based on a limited number of factors or combinations of factors. Though a strong consensus emerges on the negative environmental costs of sprawl, including toward climate change, there remain ambiguities when trying to untangle and weigh specific causes.

Keywords: climate change; urban sprawl; urban form; land use; physical planning; urban transportation; greenhouse gas emission; energy consumption

Citation: Feng, Q.; Gauthier, P. Untangling Urban Sprawl and Climate Change: A Review of the Literature on Physical Planning and Transportation Drivers. *Atmosphere* **2021**, *12*, 547. https://doi.org/10.3390/atmos12050547

Academic Editors: Hideki Takebayashi and Jihui Yuan

Received: 8 April 2021
Accepted: 22 April 2021
Published: 24 April 2021

1. Introduction

This research aimed at answering a pretty straightforward question: "what does the research say about the links between urban sprawl and climate change." The exercise stemmed from the desire of the two largest regional planning bodies of the Province of Québec to get a synthetic summary of the research approaches and key findings on the matter, in the context of the revision of their Regional Plans. The general consensus in the applied planning literature is that sprawled environments epitomized, in a North American context particularly, by post-Second World War automobile-based suburban development models, are the cause of environmental degradations while contributing to climate change. There are plenty of empirical pieces of evidence to support such tenets. Yet, upon closer examination, the issue appears more elusive than it might look at first sight. Establishing the linkages between sprawl and climate change is complicated by many factors, starting with the inherent complexities of the interactions between the natural and built environments and the difficulties to untangle the direct, indirect, mutual impacts between them, on the one hand, and difficulty of tackling the sprawl phenomenon itself, on the other hand.

The determination of the impacts of urban sprawl on climate change and their measurement can, and do, vary significantly depending on how sprawl itself is conceptualized and measured. Though the overall negative environmental impacts of sprawl are well established, seemingly inconclusive or contradictory results on their precise causes are not rare. Some of these conundrums are impeding our ability to fully understand the connections between sprawl and climate change and to articulate a proper public policy response. This paper presents key conclusions stemming from a survey of pertinent literature probing the links between urban sprawl and climate change. It sets about, firstly, to bring some conceptual clarity to the notion of sprawl, while charting, secondly, the scientific production that addresses direct and indirect links between the latter and climate change. Those links are examined through the lenses of energy consumption, and greenhouse gas (GHG), and carbon dioxide (CO_2) emissions. While tremendous efforts have already been made in exploring the relationships between urban forms, including sprawled environments, energy consumption, and climate change [1–6], it is very difficult for specialists and interested stakeholders to build a synthetic picture and to make sense of seemingly disparate research.

Following a brief discussion on the approach mobilized to review the pertinent literature and gather pertinent materials, the following sections introduce sprawl's definitions, characterizations, and quantification methods. A conceptual diagram has proposed that charts what we deem the physical planning drivers of climate change as well as their direct, indirect, and combined impacts. Key research approaches and findings on the links between sprawl and climate change are then presented according to three main streams revealed by the review of the pertinent literature. A discussion follows on the complementarities and limitations of the research programs while pointing to some ambiguities and apparent inconsistencies.

2. Materials and Methods

A two-pronged approach was adopted for identifying and analyzing the pertinent literature probing the linkages between urban sprawl and climate change. The first step centered on 7 literature reviews published between 1994 to 2015 (since expanded to include two reviews published in 2020 [7,8]) (see Table 1) on urban sprawl per se, or the environmental impacts of sprawl. The works and research themes covered in these 7 reviews laid a solid foundation for further search, by allowing, firstly, to review definitions, conceptualizations, and characterizations of sprawl, and secondly, to identify potent keywords and combinations of keywords that address the linkages between sprawl, or key aspects of it, and climate change. The second step entailed conducting a search relying on ISI's Web of Science®® database from 1979 until 2018. Various iterations of (urban sprawl OR sprawl) AND (climate change OR global warming OR greenhouse gas emission* OR CO_2) AND (transportation OR land use* OR land cover change) were performed. A preliminary assessment, including the probing of the articles' bibliographies, was followed by a more thorough examination. The analytical approach mobilized is the critical literature review. This approach does not seek exhaustivity. It is meant to document, compare and contrast contributions from different theoretical, methodological, and epistemological perspectives [9]. The exercise prioritized quantitative studies concerned with the material and spatial manifestations of sprawl in relation to outputs that contribute to climate change while excluding studies centered on economic, social, cultural, or technological factors. North American contexts have been given more weight compared to other settings. A satisfying level of saturation was reached after selecting and perusing some 220 academic contributions, including 9 literature reviews, in the fields of (by decreasing order of importance): environmental sciences, environmental studies, urban studies, ecology, water resource management, sustainable technologies, multidisciplinary geoscience, geography, and planning. Those contributions constitute the core material of this review.

Table 1. Reviews of either urban sprawl, the environmental impacts of urban sprawl, or both.

Review Author (s), Year of Publication Title of the Review	Significance of the Study
Ismael (2020) *Urban form study: the sprawling city—review of methods of studying urban sprawl*	selectively reviewed important existing and novel methods to study and measure urban sprawl from the field of urban geography.
Rubiera-Morollon and Garrido-Yserte (2020) *Recent Literature about Urban Sprawl: A Renewed Relevance of the Phenomenon from the Perspective of Environmental Sustainability*	reviewed the literature on sprawl since 2000, mainly from 2010–2020, while identifying key factors behind its renewed relevance with respect to environmental sustainability in relation to new methodological and recent theoretical advances.
Ewing and Hamidi (2015) *Compactness versus Sprawl: A Review of Recent Evidence from the United States*	revisited the debates about urban sprawl and compact city and summarized the pertinent literature on characteristics, measurements, causes, impacts, and remedies of sprawl.
Yigitcanlar and Kamruzzaman (2014) *Investigating the interplay between transport, land use and the environment: a review of the literature*	surveyed publications from database-ScienceDirect from 1990 and onwards on the latest empirical approaches and best practices worldwide to examine the interplay between transport, land use, and the environment.
Wilson and Chakraborty (2013) *The Environmental Impacts of Sprawl: Emergent Themes from the Past Decade of Planning Research*	extended and updated Johnson's (2001) work by collecting articles published since 2001 related to the environmental impacts of sprawl.
Burchell et al. (2002) *Costs of Sprawl, 2000.*	analyzed urban sprawl, its impacts on resources, personal costs of sprawl, benefits of sprawl, and ways to reduce its negative effects.
Johnson (2001) *Environmental impacts of urban sprawl: a survey of the literature and proposed research agenda*	one of the most widely cited and influential reviews associated with the environmental impacts of sprawl.
Burchell et al. (1998) *The Costs of Sprawl—Revisited*	provided "a detailed examination of most of the information that can be assembled on both sprawl and its costs . . . " (p.ii)
Ewing (1994) *Characteristics, Causes, and Effects of Sprawl: A Literature Review*	reviewed literature on definitions, characteristics, and effects of urban sprawl.

3. Results

The current cycle of climate change is attributable to natural (volcanic activity and solar output for the most part) and anthropogenic drivers. There is an overwhelming consensus to the effect that the GHG emissions caused by human activities are the most important cause of climate change [6,10–14]. Among all of the GHGs, carbon dioxide (CO_2) is the most detrimental contributor to global climate change [6,15,16]. The 2007 IPCC report identifies two primary anthropogenic drivers of increases in atmospheric CO_2: fossil fuel combustion and land-use change. Between 1970 and 2010, around 78% of the total GHG emissions increase was caused by fossil fuel combustion and industrial processes [17].

Cities are already responsible for approximately 80% of the overall GHG/CO_2 emissions [18], while urbanized populations are expected to double, as rural populations level off or decline [19] (p.3133). The Contribution of Working Group III to the Fifth Assessment Report of the Intergovernmental Panel on Climate Change identifies urban form as a significant driver for GHG emissions [17]. The term urban form designates the spatial arrangement of buildings and infrastructures in urbanized contexts.

The GHG/CO_2 emissions pertaining to urban form are predominantly attributable to the fossil energy consumption linked to transportation dynamics and for the heating and cooling in buildings [17,20,21]. Yigitcanlar and Kamruzzaman noted that "transport and land uses are the two major sectors that contribute most in emitting CO_2 in the environment" [15] (p.2121). As will be discussed, built forms, land use, land cover, and transportation respectively contribute to the GHG/CO_2 emissions on their own, while impacting one another. The following paragraphs will highlight some of those key dynamics.

3.1. Defining, Characterizing, and Measuring Sprawl

3.1.1. Defining Sprawl

The forms assumed by urbanization have evidently evolved in the course of history. For thousands of years, the forms of cities had been essentially predicated on walking. Such conditions have informed the architectural and urban configurations as well as the spatial distribution of amenities and activities. Some ancient cities have housed large populations. Thebes in Egypt reached the milestone of 100,000 inhabitants around 1000 years BCE, while Rome, Beijing, and London reached the million mark around 100 CE for the first and the turn of 19th century for the latter ones [22]. Yet, even the most populous cities displayed a spatial extension limited to a radius of 5 km or so, corresponding to an hour of walking from the periphery to the center. Such cities had to "grow from within," which entailed increased densities of populations, buildings, and activities. The introduction of trains and streetcars triggered spatial expansion and the advent of new urban configurations between 1850 to 1950. However, it's the introduction of the automobile in the 1930s, and the generalization of automobility after the Second World War, that would enable urban sprawl, i.e., the spatial dispersion of populations and activities on large territories. Sprawl has become the predominant contemporary urban form at great environmental costs. It is now seen as a "fundamental cause of unsustainability in cities" [23] (p.64).

Urban sprawl has been discussed and researched from a wide diversity of perspectives in the urban studies and planning literature. Once described as an "American zeitgeist" [24], sprawl is becoming a global phenomenon. Sprawl has been substantially altering the physical and spatial structures as well as the functioning of cities where it prevails, though the rates and patterns of sprawl vary in different parts of the world. Such variability is one of the factors that explain the difficulty to characterize the phenomenon. There is no unitary or consensual definition of sprawl in the literature, as regularly stressed by some of its most dedicated observers [25–32]. Various terms such as dispersion, suburban sprawl [33], suburbanization, suburbia and edgeless city [34] have been used in trying to qualify, contextualize or better denote sprawl, but none took hold as an appropriate alternative. The definitions of sprawl are strongly informed by the disciplinary, theoretical, and epistemological perspectives of the definers [35] (p. 3). Moreover, the term alternatively refers to a process, an outcome, or to specific material and spatial manifestations. Adding to the difficulty, each of those instances is responsive to their context and is subject to geographical and temporal variability. A collection of definitions, including from some of the most recognized experts in the field illustrate the last point (Table 2) while highlighting the difficulty to capture the complexity and multidimensional reality of sprawl in a single definition.

Table 2. The definitions of sprawl: summary of findings [36] (p. 13–14).

Authors/ Year of Publication	Definition of Urban Sprawl	Particularity of the Definition
Burchell et al., (1998)	"Sprawl refers to a particular type of suburban peripheral growth." (p. 6).	They stress that sprawl's distinguishing trait: density, should be assessed in relative terms: i.e., especially density should be set "in context", relative to localized circumstances (cultural, geographical, etc.) and relative to the sound use of the resources in that particular context.
Sierra Club, (1998)	"low-density, automobile-dependent development beyond the edge of service and employment areas".	The definition stresses some of the sprawl's spatial characteristics (density, position relative to service, etc.) and effects (automobile dependence).

Table 2. *Cont.*

Authors/ Year of Publication	Definition of Urban Sprawl	Particularity of the Definition
Nelson and Duncan, (1999)	"Unplanned, uncontrolled, and uncoordinated single-use development that does not provide for an attractive and functional mix of uses and/or is not functionally related to surrounding land uses and which variously appears as low density, ribbon or strip, scattered, leapfrog, or isolated development." (p. 1).	The definition mixes normative and affective criteria (functional, attractive), spatial attributes (scattered, isolated, etc.), and the characterization of development processes (uncontrolled, etc.).
Barners et al., (2001)	"sprawl as a pattern of land-use/land cover conversion in which the growth rate of urbanized land (land rendered impervious by development) significantly exceeds the rate of population growth over a specified time period, with a dominance of low-density impervious surfaces." (p. 4).	The definition refers to urbanization processes (land cover conversion, the ratio of land urbanized/population growth) and the resulting spatial patterns (land-use) and spatial properties (density, impervious surfaces, etc.).
Gaslter et al., (2001)	"Sprawl (n.) is a pattern of land use in a UA that exhibits low levels of some combination of eight distinct dimensions: density, continuity, concentration, clustering, centrality, nuclearity, mixed uses, and proximity." (p. 685).	Sprawl is defined in purely spatial terms, as the pattern resulting from the combination of eight properties manifested at "low-levels" of intensity. The said properties allow quantification, hence inaugurating the "first multidimensional measures of sprawl by disaggregated land-use patterns into eight different dimensions" (Ewing and Hamidi, 2014).
Jaeger et al., (2010)	"A landscape suffers from urban sprawl if it is permeated by urban development or solitary buildings." (p. 400).	Sprawl is defined in spatial and topological terms and as a gradient, which takes into consideration the developed, or "built" land cover.
Jaeger and Schwick, (2014)	"A landscape suffers from urban sprawl if it is permeated by urban development or solitary buildings and when land uptake per inhabitant or job is high". (p. 296).	Updated Jaeger et al. 2010 definition, sprawl is defined in spatial and topological terms and as a gradient, which takes into consideration the developed, or "built" land cover as well as land uptake (expressed in ratios inhabitants/land area and jobs/land area).
Ewing, Tian, and Lyons, (2018)	"sprawl is operationally defined as low density, single-use, uncentered, or poorly connected development". (p. 96).	This operational definition of sprawl centers on four spatial characters affecting the distribution of people and urban functions (land-use) and the configurational properties of the street network (connectivity).

While there is no unitary definition of sprawl in the literature, its definitions and characterizations revolve around three highly recurring aspects or characters. Sprawl manifests patterns of land development marked by low-intensity (expressed in densities of population and activities) and spatially segregated land uses (comprised of residential and other functions such as commercial, leisure, and economic production). Such spatial patterns are enabled by, and heavily dependent on, automobility. Sprawl is often contrasted with and compared (in qualitative and quantitative terms) to its polar opposite, deemed the "compact city." The latter typically refers to environments urbanized prior to the generalization of the automobile that are characterized by high densities and diversities of land-uses, and that are amenable to walking and to the deployment and use of public transit [2,6,37]. Though evocative, the compact city notion remains elusive and difficult to operationalize (and to measure, for instance) for the same reasons that affect the notion of sprawl. The difficulty stems from the relativity of the conditions that are described or probed. Whether an environment can be deemed compact or sprawled is relative. There is no way to establish criteria and thresholds empirically, let alone universal ones. At best, sprawl and compactness describe conditions at the opposite ends of the development on a continuum [2,32]. Such categorization overlooks the intermediary conditions and turns a blind eye to atypical combinatory patterns.

Feng and Gauthier propose a definition of sprawl that accounts for the context: "... the term sprawl denotes an urbanization process that produces low-intensity modes of occupation of the land. [Sprawl] is characterized by built and spatial forms that are suboptimal in serving their purposes when taking into consideration their geographical, cultural, and technological contexts and local historical precedents" [36] (p. 16). They posit further, referring to the three sustainable development pillars, that sprawl "produces a suboptimal return on investment, environmentally, socially, and economically speaking, for the community" [36] (p. 16). Feng and Gauthier's conceptualization does not preclude the possibility of measuring sprawl or some of its key components but stresses that the results need to be interpreted relative to the context [36]. While sprawl's environmental impacts can be measured in relative terms, against normative sustainable criteria and benchmarks, or in absolute terms, based on concrete GHG emissions outputs for instance. The difficulty in the latter case lays in the ability to measure accurately sprawl itself and to untangle the causes from the effects of the intertwined characteristics of the phenomenon as we will see.

3.1.2. Measuring Sprawl Key Characteristics

Sprawl is a multi-dimensional phenomenon that requires different measures for each dimension [30,38]. Pendall stated that "the measurement of sprawl is not straightforward, partly because of the variation in how sprawl is defined" [39] (p. 558). Various approaches and methods have been developed to quantify sprawl. Ewing and Hamidi classified the efforts to quantify the extent of sprawl into three stages [2]. The early research, prior to the year 2000, was crude and unidimensional, exclusively or merely focusing on density; the 2001–2010 period has featured multi-level, multi-dimensional, and multi-disciplinary approaches; the subsequent stage from 2011 aimed also to tackle changes or trends in the degrees of urban sprawl [2]. Each method has its own advantages and limitations. Some studies using different methods have delivered inconsistent, and seemingly divergent or contradictory results. As always, caution is needed when interpreting results. Sprawl is characterized by intertwined sets of characters that interact with and influence each other. Its measurement is also particularly sensitive to spatial resolution and the modifiable areal unit problem (i.e., the variability of the shape and scale of the spatial unit against which the data is aggregated) [40,41]. In recent years, major advances in geospatial technologies, including GIS, remote sensing, and photogrammetric techniques have expanded the researchers' toolbox [7,8]. They have allowed in particular to measure sprawl physical characteristics more accurately across a variety of international contexts while facilitating comparative analysis [8].

The most common categories of variables mobilized to measure properties associated with the three main characters of sprawl have been identified, by relying on inductive and deductive reasoning while probing the literature on sprawl characterization and measurement. Table 3 summarizes these findings.

Table 3. Sets of variables are used to measure sprawl.

Category	Character	Variable Name	Definition
Urban sprawl	Density	Population density	Density is most commonly defined as population/housing or employment density, which are measured per unit of analysis.
		Residential density	
	Land use	D variables, first 3 Ds the 5Ds, density, diversity, and design, destination accessibility and distance to transit)	The number of different land use in a given area (at a mesoscale: neighborhood or activity center, land use patterns are characterized by various measures of land use mix within neighborhoods and activity centers). Two land-use mix measures have become most accepted: an entropy index and a dissimilarity index.
	Transportation/automobile dependence	Commute time	Vehicle hours traveled
		Trip distance: VMT	Vehicle miles traveled (or vehicle hours traveled) "is a primary performance indicator for land use and transportation" (Ewing et al., 2014, p. 3080).
		Mode split	Probability (or percentage) of commuting by automobile, transit (rail or bus), or by non-motorized mode (walking/cycling); others also include moped, motorcycle, taxi.

Density has been used as one of the chief measurements or the sole indicator of urban sprawl in many early studies. Expressed by a number of people/housing units or jobs per geographical area unit (acre, hectare, and km^2 for instance), it is the most commonly used measurement of sprawl in the literature. Technically, density variables measure the intensity of a specific land-use category in an area of reference. They are a sub-set of land-use variables. Land use variables are routinely referred to as the so-called "D variables." They are centered on the compositional and configurational characteristics of the land allocation in a geographical area of reference, as well as on accessibility to specific amenities (jobs, transit stops) or overall accessibility by foot (relying on topological variables as proxies). Cervero and Kochelman coined the original expression "three Ds," which stands for: density, diversity (land-use composition), and design (accessibility based on a place's spatial characteristics) [42], later expanded to include destination accessibility and distance to transit, referred to as the fourth and fifth D variables [43] (p. 200). The land-use variables are typically measured at the neighborhood scale by using census data, agency data, or data that can be derived from GIS. Sprawl is almost indissociable from heavy reliance on automobiles. In the sprawling literature, auto dependence has been measured using proxies such as the modal share, and the total amount of time or the distance traveled (typically by car, expressed in vehicle miles/kms traveled—VMT/VKT). The most widely used and primary performance indicator in land use and transportation studies relies on VMT/VKT measurements [44] (p. 3079). Not surprisingly, the methods developed to analyze the main characters of urban sprawl separately or in combination, have informed the research on sprawl and climate change. Another important facet has been the analysis of the land cover, which refers to the composition of the ground surface itself.

Figure 1 illustrates the links between urban sprawl and climate change. It charts, more specifically, what we deem the physical planning (land use and land cover) and transportation drivers of change; the elements analyzed and the type of measurements; the dynamics involved; their effects and consequences, and finally the outcomes (i.e., contribution to climate change or vulnerabilities with respect to the impacts of climate change).

Figure 1. Conceptual diagram of the relationship between urban sprawl and climate change.

3.2. *Sprawl and Climate Change*

The following sections report on the three most important "streams of research" investigating the relationships between sprawl and climate change as revealed by the literature review, while summarizing some of their key conclusions. The first stream focuses on the linkages between density, fossil energy consumption, and GHG/CO_2 emissions. The second one is considering the dynamics between land use energy consumption and emissions, and between land cover change and the carbon balance. The third stream is centered on the impacts of transportation dynamics on climate change. The joined impacts of land use and transportation interactions and land use and built forms on fossil energy use and GHG/CO_2 emissions are also addressed.

3.2.1. The Impacts of Urban Density on Fossil Energy Consumption and GHG/CO_2 Emissions

Density was and still is one of the most widely used indicators to measure land-use and urban form intensity. "Density, or more specifically, low density, is one of the cardinal characteristics of sprawl" [24] (p. 6). Unsurprisingly, early research efforts to quantify the impacts of urban form on energy consumption and GHG/CO_2 emissions centered on density. Different measures of density, such as the number of inhabitants or dwellings per unit of space have been used as proxies for land-use intensity or urban form compactness. The most important line of inquiry focuses on urban density's links with fossil energy consumption and GHG/CO_2 emissions due to transportation. The second line of inquiry centers on the energy consumption and emissions of the buildings themselves (heating, cooling, etc.) depending on the density.

Early attempts to explore the links between population density on energy consumption and GHG/CO_2 emissions, which focused on the outputs of the transportation patterns associated with various land-use and urban form configurations have produced sound and valuable insights into nature and the impacts of these intertwined relationships [20,45–47]. The seminal study by Newman and Kenworthy of 32 cities from around the world shows a clear negative relationship between population density and fuel consumption [20]. Subsequent efforts have produced a deeper understanding by probing the trends and patterns of changes in the said relationships over a long period [48].

Low-density residential environments composed of single-family houses are a defining characteristic of sprawl. Many scholars investigate the relationship between residential density, the level of GHG emissions related to both transportation patterns and building energy use [49–52]. Different residential densities are typically represented by different neighborhood types—Generally classified as either traditional (from the pre-automobile era) or suburban. Living and transportation arrangements in low-density suburban contexts are compared with urban center's living, characterized by high-density apartment buildings, to measure their respective energy consumption and levels of GHG emissions. However, as Osorio et al. argued, probing the energy consumption of buildings is complicated, and even more so at the neighborhood or network of buildings scales [53]. Low-density, detached single-family buildings and urban fabrics are associated with higher energy consumption and GHG/CO_2 emissions, but specific contributing factors are difficult to untangle and measure precisely.

In general, the literature produces very strong pieces of evidence showing that population/residential densities are negatively correlated with energy consumption and GHG/CO_2 emissions and that increasing density results in lower such outputs.

Table 4 summarizes, in chronological order, the approach, context, and key findings of selected research investigating the relationships between density, energy consumption, and GHG emissions.

Table 4. The impacts of urban density on fossil energy consumption and GHG/CO_2 emissions [36] (p. 39–40).

Author(s) Year of Publication	Type of Density	Relationship Studied with Density	Geographical Context	Main Results
Newman and Kenworthy (1989)	Population density	Gasoline consumption per capita	32 global cities	Per capita gasoline consumption is negatively correlated with population density.
Norman, MacLean, and Kenned (2006)	Residential density	Energy use and GHG emissions	Toronto	CO_2 equivalent emissions are 60% less for high-density than for low-density development.
Nelson and Duncan (1999)	Residential building density	GHG emissions	Toronto	Top ten in terms of GHG emission were all located in the low-density tracts.
Andrews (2008)	Urban density	GHG emissions distribution along the rural-to-urban gradient	United States Canadian cities	Per-capita CO_2 emissions vary widely following an inverted "U" shape, with post-war suburbs at the pinnacle.
Ewing and Rong (2008)	House size and type	Housing types and energy consumption	United States	Houses located in compact counties require roughly 20% less primary energy than those in sprawling counties.

Table 4. *Cont.*

Author(s) Year of Publication	Type of Density	Relationship Studied with Density	Geographical Context	Main Results
Taniguchi, Matsunaka, and Nakamichi (2008)	Population density	Per capita automobile CO_2 emissions	38 Japanese cities	Density negatively correlated with automobile CO_2 emissions Per-capita automobile CO_2 emissions increased in all city types between 1987 and 2005.
Glaeser and Kahn (2010)	Population density	Household emissions	66 major US cities	Gasoline usage is negatively correlated with population density and positively correlated with distance from downtown.
Kim and Brownstone (2010)	Residential density	Household annual mileage traveled and fuel consumption	United States	Households residing in an area that is 1000 housing units per square mile denser drive 1500 (7.8%) fewer miles per year and consume 70 (7.5%) fewer gallons of fuel than households in the less dense areas.
Ala-Mantila, Junnila, and Heinonen (2013)	Residential types (Semi-detached and detached houses, apartment buildings)	Consumption-based carbon footprints by residential types	Finland	Low-rise lifestyle causes approximately 26% more emissions than high-rise.
Pitt (2013)	Residential types (attached, multifamily, single-family detached housing)	Residential GHG emissions and energy consumption for future housing development	United States	On average, attached homes and multi-family structures are more energy-efficient than single-family detached housing types.
Ala-Mantila, Heinonen, and Junnila (2014)	Housing and household types	Consumption-based carbon footprints by housing and household types	Finland	Rural lifestyle related to the highest GHG emissions. Emissions decrease as density increases while moving towards city centers.
Fercovic and Gulati (2016)	Population density	Average household emissions	Canadian cities	Denser cities produce fewer emissions than low-density ones. Average household emissions across all cities over time are falling.
Estiri (2016)	Households housing arrangement (city and suburban)	Energy consumption by households	United States	On average, US suburban households consume more energy in residential buildings than their city-dweller counterparts.

3.2.2. The Impacts of Land Use and Land Cover on Fossil Energy Consumption and GHG/CO_2 Emissions

The term land use refers to the composition and configurations of the land surface utilization (for housing, work, leisure, transportation, etc.). Land cover denotes the nature and the composition of the surface on the ground (such as forest land, grassland, wetlands, anthropogenic biomes of crops, or artificial infrastructures and buildings). The research probing the links between land use patterns and urban development dynamics in relation to energy consumption and GHG/CO_2 emissions touches: 1. on direct outputs and direct impacts on the carbon balance; 2. on indirect outputs linked to transportation dynamics associated with land use patterns and; 3. on increased environmental vulnerabilities.

Sprawl entails massive land cover changes involving the artificialization of natural or cultivated land. Such transformations translate into deforestation and grasslands losses; loss of valuable arable land; the creation of extensive impervious surfaces and; extensive construction of buildings and roads. Land conversion causes significant losses of biomass while altering natural habitats and ecosystems. Land cover change associated with sprawl contributes to climate change by reducing the carbon capture and storage capacities [54]. It furthers the ecosystem's vulnerabilities stemming from bioclimatic transformations induced by climate change itself and compromises these ecosystems' ability to mitigate the impacts of extreme weather events such as heavy rains [3,12].

The research investigating the relations between the composition and configurations of land utilization, energy consumption, and GHG/CO_2 emissions, usually focuses on the transportation implications of land use conditions in sprawled and compact environments respectively. The composition and spatial distributions of urban activities and functions influence people's travel behavior by "affecting decisions about how much, where, when, and how to get around" [55] (p. 2). The degree of land use mix not only correlates with VMT/VKT but exerts also an influence on the choice of the mode of transportation.

A majority of empirical studies surveyed herein conclude that more mixed land uses and compact urban forms that are complemented by a good public transit system and a well-connected and easily accessible street network are associated with fewer VMT/VKT, lower levels of GHG/CO_2 emissions, and energy consumption, as well as a lesser dependence to the automobile when compared to sprawled contexts [6,56–58].

Table 5 summarizes, in chronological order, the contexts, methods, and main findings of studies probing the impacts on energy consumption and GHG/CO_2 emissions of land use/land cover change, composition, and configurations.

3.2.3. Transportation, Automobile Dependence, Energy Consumption, and GHG/CO_2 Emissions

The transportation sector consumed more than half of the oil used globally in 2015 [59] and has been identified as the largest emitter of CO_2, outpacing other sectors [19]. Energy consumption and GHG/CO_2 emissions related to transportation, and road transportation, in particular, have seen sharp increases worldwide [6,12,55,60]. Given the fact that the vast majority of vehicles are powered by combustion engines using fossil energy, the road transport sector contributes greatly to climate change through GHG/CO_2 emissions.

Sprawl would have been impossible at its current scale without heavy reliance on the automobile. The general consensus in the literature is that sprawled urban forms generate more car travel and entails greater energy consumption and more GHG/CO_2 emissions as a consequence [4,6,19]. The research reaches the same conclusions when such outcomes are measured at the local, or neighborhood scale, or the regional level. In other words, low-density suburban environments generate more emissions than compact environments in the same city, and the more a city is marked by sprawl overall, the more emissions it generates.

Table 5. The impacts of land use and land cover on fossil energy consumption and GHG/CO_2 emissions [36] (p. 47–48).

Author(s) Year of Publication	Scope and Location	Main Method(s)	Data/Time Frame	Land Use Factors	Main Findings
Bart (2010)	EU Member States	A simple linear multiple regression analysis	CORINE database between 1990 and 2000	Increase in artificial land area	Sprawling development is strongly associated with increases in transport-related emissions and is the most important driver of emission growth.
Stone, Hess, and Frumkin (2010)	Metropolitan regions in the U.S.	Applying a widely used sprawl index	Urban form in 2000. Extreme Heat Events between 1956 and 2005	Sprawl index, frequency of EHEs	"The rate of increase in the annual number of EHEs in the most sprawling metropolitan regions is more than twice the rate of increase observed in the most compact metropolitan regions" (p. 1425).
Bereitschaft and Debbage (2013)	86 U.S. metropolitan areas	A series of linear regression models have been applied	Air pollutants data collected based on 2000 census	5 pre-existing urban sprawl indexes were selected	After controlling other variables, higher levels of urban sprawl or sprawled urban form are closely linked with a higher level of air pollution and CO_2 emissions.
Kim, Lee, and Choi (2014)	Los Angeles Metropolitan Area (LAMA) vs. Seoul Metropolitan Area (SMA)	Comparative approach by employing the Cobb-Douglas functions	Data were collected based on the status quo from 2008	Distinctive land-use density: an auto-centric area vs. dense, intensive land-use area	Reduction of CO_2 emissions in both areas can be achieved by the public transit mode share adjustment without weakening existing mobility levels. However, the amount of CO_2 reduction of the SMA is much more significant than that of the LAMA.

Table 5. *Cont.*

Author(s) Year of Publication	Scope and Location	Main Method(s)	Data/Time Frame	Land Use Factors	Main Findings
Adeyemi et al. (2015)	Tshwane metropolis, Gauteng Province, South Africa	a correlation analysis to test the relationship between Land Surface Temperature, Normalized Difference Vegetation Index, and Normalized Difference Built-up Index	Landsat 8 LCDM, 2003, and Landsat 7 ETM+, 2013	Vegetation cover and impervious surface area	LST has a positive relationship with NDBI, while has a negative relationship with NDVI.
Wang, Li, and Yang (2015)	Southern China	A structural equation model	1988 and 2005	Vegetation, urban and surrounding area, and other	"Adding vegetation area is the main method to mitigate regional climate change" (p. 1).
Iwata and Managi (2016)	Japanese cities (1750)	Linear model	City-level data from 1990 to 2007	Impacts of different land-use strategies	Different urban planning instruments impact the level of vehicular CO_2 emissions differently. Some methods are more effective in low-density cities, while others work better in high-density cities.
Emadodin, Taravat, and Rajaei (2016)	Tehran, Iran	MLP neutral network has been used; more detailed presentation sees p. 233.	Satellite images: every 5 years from 1975 to 2015; Local climatic data: 1990 to 2010.	IDM has been used to measure changes in aridity between 1990–2000 and 2001–2010.	Between these two time periods, the average temperature has increased from 17.43 to 18.31. More arid area has experienced greater temperature increase.
Lu and Liu (2016)	287 Chinese cities: four provincial-level cities and 283 prefecture-level cities	A geographically weighted regression (GWR) model	NO2 data from 2008; SO2 data from 2007	Urban form indexes: the compact ration index, the fractal dimension index, and the Boyce–Clark shape index	Urban form characteristics significantly affect urban air quality in China.

Table 5. *Cont.*

Author(s) Year of Publication	Scope and Location	Main Method(s)	Data/Time Frame	Land Use Factors	Main Findings
Cai et al. (2017)	Chinese and American cities	Compare and quantify the correlation among nighttime light intensity, surface thermal changes, and city size	MODIS LST and DMSP/OLS Nighttime light data sets 2001–2012	Spatiotemporal changes of the urbanization process	In general, despite the spatial heterogeneities, light intensity increases with increasing city size.
Moradi and Tamer (2017)	Bursa City	Paired Samples t-Test; Holdren Model	1984 to 2014	The growth of the urban settlement, the growth of urban population Emissions decrease as density increases while moving towards city centers.	During 1995 to 2003, urban growth was ascribed to 65% of urban sprawl, accompanied by a loss of forests and agricultural land, and an increase of 1.36 °C monthly minimums temperature (p. 26).

As was mentioned before, low densities and other land use characteristics of sprawl impact the transportation negative outputs in such contexts. But studies focused specifically on transportation dynamics point also to modal share and trip characteristics in sprawled versus more compact environments. Relative to more compact environments, sprawl fosters higher rates of automobile ownership. It is less amenable to public transit deployment and use. It is associated with lower levels transit ridership, is generating longer commuting times, and is increasing VMT/VKT [61,62]. Such conditions are associated with a number of other externalities, among which is the amount of space dedicated to car infrastructure itself, in the form of roads, highways, parking lots, etc., all impervious surfaces that alter the water cycle and contribute to higher ground-level temperatures.

3.3. *On Some Gaps, Limitations, and Ambiguities in the Literature*

While the consensus is strong about the environmental costs of sprawl and its contribution to climate change, there exists a number of gaps, limitations, and seemingly diverging interpretations on a limited number of specific aspects. Differences in the conceptualization of sprawl, and consequently in the measurements of its various dimensions and overall configurational patterns, affect the ability to analyze more precisely the impacts of specific urban form and land use attributes on transportation patterns and their associated GHG/CO_2 emissions. In addition, as already mentioned, the measurement of sprawl or some of its attributes, such as low densities, are particularly sensitive to the modifiable areal unit problem, or MAUP, according to which seemingly discordant results are due to the spatial partitioning used, or on the spatial resolution at which the analysis is conducted [40,41].

Another issue stems from entangled factors and conditions in sprawled environments (or their polar opposite the compact city). Higher urban density for instance is generally accompanied by mixed land uses and better public transit, so that variables measuring those aspects tend to correlate with one another [55], and with fossil energy use and GHG/CO_2 emissions. However, it is unlikely that there exists a simple relationship between urban form and travel behavior for instance. A majority of studies surveyed herein conclude that urban form exerts a significant influence on people's travel behavior mainly through influencing vehicle VMT/VKT, modal choice (public transit versus car), and modal split (between automobile, public and active transportation modes), but it is far less clear how urban design and specific land use characteristics influence people's travel [63,64]. The latter uncertainties do not invalidate or weaken the general conclusions that sprawl is associated with greater emissions levels, but they limit considerably the ability of interested parties to intervene efficiently to retrofit existing environments to reduce their environmental footprint as it is impossible to alter all aspects at once.

4. Discussion

Any serious attempts to measure the impacts of urbanized habitats on the environment, or to intervene on such issues with the aims of reducing their environmental footprints or to build-up resilience, require a deep understanding of the urban material and spatial forms that are manifested in the contemporary city. Proper theorization and characterization of the urban built environments constitute an essential facet of any such research effort. The conceptual ambiguities pertaining to the notion of sprawl, or the lack of unified normative or operational definitions have hindered researchers' abilities to engage with the multidimensionality of sprawl and to analyze more accurately its environmental costs.

Firstly, lacking proper theorization of urban form often leads to a crudely approximative characterization and quantification of spatial conditions (e.g., the widespread use of density indicators that does not account for the variability of spatial composition and configuration in which the same density can be manifested). Many studies compare people's travel behavior between different types of neighborhood. In most cases, a dichotomous classification is employed to compare and contrast internally homogeneous neighborhoods that are either "traditional" i.e., compact, or suburban, i.e., sprawled. There are several problems with this categorization [65]. Bagley, Mokhtarian, and Kitamura criticized that a "binary designation of a residential neighborhood as either traditional or suburban is a distortion of reality, since some locations may have some characteristics of both types" [66] (p. 689) or present intermediary conditions on some or all aspects.

Secondly, sprawl should be and can only be fully understood in relative terms spatially, geographically, culturally, and temporarily. Sprawl can assume very different meanings in different geographical and cultural contexts, or in different urbanization development phases. Cities and regions sprawl differently by presenting varying patterns, rates, extents, and trends of sprawl. Local realities should be carefully taken into account when referring to thresholds used in other contexts. Cities are constantly being transformed and rebuilt upon themselves, often entailing densification and reshuffling of land utilization. Any stage displays conditions that are the temporary results of ongoing processes. As a consequence, sprawl must be conceived in both their space and time contexts.

Thirdly, the terms "sprawl" and "compact city" have been used to represent polar opposites situated at the ends of a spectrum. This raises several complex theoretical and methodological questions and poses significant challenges for the operationalization of these concepts, both for analytical and applied purposes. Morphologically speaking, such a dichotomic representation is fallacious. A city's spatial expansion over time produces a variety of urban configurations and combinatory patterns. Those various parts coexist in the same city. They are connected with one another and to the city as a whole. When a city expands, the properties in the new areas alter the spatial conditions and transportation dynamics of the whole city.

5. Conclusions

This review has shown that there is a significant amount of literature analyzing the relationships between sprawl and climate change and that there is a renewed interest in the topic [8]. These research efforts address different aspects of those relationships by pointing to a variety of direct and indirect links, depending on the factors and combination of factors considered. The fragmentation of the research landscape can challenge one's ability to build a synthetic picture. Highlighting three main streams of research contributes to bringing some clarity. Those coincide roughly with the three main characteristics of sprawl highlighted in the literature centered on that phenomenon. Yet, as seen, there remains significant explicit and implicit overlapping between the streams. This is partly due to ambiguities in the conceptualization of sprawl itself, as discussed in this paper, as well as the difficulty to untangle its defining characteristics when probing it quantitatively, or by extension, measuring its impacts. The latter considerations produce some "noise" and seemingly incongruent results, as variables used to measure land use, including density, urban form properties, and transportation tend to correlate with one another in "pure" sprawled, or compact, contexts, but express more elusive interrelations in less archetypical configurations, or when measured against different spatial partitioning (scale and spatial boundaries). In spite of such limitations, much of the evidence gathered in this exercise demonstrates that sprawl is associated with indisputable environmental costs, including climate change. What is at stake is not determining whether sprawl is sustainable, it is not, nor what factors make it suboptimal, those are known, but rather what combinations of factors are more potent relative to the outcomes. There is a growing consensus on the need to develop combined strategies that consider transportation, land-use, and urban form synergies [6,67,68]. Based on the literature reviewed, this is the way to go to cope with the inherent complexities of sprawl. Yet, the operationalization of such approaches remains highly challenging. For, research has been fragmented. It has yet to produce a cohesive framework clarifying what could be realistically expected from retrofitting urban form, land-use, and transportation systems, considered separately and, more importantly, in combination, to foster modal shifts toward public transportation and the reclaiming of the space lost to automobility.

Funding: This research was funded by Ouranos via the Plan d'action 2013-2020 sur les changements climatiques du Gouvernement du Québec.

Conflicts of Interest: The authors declare no conflict of interest.

References

1. Baklanov, A.; Molina, L.T.; Gauss, M. Megacities, air quality and climate. *Atmos. Environ.* **2016**, *126*, 235–249. [CrossRef]
2. Ewing, R.; Hamidi, S. Compactness versus Sprawl: A Review of Recent Evidence from the United States. *J. Plan. Lit.* **2015**, *30*, 413–432. [CrossRef]
3. Baur, A.H.; Förster, M.; Kleinschmit, B. The spatial dimension of urban greenhouse gas emissions: Analyzing the influence of spatial structures and LULC patterns in European cities. *Landsc. Ecol.* **2015**, *30*, 1195–1205. [CrossRef]
4. Stone, B.; Hess, J.J.; Frumkin, H. Urban Form and Extreme Heat Events: Are Sprawling Cities More Vulnerable to Climate Change Than Compact Cities? *Environ. Health Perspect.* **2010**, *118*, 1425–1428. [CrossRef]
5. Stone, J.B. Urban sprawl and air quality in large US cities. *J. Environ. Manag.* **2008**, *86*, 688–698. [CrossRef] [PubMed]
6. Ewing, R.; Bartholomew, K.; Winkelman, S.; Walters, J.; Chen, D. *Growing Cooler: The Evidence on Urban Development and Climate Change*; Urban Land Institute: Washington, DC, USA, 2008.
7. Ismael, H.M. Urban form study: The sprawling city—Review of methods of studying urban sprawl. *GeoJournal* **2020**. [CrossRef]
8. Rubiera-Morollón, F.; Garrido-Yserte, R. Recent Literature about Urban Sprawl: A Renewed Relevance of the Phenomenon from the Perspective of Environmental Sustainability. *Sustainability* **2020**, *12*, 6551. [CrossRef]
9. Grant, M.J.; Booth, A. A typology of reviews: An analysis of 14 review types and associated methodologies. *Health Inf. Libr. J.* **2009**, *26*, 91–108. [CrossRef] [PubMed]
10. IPCC. *Climate Change 2001: Synthesis Report. A Contribution of Working Groups I, II, and III to the Third Assessment Report of the Integovernmental Panel on Climate Change*; Watson, R.T., Core Writing Team, Eds.; Cambridge University Press: Cambridge, UK; New York, NY, USA, 2001.
11. Hamin, E.M.; Gurran, N. Urban form and climate change: Balancing adaptation and mitigation in the U.S. and Australia. *Habitat Int.* **2009**, *33*, 238–245. [CrossRef]
12. Bart, I.L. Urban sprawl and climate change: A statistical exploration of cause and effect, with policy options for the EU. *Land Use Policy* **2010**, *27*, 283–292. [CrossRef]
13. Bulkeley, H.; Betsill, M. Rethinking Sustainable Cities: Multilevel Governance and the 'Urban' Politics of Climate Change. *Environ. Politi* **2005**, *14*, 42–63. [CrossRef]
14. Hennicke, P. Long term scenarios and options for sustainable energy systems and for climate protection: A short overview. *Int. J. Environ. Sci. Technol.* **2005**, *2*, 181–191. [CrossRef]
15. Yigitcanlar, T. Kamruzzaman Investigating the interplay between transport, land use and the environment: A review of the literature. *Int. J. Environ. Sci. Technol.* **2014**, *11*, 2121–2132. [CrossRef]
16. While, A.; Jonas, A.E.G.; Gibbs, D. From sustainable development to carbon control: Eco-state restructuring and the politics of urban and regional development. *Trans. Inst. Br. Geogr.* **2010**, *35*, 76–93. [CrossRef]
17. Intergovernmental Panel on Climate Change (IPCC). *Climate Change 2014: Mitigation of Climate Change, Contribution of Working Group III to the Fifth Assessment Report of the Intergovernmental Panel on Climate Change*; Edenhofer, O., Pichs-Madruga, R.Y., Sokona, E., Farahani, S., Kadner, K., Seyboth, A., Adler, I., Baum, S., Brunner, P., Eickemeier, B., et al., Eds.; Cambridge University Press: Cambridge, UK; New York, NY, USA, 2014; Available online: http://ipcc.ch/report/ar5/wg3/ (accessed on 11 March 2018).
18. Iwata, K.; Managi, S. Can land use regulations and taxes help mitigate vehicular CO_2 emissions? An empirical study of Japanese cities. *Urban Policy Res.* **2016**, *34*, 356–372. [CrossRef]
19. Marshall, J.D. Energy-Efficient Urban Form. *Environ. Sci. Technol.* **2008**, *42*, 3133–3137. [CrossRef]
20. Newman, P.; Kenworthy, J. Gasoline consumption and cities: A comparison of U.S. cities with a global survey. *J. Am. Plan. Assoc.* **1989**, *55*, 24–37. [CrossRef]
21. Seto, K.C.; Dhakal, S.; Bigio, A.; Blanco, H.; Delgado, G.C.; Dewar, D.; Huang, L.; Inaba, A.; Kansal, A.; Lwasa, S.; et al. Human Settlements, Infrastructure and Spatial Planning. In *Climate Change 2014: Mitigation of Climate Change. Contribution of Working Group III to the Fifth Assessment Report of the Intergovernmental Panel on Climate Change*; Edenhofer, O., Pichs-Madruga, R.Y., Sokona, E., Farahani, S., Kadner, K., Seyboth, A., Adler, I., Baum, S., Brunner, P., Eickemeier, B., et al., Eds.; Cambridge University Press: Cambridge, UK; New York, NY, USA, 2014.
22. Reba, M.; Reitsma, F.; Seto, K.C. Spatializing 6,000 years of global urbanization from 3700 BC to AD 2000. *Sci. Data* **2016**, *3*, 160034. [CrossRef]
23. Newman, P.; Kenworthy, J. *Sustainability and Cities: Overcoming Automobile Dependence*; Island Press: Washington, DC, USA, 1999.
24. Burchell, R.W.; Shad, N.A.; Listokin, D.; Phillips, H.; Downs, A.; Seskin, S.; Davis, J.S.; Moore, T.; Helton, D.; Gall, M. *The Costs of Sprawl Revisited: Transportation Research Board Report 39*; National Academy Press: Washington, DC, USA, 1998.
25. Hamidi, S.; Ewing, R.; Preuss, I.; Dodds, A. Measuring Sprawl and Its Impacts. *J. Plan. Educ. Res.* **2014**, *35*, 35–50. [CrossRef]
26. Zeng, C.; Liu, Y.; Liu, Y.; Qiu, L. Urban sprawl and related problems: Bibliometric analysis and refined analysis from 1991 to 2011. *Chin. Geogr. Sci.* **2013**, *24*, 245–257. [CrossRef]
27. Torrens, P.M. A Toolkit for Measuring Sprawl. *Appl. Spat. Anal. Policy* **2008**, *1*, 5–36. [CrossRef]

28. Cutsinger, J.; Towns, D.; Galster, G.; Wolman, H.; Hanson, R. Verifying the Multi-Dimensional Nature of Metropolitan Land Use: Advancing the Understanding and Measurement of Sprawl. *J. Urban Aff.* **2005**, *27*, 235–259. [CrossRef]
29. Wilson, E.H.; Hurd, J.D.; Civco, D.L.; Prisloe, M.P.; Arnold, C. Development of a geospatial model to quantify, describe and map urban growth. *Remote Sens. Environ.* **2003**, *86*, 275–285. [CrossRef]
30. Galster, G.; Cutsinger, J. Racial Settlement and Metropolitan Land Use Patterns: Does Sprawl Abet Black-White segregation? *Urban Geogr.* **2007**, *28*, 516–553. [CrossRef]
31. Johnson, M.P. Environmental Impacts of Urban Sprawl: A Survey of the Literature and Proposed Research Agenda. *Environ. Plan. A* **2001**, *33*, 717–735. [CrossRef]
32. Ewing, R. Is Los Angeles-Style Sprawl Desirable? *J. Am. Plan. Assoc.* **1997**, *63*, 107–126. [CrossRef]
33. Whyte, W. *The Exploding Metropolis*; Doubleday: Garden City, NY, USA, 1958.
34. Garreau, J. *Edge City: Life on the New Frontier*; Anchor Books: New York, YN, USA, 1992.
35. Barnes, K.B.; Morgan, J.; Roberge, M.C.; Lowe, S. *Sprawl Development: Its Patterns, Consequences and Measurement*; Retrieved 11 March 2021 from SPRAWL DEVELOPMENT: (towson.edu); Towson University: Towson, MD, USA, 2001.
36. Feng, Q.; Gauthier, P. *Urban Sprawl and Climate Change: A Survey of the Pertinent Literature on Physical Planning and Transportation Drivers-Report*; Financed by OURANOS and MITACS; OURANOS: Montreal, QC, Canada, 2018.
37. Breheny, M. The Compact City and Transport Energy Consumption. *Trans. Inst. Br. Geogr.* **1995**, *20*, 81. [CrossRef]
38. Galster, G.; Hanson, R.; Ratcliffe, M.R.; Wolman, H.; Coleman, S.; Freihage, J. Wrestling Sprawl to the Ground: Defining and measuring an elusive concept. *Hous. Policy Debate* **2001**, *12*, 681–717. [CrossRef]
39. Pendall, R. Do land-use controls cause sprawl? *Environ. Plan. B* **1999**, *26*, 555–571. [CrossRef]
40. Wong, D. *The Modifiable Area Unit Problem (MAUP), In The SAGE Handbook of Spatial Analysis*; Fortheringham, A.S., Rogerson, P.A., Eds.; SAGE Publications, Ltd.: Newbury Park, CA, USA, 2009. [CrossRef]
41. Flowerdew, R. *Understanding the Modifiable Areal Unit Problem*. 2009. Available online: https://www.ed.ac.uk/files/imports/fileManager/RFlowerdew_Slides.pdf (accessed on 28 November 2017).
42. Cervero, R.; Kockelman, K. Travel demand and the 3Ds: Density, diversity, and design. *Transp. Res. Part D* **1997**, *2*, 199–219. [CrossRef]
43. Boarnet, M.G. A Broader Context for Land Use and Travel Behavior, and a Research Agenda. *J. Am. Plan. Assoc.* **2011**, *77*, 197–213. [CrossRef]
44. Ewing, R.; Hamidi, S.; Gallivan, F.; Nelson, A.C.; Grace, J.B. Structural equation models of VMT growth in US urbanised areas. *Urban Stud.* **2014**, *51*, 3079–3096. [CrossRef]
45. Cervero, R.; Murakami, J. Effects of Built Environments on Vehicle Miles Traveled: Evidence from 370 US Urbanized Areas. *Environ. Plan. A: Econ. Space* **2010**, *42*, 400–418. [CrossRef]
46. Glaeser, E.L.; Kahn, M.E. The greenness of cities: Carbon dioxide emission and urban development. *J. Urban Econ.* **2010**, *67*, 404–418. [CrossRef]
47. Karathodorou, N.; Graham, D.J.; Noland, R.B. Estimating the effect of urban density on fuel demand. *Energy Econ.* **2010**, *32*, 86–92. [CrossRef]
48. Fercovic, J.; Gulati, S. Comparing household greenhouse gas emissions across Canadian cities. *Reg. Sci. Urban Econ.* **2016**, *60*, 96–111. [CrossRef]
49. Norman, J.; MacLean, H.L.; Kennedy, C.A. Comparing high and low residential density: Life-cycle analysis of energy use and greenhouse gas emissions. *J. Urban Plan. Dev.* **2006**, *132*, 10–21. [CrossRef]
50. Nichols, B.G.; Kockelman, K. Urban form and life-cycle energy consumption: Case studies at the city scale. *J. Transp. Land Use* **2015**, *8*, 115–128. [CrossRef]
51. Nichols, B.G.; Kockelman, K.M. Life-cycle energy implications of different residential settings: Recognizing buildings, travel, and public infrastructure. *Energy Policy* **2014**, *68*, 232–242. [CrossRef]
52. Senbel, M.; Giratalla, W.; Zhang, K.; Kissinger, M. Compact development without transit: Life-cycle GHG emissions from four variations of residential density in Vancouver. *Environ. Plan. A* **2014**, *46*, 1226–1243. [CrossRef]
53. Osorio, B.; McCullen, N.; Walker, I.; Coley, D. Understanding the Relationship between Energy Consumption and Urban Form. *Athens J. Sci.* **2017**, *4*, 115–142. [CrossRef]
54. Tomlinson, R.W. *CLIMATE CHANGE—Land Use, Land-Use Change and Carbon Stocks—Synthesis Report (ENVIRONMENTAL RTDI PROGRAMME 2000–2006)*; Environmental Protection Agency: Johnstown Castle, Ireland, 2006.
55. United States Environmental Protection Agency (EPA). *Our Built and Natural Environments: A Technical Review of the Interactions between Land Use, Transportation, and Environmental Quality*, 2nd ed.; 2013. Available online: https://www.epa.gov/smartgrowth/our-built-and-natural-environments (accessed on 17 March 2018).
56. Hankey, S.; Marshall, J.D. Impacts of urban form on future US passenger-vehicle greenhouse gas emissions. *Energy Policy* **2010**, *38*, 4880–4887. [CrossRef]
57. Brownstone, D.; Golob, T.F. The impact of residential density on vehicle usage and energy consumption. *J. Urban Econ.* **2009**, *65*, 91–98. [CrossRef]
58. Taniguchi, M.; Matsunaka, R.; Nakamichi, K. A time-series analysis of the relationship between urban layout and automobile reliance: Have cities shifted to integration of land use and transport? *Urban Transp. XIV* **2008**, *101*, 415–424. [CrossRef]

59. Poumanyvong, P.; Kaneko, S.; Dhakal, S. Impacts of urbanization on national transport and road energy use: Evidence from low, middle and high income countries. *Energy Policy* **2012**, *46*, 268–277. [CrossRef]
60. Grazi, F.; Bergh, J.C.V.D.; Van Ommeren, J.N. An Empirical Analysis of Urban Form, Transport, and Global Warming. *Energy J.* **2008**, *29*, 97–122. [CrossRef]
61. Holcombe, R.G.; Williams, D.W. Urban Sprawl and Transportation Externalities. *Rev. Reg. Stud.* **2010**, *40*, 257–273. [CrossRef]
62. Ewing, R.; Pendall, R.; Chen, D. Measuring Sprawl and Its Transportation Impacts. *Transp. Res. Rec. J. Transp. Res. Board* **2003**, *1831*, 175–183. [CrossRef]
63. Boarnet, M.G.; Sarmiento, S. Can Land-use Policy Really Affect Travel Behaviour? A Study of the Link between Non-work Travel and Land-use Characteristics. *Urban Stud.* **1998**, *35*, 1155–1169. [CrossRef]
64. Crane, R. The Influence of Urban Form on Travel: An Interpretive Review. *J. Plan. Lit.* **2000**, *15*, 3–23. [CrossRef]
65. Bagley, M.N.; Mokhtarian, P.L. The impact of residential neighborhood type on travel behavior: A structural equations modeling approach. *Ann. Reg. Sci.* **2002**, *36*, 279–297. [CrossRef]
66. Bagley, M.N.; Mokhtarian, P.L.; Kitamura, R. A Methodology for the Disaggregate, Multidimensional Measurement of Residential Neighbourhood Type. *Urban Stud.* **2002**, *39*, 689–704. [CrossRef]
67. Bertolini, L.; Le Clercq, F.; Kapoen, L. Sustainable accessibility: A conceptual framework to integrate transport and land use plan-making. Two test-applications in the Netherlands and a reflection on the way forward. *Transp. Policy* **2005**, *12*, 207–220. [CrossRef]
68. Barbour, E.; Deakin, E.A. Smart Growth Planning for Climate Protection. *J. Am. Plan. Assoc.* **2012**, *78*, 70–86. [CrossRef]

atmosphere

MDPI

Article

Urban Resilience of Shenzhen City under Climate Change

Weiwei Shao [1], Xin Su [1], Jie Lu [2], Jiahong Liu [1,*], Zhiyong Yang [1], Chao Mei [1], Chuang Liu [1] and Jiahui Lu [1]

1 State Key Laboratory of Simulation and Regulation of Water Cycle in River Basin, China Institute of Water Resources and Hydropower Research, Beijing 100038, China; shaoww@iwhr.com (W.S.); suxin199109@gmail.com (X.S.); yangzy@iwhr.com (Z.Y.); meichao@iwhr.com (C.M.); liuchuang1619@gmail.com (C.L.); lujiahui1215@gmail.com (J.L.)

2 School of Geographical Sciences, Liaoning Normal University, Dalian 116029, China; lj895232175@gmail.com

* Correspondence: liujh@iwhr.com

Abstract: The Chinese government attaches great importance to climate change adaptation and has issued relevant strategies and policies. Overall, China's action to adapt to climate change remains in its infancy, and relevant research needs to be further deepened. In this paper, we study the future adaptive countermeasures of Shenzhen city in the Pearl River Delta in terms of climate change, especially urban flood risk resilience. Based on the background investigation of urban flood risk in Shenzhen, this paper calculates the annual precipitation frequency of Shenzhen from 1953 to 2020, and uses the extreme precipitation index as a quantitative indicator to analyze the changes in historical precipitation and the impact of major flood disasters in Shenzhen city in previous decades. Based on the six kinds of model data of the scenario Model Inter-comparison Project (MIP) in the sixth phase of the Coupled Model Inter-comparison Project (CMIP6), uses the Taylor diagram and MR comprehensive evaluation method to evaluate the ability of different climate models to simulate extreme precipitation in Shenzhen, and the selected models are aggregated and averaged to predict the climate change trend of Shenzhen from 2020 to 2100. The prediction results show that Shenzhen will face more severe threats from rainstorms and floods in the future. Therefore, this paper proposes a resilience strategy for the city to cope with the threat of flood in the future, including constructing a smart water management system and promoting the development of a sponge city. Moreover, to a certain extent, it is necessary to realize risk transfer by promoting a flood insurance system.

Keywords: extreme precipitation; temporal change; CMIP6; prediction; urban resilience; sponge city

Citation: Shao, W.; Su, X.; Lu, J.; Liu, J.; Yang, Z.; Mei, C.; Liu, C.; Lu, J. Urban Resilience of Shenzhen City under Climate Change. *Atmosphere* **2021**, *12*, 537. https://doi.org/10.3390/atmos12050537

Academic Editors: Hideki Takebayashi and Jihui Yuan

Received: 20 March 2021
Accepted: 18 April 2021
Published: 22 April 2021

1. Introduction

Climate change is currently a global issue that affects human survival and sustainable societal development. In the past 100 years, the global climate has been characterized by warming [1]. The Fifth Assessment Report of the International Intergovernmental Panel on Climate Change (IPCC) states that the average global temperature increased by 0.65–1.06 °C from 1880 to 2012, and that the global temperature will continue to increase in the future [2,3]. In the context of continued global warming, changes in climate variability (such as monsoon, precipitation, and atmospheric circulation) will not only have significant regional hydrothermal impacts but will also further increase the risk of extreme climate events [4,5]. Although extreme climate events are low probability events, they are extremely sudden and destructive, and are difficult to predict [6]. In recent years, with the development of urbanization, problems have arisen that are associated with the increase in the area of hardened urban subsurfaces, the proliferation of impervious facilities, imperfect construction of drainage networks and other infrastructure, and the uncoordinated development of cities themselves. Among the challenges, heavy rainfall and flooding caused by extreme precipitation tend to result in internal flooding in cities with insufficient drainage capacity, increasing the frequency and risk of urban natural disasters and affecting regional infrastructure and economic development [7]. Therefore, the quantitative evaluation description, simulation, and early warning assessment of extreme precipitation events in cities has become a current research hotspot in the field of climate change and urban construction.

Domestic and international scholars have carried out a considerable amount of research on extreme climate change, simulation prediction, and urban response [8–10], mainly focusing on three aspects. Firstly, the quantitative research on extreme climate change, including the characterization definition, spatial and temporal evolution patterns, and the causes of extreme climate indices [11,12]. The Expert Team on Climate Change Detection and Indices (ETCCDI) of the World Meteorological Organization (WMO) has defined 27 extreme climate indices in terms of intensity, frequency, and duration using statistical methods such as the percentage method and threshold determination. These indices are widely used in the quantitative analysis of extreme climate change [13–16]. Secondly, the application of global climate model data to simulate and predict future extreme climate indices. The World Climate Research Program has developed the Coupled Model Inter-comparison Project (CMIP) as a database for making such predictions [17]. The latest version of the current CMIP series is CMIP6, which presents new prognostic scenarios using six integrated assessment models (IAMs), based on different shared socioeconomic pathways (SSPs) and the latest anthropogenic emission trends. The new scenarios not only include future social and economic changes (such as population, economic development, ecosystems, resources, systems, and social factors), but also future efforts to mitigate, adapt, and respond to climate change, with better resolution [18–20]. Many scholars have attempted to use the CMIP6 outputs to evaluate the characteristics of extreme rainfall under different scenarios in the future [21–23]. The result showed that extreme temperature and precipitation have shown increasing trends in this century overall, and these changes are more obvious in the CMIP6 simulation than in the CMIP5 simulation. However, not all models exhibit a better performance in a certain region. Therefore, the best choice is an optimal model prediction that can better reduce the uncertainty of the simulation. With the application of global climate model data, a large number of researchers have found that a multi-model ensemble (MME) has better evaluation performance than a single model [24,25]. The third aspect is that many scholars are committed to proposing an effective measure to enable cities to better cope with extreme climate change [26–29]. Therefore, many new concepts have been proposed, such as sponge city, Smart Water Management (SWM), flood insurance, etc.

Shenzhen is one of the most innovative cities with the highest degree of marketization, the most perfect market system and the most important economy in China. It plays an irreplaceable role in promoting the rapid development of the national economy. In recent years, with the acceleration of urban modernization and the continuous improvement of urban function positioning, Shenzhen has become increasingly sensitive to flooding disasters, and climate risks have also increased. However, the meteorological disasters caused by extreme rainfall cause serious loss of life and property almost every year, which greatly restricts economic and social development. For example, the 2008 typhoon "Fengshen" was the strongest typhoon encountered in Shenzhen. There were more than 90 waterlogged sites in the city, and at least 18 sites with water depths of more than 1 m; traffic jams in many road sections in the city were serious; in addition, the heavy rain caused landslides in 10 places in the city, the most serious being on the side of the Buji River. The slope of Shenhui Road collapsed more than 70 m, and nearly one million people were affected by the heavy rain in the city, causing a direct economic loss of about 490 million yuan [30]. Improving our understanding of the characteristics of flood disasters and optimizing flood control and disaster reduction countermeasures is the main subject and overall goal of Shenzhen's future urban flood control work.

At present, there is a lack of research using the CMIP6 outputs to predict extreme precipitation changes at the urban scale [31]. Meanwhile, there is a lack of systematic measures and suggestions on how cities should deal with climate change. Therefore, based on the simulation capabilities of different CMIP6 outputs, this paper reveals the evolutionary laws and characteristics of extreme climate events, and proposes countermeasures to manage extreme precipitation and urban waterlogging, with a view to minimizing disaster losses and improving the city ability to deal with climate risks.

2. Literature Review

2.1. Resilient City

As a complex giant system, cities are becoming more and more powerful at the same time they are becoming more and more vulnerable, such as floods caused by extreme climates. Therefore, how a city can respond to various changes and maintain its own vitality in the face of numerous challenges and crises has become an important issue that needs to be resolved.

The concept of elasticity originated from ecology and was put forward by the American scholar Holling in the 1970s [32]. With the development of research, the concept of resilience began to be combined with disciplines other than ecology. In the field of urban planning, Albert [33] proposed the concept of "resilient city", which is defined as the ability and degree of a city to dissolve and absorb changes before structural and process restructuring changes. The Resilience Alliance defines a resilient city as: the ability of a city or urban system to absorb and absorb external disturbances and maintain the original main features, structure, and key functions [34]. However, in fact, a resilient city includes not only the ability of the city system to adjust itself to respond to various negative uncertainties and sudden attacks, but also the ability to effectively transform those positive opportunities into capital [35].

2.2. Resilient Strategies

Frequent water disasters have caused people to reflect on whether the single-target engineering resistance strategy is reasonable, such as raising flood dikes and diverting water, and actively looking for other solutions. In this context, resilient strategies have received attention and development [36]. Resilient strategies are more resilient than resistance strategy and more adaptable to various uncertain changes [37]. Resilience strategies include structural measures and non-structural measures [38]. Structural measures include river net flow management, flood adaptation, hanging water and construction measures, while non-structural measures are reflected in flood policies and management.

China's 13th Five-Year Plan proposes that "sponge cities" are an important direction for the development of new urbanization. "Sponge city" is a new generation of urban stormwater management concept [39]. It means that the city can be like a sponge and has good flexibility in adapting to environmental changes and coping with natural disasters caused by rainwater. It can also be called a "water resilient city" [40]. The "sponge city" has the flexibility to adapt to environmental changes and respond to rain and flood disasters. The construction of "sponge city" will help solve the problem of urban waterlogging, improve the urban ecological environment, and improve the quality of life of the people [41].

At present, it is also a hot issue to promote smart water management with the help of emerging information technologies such as the Internet, cloud computing, big data, and artificial intelligence [42,43]. The urban flood control system mainly consists of a source monitoring system, a river management system, a flood forecasting system, an alarm system, a flood risk assessment system, and a flood-related database. The source flood level and flow are monitored through the source monitoring system, and then the flood level is forecasted through the flood forecasting system. The flood was assessed through the flood risk assessment system. Through these non-engineering measures, flood attacks can be avoided, prevented, or reduced, adapt to changes in various types of floods, and better utilize the benefits of flood control projects, thereby reducing flood losses [44].

In addition, in many developed countries, flood insurance is often an important part of disaster prevention laws, and catastrophe insurance plays a vital role in disaster relief and protection of people's lives and property [45,46]. In the 20th century, although the American flood insurance measures were unable to restrain the upward trend of flood losses, they effectively reduced the ratio of flood losses to GNP and restricted flood risks to acceptable limits [47]. As an important non-engineering measure for flood control, will play an increasingly important role in transferring flood risk [48].

3. Materials and Methods

3.1. Study Area

Shenzhen is located between 113°43′–114°38′E and 22°24′–22°52′N (Figure 1). It is a coastal city in the south of China with an area of 1997.47 km^2. Shenzhen experiences a southern subtropical monsoon climate with variable weather in spring, prevailing easterly winds, a long summer, and a short winter. The climate is mild, with mean annual temperature of 22.4 °C. Shenzhen is rich in precipitation resources. Annual rainfall of 1933.3 mm occurs during the rainy season from April to September each year. The prevailing winds are southeasterly and easterly, with an influence from tropical cyclones 4–5 times per year on average [49].

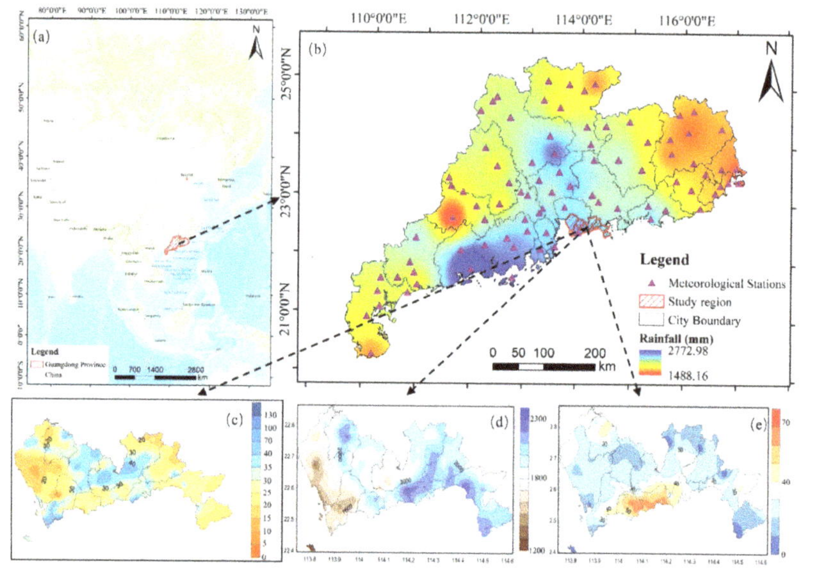

Figure 1. Study area location and rainfall characteristics. (**a**) Location of Guangdong Province; (**b**) spatial distribution of the mean annual rainfall in Guangdong Province from 2000 to 2019; (**c**) distribution map of maximum 1 h rainfall in Shenzhen from 19–22 May 2020; (**d**) distribution of accumulated rainfall in Shenzhen throughout 2019; (**e**) maximum 30 min rainfall distribution in Shenzhen from 21:00 to 23:00 on 11 April 2019. Where (**c**–**e**) were derived from the Shenzhen Climate Bulletin.

As of the end of 2019, the city had a built-up area of 927.96 km^2, an urban population of 13.438 million, and an urbanization rate of 100%. It is the first fully urbanized city in China. Recently, Shenzhen has become a national economic center and a national innovation city with rapid urban development. However, due to its unique geographical location, population density, and social conditions, severe weather such as strong convection, low temperature and rainfall, tropical cyclones, thunderstorms, and heavy rain frequently occur, causing serious economic losses.

As a result of climate change, the rainfall pattern in Shenzhen has changed considerably, the frequency of heavy rainfall has increased, and multiple climate risks such as heavy rains, floods, and typhoons have been superimposed, which will increase the risk of future urban floods. In 2019, the temporal and spatial distribution of precipitation is extremely uneven, the precipitation is concentrated in stages, and the rain is strong. The city's annual average rainfall is 1882.9 mm. Figure 1c,d respectively display two typical rainstorm events. The heavy rain on 11 April 2019, lasted just one hour and 42 min, with a maximum rainfall of 65.5 mm. The maximum one-hour rainfall on 22 May 2020 reached 153.7 mm, setting a new record [30]. Shenzhen is a typical and representative city, so it was selected as a study area for assessment of flood risk impacts under climate change, which will be of great reference significance for other cities in China to cope with climate change.

3.2. Data Sources

CMIP6 is the sixth stage of CMIP. It has the largest number of experimental models, the most complete experimental design, and the largest amount of data simulation since the implementation of the CMIP plan [50]. This study used six global climate models from ScenarioMIP, i.e., the scenario model comparison program of CMIP6 (Table 1), with different spatial resolutions for each model; more details can be found at https://esgf-node.llnl.gov/search/cmip6/, accessed on 20 April 2021.

Table 1. Global climate model information.

Serial Number	Model Name	Country	Institution	Resolution
1	BCC-CSM2-MR	China	Beijing Climate Center (BCC)	$1.125^\circ \times 1.125^\circ$
2	CanESM5	Canada	Canadian Centre for Climate modelling and analysis (CCCma)	$2.81^\circ \times 2.81^\circ$
3	CMCC-CM2-SR5	Italy	Euro-Mediterranean Center on Climate Change (CMCC) Foundation	$1^\circ \times 1^\circ$
4	FGOALS-g3	China	Chinese Academy of Sciences (CAS)	$2.3^\circ \times 2^\circ$
5	IPSL-CM6A-LR	France	Institut Pierre Simon Laplace (IPSL)	$1.26^\circ \times 2.5^\circ$
6	MPI-ESM1-2-LR	Germany	Max Planck Institute for Meteorology (MPI-M)	$1.5^\circ \times 1.5^\circ$

The data in this study were historical experimental data from 1953 to 2014 and simulated data from 2020 to 2100 under four combined scenarios (SSP1-2.6, SSP2-4, SSP3-7.0, and SSP5-8.5), where SSP1-2.6 was the updated CMIP5 RCP2.6 scenario in CMIP6, which represents the combined effects of low vulnerability, low mitigation pressure, and low radiative forcing; SSP2-4.5 is the updated CMIP5 RCP4.5 scenario in CMIP6, which represents a combination of moderate social vulnerability and moderate radiative forcing. SSP3-7.0 was a new radiative forcing scenario in CMIP6, representing a combination of high social vulnerability and relatively high anthropogenic radiative forcing, which is important for IAM and climate change impact, mitigation, and adaptation (IAV) studies; SSP5-8.5 is the updated CMIP5 RCP8.5 scenario in CMIP6 and was the only shared socioeconomic pathway to achieve an anthropogenic radiative forcing of 8.5 W/m^2 by 2100.

In order to assess the simulation capability of climate variables output from the climate model for the historical reference period (1953–2014), the measured daily-scale precipitation data of Shenzhen Station from 1953 to 2020, which were derived from the China Meteorological Data Network (http://data.cma.cn/site/index.html, accessed on 20 April 2021), were used as the benchmark in this study.

3.3. Analysis Method

3.3.1. Extreme Precipitation Index Method

We selected 11 of the 15 extreme precipitation index indexes determined by the Expert Team on Climate Change Detection and Indices (ETCCDI) [51] (http://etccdi.pacificclimate.org/docs/ETCCDMIndicesComparison1.pdf, accessed on 20 April 2021), which are described in detail in Table 2. The extreme precipitation index was calculated based on the RclimDex 1.0 model, and the linear trend analysis method and the P test method were used to calculate the inter-annual change trends and to determine the significance of each index, the change characteristics of the extreme climate index were then analyzed.

Table 2. Definitions and classifications of extreme precipitation indicators.

Type	Index Code	Index Name	Definition	Unit
Intensity Index	R95p	Very wet days	Annual total PRCP when RR > 95th percentile	Mm
	R99p	Extremely wet days	Annual total PRCP when RR > 99th percentile	mm
	RX1day	Max 1-day precipitation amount	Monthly maximum 1-day precipitation	Mm
	Rx5day	Max 5-day precipitation amount	Monthly maximum consecutive 5-day precipitation	Mm
	PRCPTOT	Annual total wet-day precipitation	Annual total PRCP in wet days (RR $\geq$ 1 mm)	mm
	SDII	Simple daily intensity index	Annual total precipitation divided by the number of wet days (defined as PRCP $\geq$ 1.0 mm) in the year	Mm/day
Duration Index	CDD	Consecutive dry days	Maximum number of consecutive days with RR < 1 mm	Days
	CWD	Consecutive wet days	Maximum number of consecutive days with RR $\geq$ 1 mm	Days
Frequency Index	R10	Number of heavy precipitation days	Annual count of days when PRCP $\geq$ 10 mm	Days
	R20	Number of very heavy precipitation days	Annual count of days when PRCP $\geq$ 20 mm	Days
	R25	Number of days above 25 mm	Annual count of days when PRCP $\geq$ 25 mm, 25 is user defined threshold	Days

3.3.2. Evaluation of Climate Model Simulation Results

Taylor graph method [52]: The Taylor graph method is a method proposed by Taylor et al. to evaluate the similarity between two datasets. It also considers the correlation coefficient (COR), relative standard deviation (RSTD), and standard root mean square error (RMSD) between the two datasets. Furthermore, because these three values have a mathematical transformation relationship, they can be comprehensively displayed in the same graph, so as to compare the similarity between the two datasets more intuitively. In this paper, the simulation capabilities of different climate models were described by the Taylor diagram. For the set of simulated climate variables, X, and the set of observed climate variables, Y, the calculation methods of each statistic are as follows.

$$COR(X,Y) = \frac{\sum_{i=1}^{n}(X-\overline{X})(Y-\overline{Y})}{\sqrt{\sum_{i=1}^{n}(X-\overline{X})^2\sum_{i=1}^{n}(Y-\overline{Y})^2}} \quad (1)$$

$$RSTD(X,Y) = \sqrt{\sum_{i=1}^{n}\frac{(X-\overline{X})^2}{(Y-\overline{Y})^2}} \quad (2)$$

$$RMSD(X,Y) = \sqrt{RSTD(X,Y) - 2 \times RSTD(X,Y) \times COR(X,Y) + 1} \quad (3)$$

If the RSTD of simulated and observed values of climate variables is smaller and the COR is larger, the RMSD is smaller, thereby indicating a better fitting ability of the climate model.

MR (Metrics Rating) comprehensive evaluation [53]: In order to further evaluate the comprehensive simulation capability of each climate model, the COR, RSTD, and RMSD of the simulated and observed values of each extreme precipitation index were counted separately, with a total of 33 indicators, and the calculated comprehensive simulation capability of each climate model was ranked.

$$MR_j = rank\left(1 - \frac{\sum_{i=1}^{n} r_{ij}}{n \times m}\right) \quad (4)$$

Here, r_{ij} is the descending ranking of the simulation ability of climate models for individual indicators; m is the number of models (taken as 26); n is the number of evaluation indicators (taken as 18); and *rank* is the descending ranking of the comprehensive simulation ability of climate models. Note that the higher the ranking, the stronger the comprehensive simulation ability of extreme precipitation indices.

Comprehensive simulation scoring [54]: Based on the comprehensive evaluation results of climate models, the optimal MME average method was adopted to reduce the uncertainty of the simulation results of a single climate model. In order to determine the optimal number of models, a comprehensive simulation scoring index (CSS) that also considers the COR and RSTD between the simulated and observed extreme precipitation index values was adopted. The formula is as follows.

$$CSS = \frac{(1 + COR)^2}{\left(RSTD + \frac{1}{RSTD}\right)^2} \quad (5)$$

The larger the CSS, the better the performance of the climate model, and vice versa.

In summary, based on the background investigation of the urban flood risk in Shenzhen, this paper analyzed the changes in historical precipitation and the impacts of major floods in Shenzhen during the past 68 years. In addition, based on CMIP6 data, we comprehensively evaluated the simulation capability of different climate models for extreme precipitation in Shenzhen. On this basis, we preferably selected models for ensemble averaging, predicted the change characteristics of extreme precipitation in Shenzhen in the 21st century (2020-2100), and proposed a resilience strategy for integrated urban flood risk management. Figure 2 presents the technical roadmap of this paper.

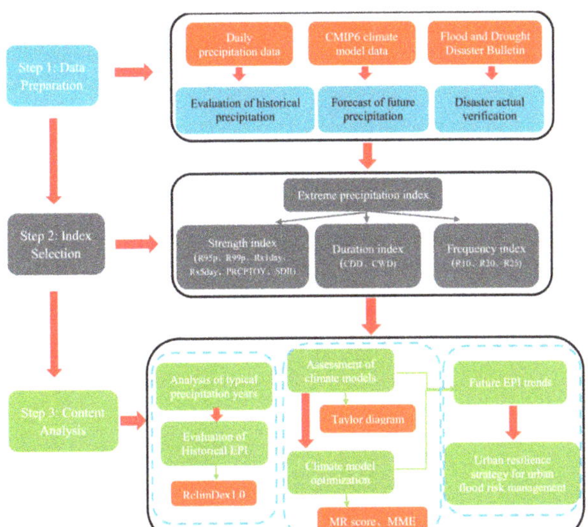

Figure 2. Overall research framework (see Table 2 for definitions and classifications of extreme precipitation indicators).

4. Results and Discussion

4.1. Analysis of Historical Precipitation Evolution in Shenzhen

4.1.1. Changes in Measured Precipitation in Shenzhen

Based on the measured precipitation data of the Shenzhen Station, annual and monthly precipitation in Shenzhen from 1953 to 2020 were determined. The change in precipitation is shown in Figure 3. The redder the color, the more rainfall, and the bluer the color, the less rainfall. The mean monthly rainfall in Shenzhen was 160.8 mm, and the maximum monthly rainfall was 1395.3 mm, with rainfall mainly concentrated in April-September (accounting for 85.11% of the annual rainfall). Extreme rainfall mainly occurred in June.

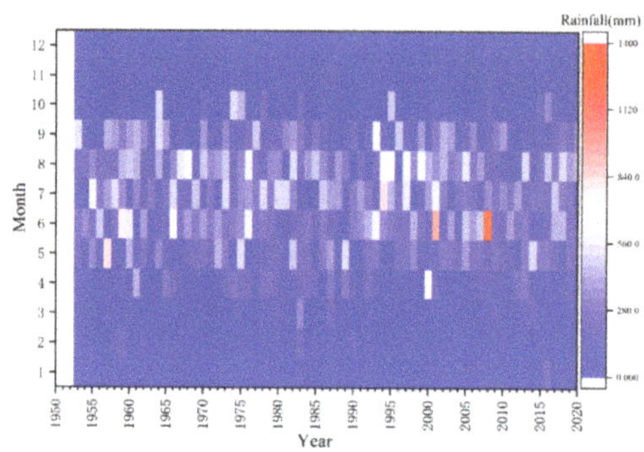

Figure 3. Interannual variation of precipitation in Shenzhen.

To better understand the evolutionary characteristics of extreme precipitation in Shenzhen, the annual precipitation in Shenzhen from 1953 to 2020 was calculated by ranking frequency, as shown in Figure 4. Statistically, 50% of the extremely wet and extremely dry years (i.e., years with an empirical frequency < 10% and an empirical frequency > 90%, respectively) occurred after 2000, indicating that extreme precipitation events have not only intensified in recent years but that their frequency has also increased and humans are facing more extreme climate challenges. In addition, three typical years were selected according to the cumulative frequency, namely, dry, normal, and humid years, which were 1963, 2002, and 2001 (empirical frequencies were 98.55%, 50.0%, and 1.45%, respectively). The accumulated precipitation in a drought year was 911.9 mm (i.e., 51.58% less than a normal year) and the lowest precipitation occurred in winter. Therefore, attention should be paid to the prevention and control of winter droughts. The cumulative precipitation in the wet year was 2747 mm (i.e., 45.9% more than in a normal year) and the highest daily precipitation in 2001 occurred on 27 June. Due to the influence of the low pressure trough in June, as of 27 June 2001, the cumulative rainfall in Shenzhen in June had reached 925.2 mm, breaking not only the 50-year historical record of 790.9 mm in June, but also the 50-year historical record of 826.2 mm of cumulative rainfall in a single month, with repeated heavy rainfall events [55].

The highest precipitation occurred in 2001 and 2008; however, the social and economic losses from heavy rains and floods were quite different. The empirical frequency of precipitation in 2001 was 1.45% and the second extremely wet year was 2008 with an empirical frequency of 2.90%. Although slightly more precipitation fell in 2001 than in 2008, flood damage was much greater in 2008 than in 2001, as shown in Table 3. From 13–14 June 2008, there was continuous heavy rainfall in Shenzhen. The rainfall was concentrated, with a wide range and high intensity. The recurrence period of heavy rain exceeded 1-in-100 years. More than 1000 waterlogging or flooding events of various degrees occurred throughout the city. The heavy rain caused more than 70 houses to collapse, 5 people were killed, 3 people went missing, and the direct economic loss equated to approximately 500 million yuan [56]. From 29–31 August 2008, there was an historically rare period of three consecutive days of extremely heavy localized rainfall, during which the maximum 12 h rainfall (333.9 mm), the maximum 24 h rainfall (417.2 mm), and the maximum 48 h rainfall (486.9 mm) all exceeded the previous August rainfall records. The city saw several reservoirs exceeding the flood control limit, several severe waterlogging, many traffic and power supply disruptions, landslides, and other dangerous situations [57]. This may be due to the fact that more extreme precipitation events occurred in 2008 than in 2001, in addition to the fact that the natural water circulation system was altered by the urbanization of Shenzhen, resulting in poor drainage and the occurrence of urban flooding problems with severe losses. This phenomenon also reflects the need to study extreme precipitation events.

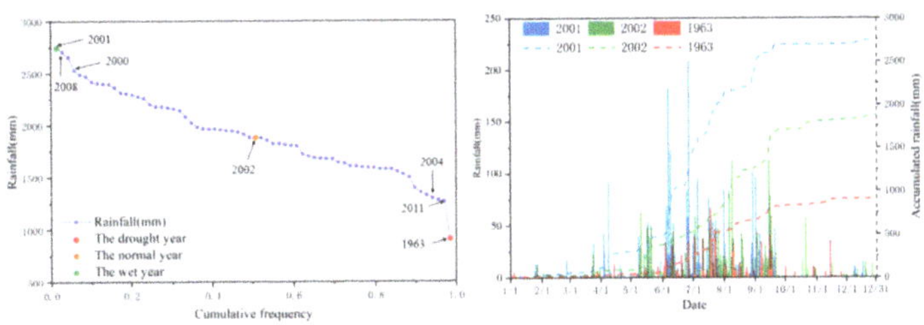

(a) (b)

Figure 4. Empirical frequency ranking of precipitation in Shenzhen (**a**) and daily and cumulative precipitation in extreme dry, normal, and extreme wet years (**b**).

Table 3. Comparison of heavy rainfall disasters in Shenzhen in 2001 and 2008 [58].

Disasters Caused by Floods	2001	2008
Affected population/million people	0.050	37.873
Number of dead (missing)	1	21
Collapsed house/room	0	88
Direct economic loss/billion yuan	0.3050	12

4.1.2. Time Series Evaluation of Historical Extreme Precipitation Indices in Shenzhen

Based on the meteorological data of Shenzhen Station from 1953 to 2020, the extreme precipitation index was calculated through RclimDex1.0 software using the least squares and local weighted regression method to analyze the time series changes of 11 extreme precipitation indices at Shenzhen Station in the past 68 years. As can be seen from Figure 5, all indices showed similar fluctuation trends in the time series, with biases in the 1960s and early 21st century. Among the 11 precipitation indices, PRCPTOT, SDII, CDD, R10, and R20 showed increasing trends; however, these trends were not significant, indicating that the extreme precipitation at Shenzhen Station did not change considerably during 1953–2020, and that precipitation showed a weakly increasing trend. Although the intensity of precipitation decreased, the number of consistently dry days decreased, the number of consistently wet days increased, and the frequencies of light and heavy rainfall also increased, indicating that extreme wet events were became more frequent during the 68 years from 1953 to 2020, leading to urban flooding, flash floods, mudslides, inundation of factory farmland, etc. The disasters caused by extremely heavy precipitation are often severe, and the risk of flash flooding has further increased.

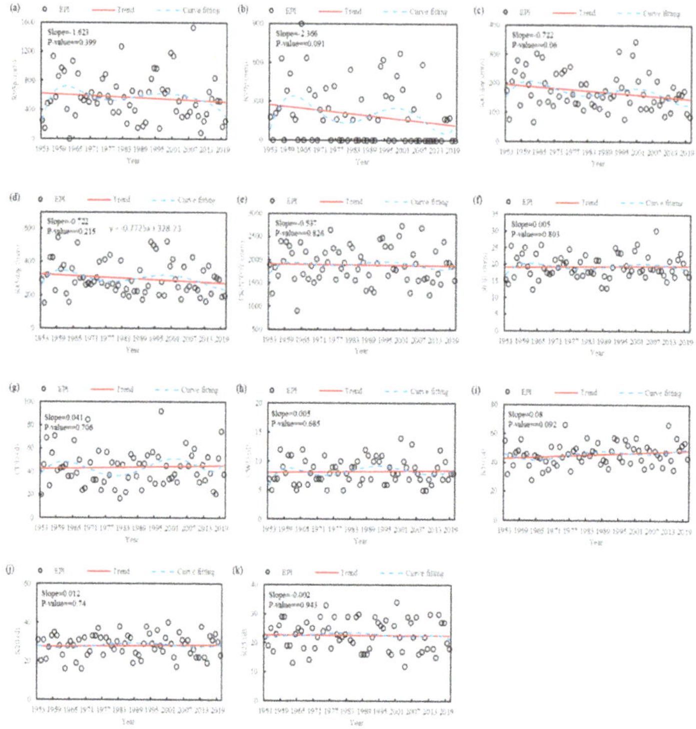

Figure 5. Linear trend and curve fitting of the extreme precipitation index (EPI) in Shenzhen. (**a**) R95p; (**b**) R99p; (**c**) Rx1day; (**d**) Rx5day; (**e**) PRCPTOT; (**f**) SDII; (**g**) CDD; (**h**) CWD; (**i**) R10; (**j**) R20; and (**k**) R25 (see Table 2 for definitions and classifications of extreme precipitation indicators).

4.2. Preferred Climate Model for Shenzhen and Prediction of Future Trend of Extreme Precipitation Index

4.2.1. Taylor Diagram-Based Simulation Evaluation

In order to objectively evaluate the simulating capabilities of the six climate models for Shenzhen's extreme precipitation index, in this paper we counted the 11 extreme precipitation indexes simulated by each climate model during the climate reference period and compared them with actual observations. The horizontal and radian axes of the figure indicate the RSTD and COR of the simulated values relative to the measured values, respectively, while the concentric circles surrounded by green dashed lines indicate the standard RMDS of the simulated values relative to the measured values, and the red dots indicate the fitting ability of different climate models for extreme precipitation, respectively.

From Figure 6, it can be seen that in terms of fitting ability, all the models had poor fitting ability for different extreme precipitation indices. Specifically, for the 11 extreme precipitation indices, the temporal CORs of most of the models were less than 0.3. In terms of RSTDs, the standard deviations and mean deviations of the six climate models were relatively large, which indicates that the interannual fluctuations of the 11 extreme precipitation indices were overestimated to different degrees by the six climate models, thus making the fitting deviations significantly larger. From the comparison of the fitting ability of different models, BCC-CSM2-MR and MPI-ESM1-2-LR were superior for each index, with relatively large CORs; however, the deviation of BCC-CSM2-MR remained large, and CanESM5 had the smallest COR, a large RSTD, and the worst simulation effect. The fitting ability of the other different models for each index was relatively close, at a medium level. Considering that the average of MME simulations works more effectively for most of the individual model results, this study used MR ranking and MME averaging for model preference for extreme temperature prediction, which has been widely accepted in other related studies [59,60] and can further improve its fitting ability.

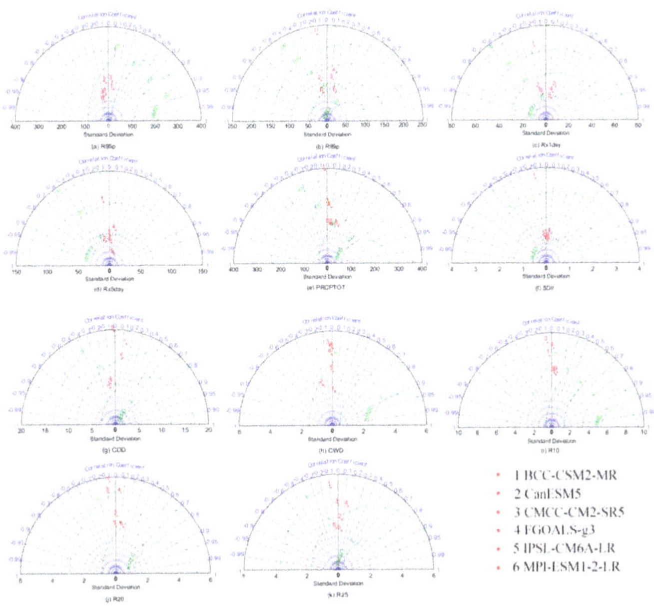

Figure 6. Taylor diagram of extreme precipitation index simulated by different climate models. (**a**) R95p; (**b**) R99p; (**c**) Rx1day; (**d**) Rx5day; (**e**) PRCPTOT; (**f**) SDII; (**g**) CDD; (**h**) CWD; (**i**) R10; (**j**) R20; and (**k**) R25 (see Table 2 for definitions and classifications of extreme precipitation indicators).

4.2.2. Preferred Extreme Precipitation index Model in Shenzhen

Optimal MME averaging can improve the simulation capability of climate models by offsetting the errors between different models and reducing the uncertainty of the simulation results of individual climate models [61,62]. In order to determine the optimal models set, this paper statistically ranked the scores of six climate models for different extreme precipitation index time fitting abilities from 1953 to 2020 based on COR, RSTD, and RMSD, and plotted the ranking results of different climate models, as shown in Figure 7 (note: For a certain extreme precipitation index, the COR, RSTD, and RMSD are indicated.) The final ranking of the best to worst models for the integrated simulation of extreme precipitation indices was BCC-CSM2-MR > PI-ESM1-2-LR > GOALS-g3 > PSL-CM6A-LR > MCC-CM2-SR5 > CanESM5. In addition, despite the differing sets of better and worse models for different indices, there remains a strong similarity, indicating that the assessment results of the climate models obtained by the MR composite score can be applied to each extreme precipitation index.

Finally, we used the comprehensive simulation scoring index (CSS) to confirm the number of optimal modes to be selected for MME; the larger the comprehensive simulation scoring index (CSS), the stronger the simulation ability. As can be seen from Figure 8, with the increase in the number of models, the overall simulation ability of MME for different extreme precipitation indices showed a trend of initially increasing and then decreasing. Therefore, according to the ranking of the comprehensive simulation capabilities of each model, we finally selected the optimal number of models for the arithmetic average, and obtained the result of the MME average. That is, for one mode that performs well, we selected the BCC-CSM2-MR modes for evaluation, such as R95p, R99p, Rx1day, Rx5day, PRCPTOT, SDII, R10, R20, and R25 indices. For the three modes that performed well, we selected the pooled mean of the BCC-CSM2-MR, MPI-ESM1-2-LR, and FGOALS-g3 modes, such as the CDD and CWD indices from Table 4, and it can be seen that the MME averaging results eventually improved the fitting ability of the extreme precipitation indices to some extent.

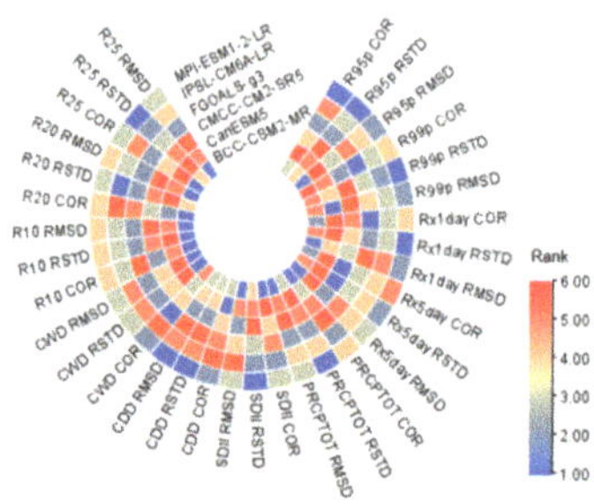

Figure 7. MR score ranking of climate model simulated extreme precipitation index (see Table 2 for definitions and classifications of extreme precipitation indicators).

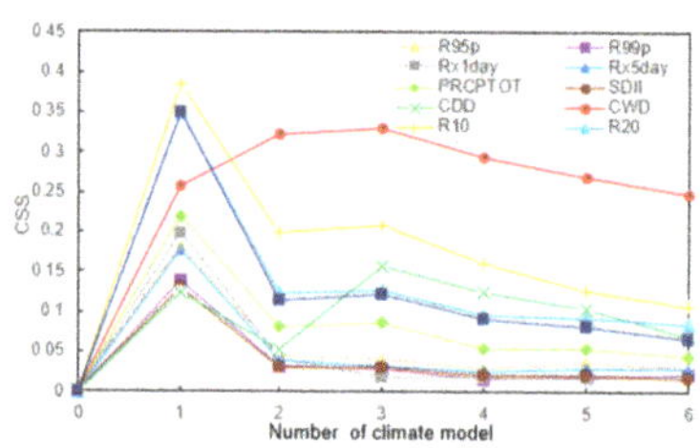

Figure 8. Trend of fitting effect of MME average simulated extreme precipitation index (see Table 2 for definitions and classifications of extreme precipitation indicators). CSS—comprehensive simulation scoring.

Table 4. Statistical characteristics of the fitting effect of the MME average simulated extreme precipitation index (see Table 2 for definitions and classifications of extreme precipitation indicators).

	R95p	R99p	Rx1day	Rx5day	PRCPTOT	SDII	CDD	CWD	R10	R20	R25
Correlation coefficient	0.15	−0.04	0.20 *	0.26 *	0.30 *	0.26 *	0.05	0.16	0.30 *	0.31 *	0.33
Standard deviation	0.44	0.47	0.44	0.38	0.43	0.31	0.46	0.89	0.74	0.62	0.60
Standard root mean square error	1.02	1.12	1.01	0.97	0.96	0.96	1.07	1.23	1.05	0.99	0.98

Note: * indicates that the data passed the $p < 0.05$ significance level test.

4.2.3. Trend Analysis of Future Extreme Precipitation Indices in Shenzhen

Figure 9 shows the relative changes of 11 extreme precipitation indices in Shenzhen under different future scenarios. Relative to the mean value of the indices in the base period of 1953–2020, very wet days (R95p) show a significant decreasing trend in the future; however, the extremely wet days (R99p) increase by means of 17.24%, 15.01%, 25.18%, and 31.94%, under the four scenarios, respectively. The occurrence of flooding is closely related to the precipitation intensity, and an increase in extremely wet days (R99p) in the future is likely to lead to an increase in flood risk and geological hazards. Among the 11 indexes, the maximum 1 d precipitation amount (Rx1day) and the maximum 5 d precipitation amount (Rx5day) have the largest increases, indicating that the extreme precipitation values in the future reflect climate change most significantly. The relative changes in the three frequency indices show that R25 > R20 > R10, that the number of heavy and stormy days will increase significantly in the future, and that the more concentrated the precipitation, the greater the possibility that flooding will occur. In addition, both CDD and CWD indices show insignificant trends in the future; however, they will increase by 0.31% and 0.10%, respectively, under the ssp5-8.5 scenario, indicating that droughts and floods may increase simultaneously in the future under the high discharge scenario. In addition, the annual total wet-day precipitation and simple daily intensity index are both positive, and the overall trend is weakly increasing. Comparing the relative rate of change of each index under the four emission scenarios, the relative change increases sequentially from SSP1-2.6 to SSP5-8.5. The increase of each index is larger in the high emission scenario, and the risk of socio-economic disasters caused by extreme rainfall in the future will increase accordingly.

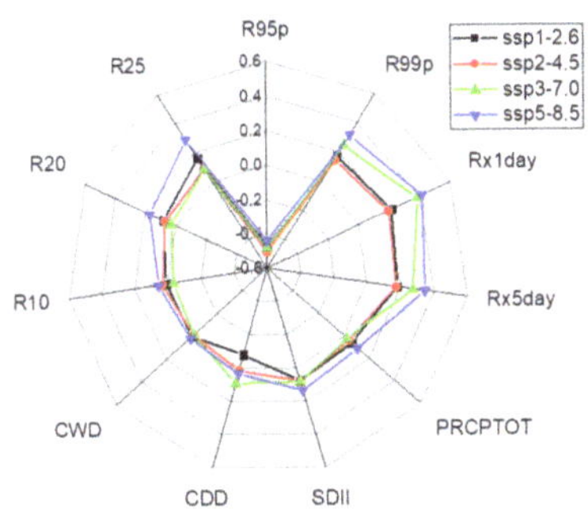

Figure 9. Relative changes in the extreme precipitation indices in Shenzhen under different scenarios (see Table 2 for definitions and classifications of extreme precipitation indicators).

Figure 10 shows the change series of 11 extreme precipitation indices under four future scenarios. It can be seen that the extreme precipitation indices in Shenzhen will show fluctuating changes in the future, among which the fluctuations in the number of extremely wet days, maximum daily precipitation, and heavy rainstorm days are large, indicating that extreme precipitation will tend to be unstable in the future. Unlike temperature, the precipitation elements do not change significantly under different future scenarios. For Shenzhen, R99p, Rx1day, Rx5day, R20, and R25 will be the most important influencing factors for the occurrence of flooding events in the future, especially R99p, Rx1day, and Rx5day, with maximum values of 1166.59 mm, 792.25 mm, and 992.53 mm, respectively. The Rx1day peak does not necessarily correspond to the Rx5day peak. It indicates that the impact of maximum 1 d precipitation amount on the overall maximum 5 d precipitation amount total is small, and that the increased occurrence of heavy precipitation events in the future and the short confluence time of floods caused by heavy precipitation will likely trigger flash floods, damage roads, inundate farmland, and cause considerable economic losses due to flooding. By the end of the 21st century, the numbers of days of moderate and heavy rain under the SSP5-8.5 scenario will increase by 16.93% and 29.33%, respectively, indicating that future rainfall will increase, mainly exhibiting increases in moderate and heavy rain, and that the growth trend is more significant than the growth trend of daily maximum precipitation. The extreme precipitation index does not change significantly under the SSP2-4.5 scenario, while there is a more pronounced overall upward trend relative to SSP5-8.5, indicating a greater likelihood of future extreme precipitation events under this scenario. Although the CDD, CWD, PRCPTOT, and SDII indices do not have a wide range of fluctuations, they all show overall increasing trends. In summary, extreme precipitation events will increase in the future, and the possible precipitation risk in Shenzhen will also increase. In this context, there is an urgent need to improve the flood warning system in Shenzhen to reduce flood risks and losses, to enable the city to better cope with climate change.

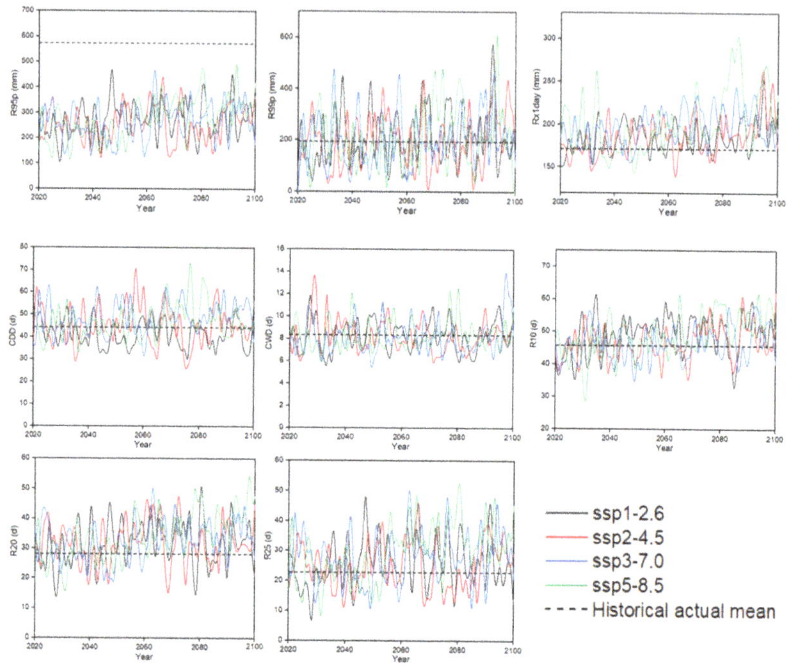

Figure 10. Variation series of extreme precipitation indicators in Shenzhen under four scenarios (see Table 2 for definitions and classifications of extreme precipitation indicators).

4.3. *Main Measures to Cope with Climate Change Risks in Shenzhen*

With the intensification of climate change, extreme weather events will be more frequent in the future. If Shenzhen is negatively affected by heavy rainfall and flooding for a long time period, it will not only cause huge economic losses to the city, but will also seriously threaten urban security. In the face of the increasingly serious flood situation, especially in high-density urbanized areas, flood prevention and drainage should be further applied as rigid constraints for urban construction based on the rigid water resource constraints. The key to integrated urban flood risk management is sponge city construction and urban resilience enhancement. In order to better cope with climate change and reduce climate risks and flood losses, on the basis of systematic assessment of climate risks in Shenzhen combined with the current situation of extreme precipitation risks and future prediction results, we propose an integrated urban flood risk management resilience strategy in terms of smart city construction, smart water utilities, sponge city construction, and system improvements.

4.3.1. Strengthening Smart City and Smart Water Utilities Construction

From the predicted results of the extreme precipitation index in Shenzhen, it is expected that the intensity of heavy rainfall will continue to reach new maximum levels, relying on natural ecological storage and purification methods is difficult to "absorb" the rain. Urban construction in Shenzhen should take into account the whole process of the natural-social water cycle. Not only can it effectively mitigate the impact of urban flood disasters but it can also consider water ecological protection and flood resource reuse. Spatial planning and land-use adjustment should enhance the spatial flood resilience of the city and realize the resource utilization of rainwater [63].

On 12 July 2018, the Shenzhen Municipal People's Government approved and issued the "Overall Plan for the Construction of a New Type of Smart City in Shenzhen", which specifies the working ideas and overall objectives, general framework, construction principles, implementation mechanism, and key projects, and guarantee measures for the construction of a smart city [64]. The Overall Plan states that Shenzhen should construct a new integrated smart city and develop a new national type of smart city benchmark city of world-class standard.

The construction of smart cities plays a crucial role in a cities' response to climate change [65–67]. In the face of urban flooding, digital means are used to empower urban drainage situation management, comprehensive scheduling and control of road traffic, and management of emergencies. This allows the realization of the prediction of crises, the management of emergency scheduling, and the rapid formulation of response measures to minimize disaster losses and improve a cities' response to climate risks. Smart water utilities can improve the lack of information perception in water security, build business systems (such as smart basin management, smart reservoir area, water diversion project management, whole process management of water projects, water administration and law enforcement, and a joint scheduling model of multiple water sources, etc.). This will enable realization of the intelligent support provided by information technology for water security and other businesses in Shenzhen and provide information security for the completion of key water services such as water quality control and river chief system management [68].

4.3.2. Promotion of Sponge City Construction

Due to the increased frequency of extreme precipitation events and the increased risk of urban flooding in the future, urban construction in Shenzhen faces additional requirements to cope with extreme weather. In view of this, the implementation of the sponge concept and resilience strategy and the construction of a sponge city that integrates water system management and flood control functions will greatly improve the current waterlogging prone situation of urban flooding and the urban ecological environment [69]. Sponge cities is a new generation of urban rainwater management concept, and refers to constructing cities to act like a sponge. This is achieved through strengthening urban planning and construction management, make the building, road and green space, water system and other ecosystems produce effect on rainwater absorption, storage, and slow release, and effective control of rainwater runoff, to realize the natural accumulation, infiltration and purification of rainfall in urban areas. Sponge cities are resilient in adapting to environmental changes and responding to natural disasters caused by rainwater and can also be referred to as "water-resilient cities" [70].

Shenzhen became one of the second batch of national sponge city pilot cities in 2016, and in January 2019, the Shenzhen Planning and National Resources Committee released the "Shenzhen Sponge City Construction Special Planning and Implementation Plan (Optimization)". The plan specifies that the overall goal is to minimize the impact of urban development and construction on the ecological environment through the construction of sponge cities and the comprehensive adoption of measures such as "seepage, retention, storage, purification, use, and discharge". In this context, the use of the internet, cloud computing, big database, artificial intelligence, and other emerging information technologies can also promote information regarding sponge cities. On the basis of the existing various types of water-related information management system, combining remote sensing big data and a spatio-temporal intelligence model, which will be integrated into the wisdom of the city cloud computing center, disaster risk warning, rainfall and flood storage, ecological environment monitoring, and other multi-functional integration, play a comprehensive and integrated ecological wisdom system services [71–73].

4.3.3. Engineering and Non-Engineering Measures

Cheng et al. conducted research in the Taihu Lake Basin and found that the urban flood risk could, in theory, be mitigated by improving flood control and drainage standard, such as the construction of flood control projects [74]. However, there is a threshold of the flood control and drainage standard, and when the standard is greater than this threshold, the cost will increase significantly; however, the flood control benefit is lower than the input cost [75,76]. Therefore, according to this feature, the present study proposes the following two measures:

1. Flood standard $\leq$ once in 50 years: Risk control as the main measure.

Improving urban flood resilience must be achieved by taking engineering and non-engineering measures to resist, absorb, and adapt to flood risks. Furthermore, in order to maintain urban structure and function, engineering measures (such as strengthening regional pumping capacity) and non-engineering measures (such as improving flood control planning) must be taken to enhance the responsiveness of cities to resist, absorb, and adapt to flood risks.

Engineering measures: These mainly rely on the construction or improvement of the engineering standards of urban drainage systems and flood control and drainage systems to treat and remove urban rainwater (mainly including urban rivers, municipal drainage pipelines, drainage pumping stations, dikes, sluices, reservoirs, etc.). It is also possible to increase the water storage capacity of cities by building sponge cities.

Non-engineering measures: These mainly include strengthening risk management, improving forecasting and early warning capabilities, improving flood control standards and related regulations, raising public awareness, and building smart water services and smart cities. Among them, strengthening risk management requires governments at all levels to improve the mechanism and system, and to improve the level of risk management in terms of three aspects: (1) Strengthening the government's comprehensive coordination and command, promoting departmental collaboration and linkage, and encouraging social forces to participate; (2) improving forecasting and early warning capabilities by establishing an independent flood forecasting and early warning system to forecast the flood characteristics of urban rivers based on the rainfall and water conditions in the upstream basin, and make scientific decisions; (3) improving flood control and drainage standards and related laws and regulations, i.e., unifying urban drainage standards and water conservancy drainage standards, and restricting and sanctioning economic and social activities that are not conducive to flood control and disaster mitigation through mandatory codes of conduct; (4) finally, the flood risk awareness and flood risk prevention capabilities of the society should be improved, and public participation in flood risk management should be strengthened. Redesigning the organizational structure enhances the flexibility and adaptability of the organization, and improves the flexibility of the system to adapt to various uncertain disturbances.

2. Flood standard > once in 50 years: Consider risk transfer.

There is a threshold value for engineering measures to reduce climate risk losses, and the cost effectiveness of engineering measures decreases sharply when the standard of flood control reaches > 50 years, which requires transferring flood risks through market-based means such as catastrophic insurance (i.e., flood insurance). Although insurance itself cannot reduce disaster losses, on the one hand it can relieve the government's economic pressure on flood relief, and on the other hand it can indirectly play a role in regulating urban flood prevention and mitigation [77,78].

Urban flood insurance is one of the main measures of urban flood risk management. The risks it bears mainly have three characteristics, as follows: (1) Non-eliminability—the risk of flooding can be reduced at a limited cost, but the risk is not completely eliminated; (2) relative predictability—compared with catastrophes such as earthquakes and tsunamis, flood disasters occur with a certain frequency and are, therefore, more controllable; (3) catastrophic characteristics—although urban flood disasters themselves have certain controllability, urban flood disasters still have the common characteristics of catastrophes with concentrated risk, high unpredictability, and huge losses [79].

At the same time, urban flood insurance also has two major attributes [80]: Insurability: region-specific and frequent flood risks are not strictly insurable risks. Therefore, flood insurance must be based on the regional flood control facilities reaching certain standards and ensuring a large insurance coverage in order to better comply with the accomplishment of large numbers of insurance and have a strong insurability; quasi-public product attributes: urban flood insurance is a product with obvious public welfare and high social benefits, but which also has a private product nature, i.e., it is a quasi-public product. This characteristic determines that flood insurance cannot be carried out by commercial insurance companies alone, and that its implementation must implement a government-led or government and market combined mechanism.

5. Conclusions

Based on the background investigation of the flood disaster risk in Shenzhen city, this paper analyzed the historical precipitation changes and the impact of major flood disasters in Shenzhen in recent decades. Using the six kinds of model data of the sixth stage scenario MIP of the CMIP6, we predicted the climate change trend of Shenzhen from 2020 to 2100. Further, on this basis, we proposed measures for cities to deal with future flood threats. The main findings were as follows:

1. The mean monthly rainfall in Shenzhen is 160.8 mm, and the maximum monthly rainfall is 1395.3 mm. Rainfall is mainly concentrated from April to September, during which the rainfall in Shenzhen Station accounts for 85.11% of the annual rainfall. Extreme rainfall mainly occurs in June. During the period from 1953 to 2020, extreme precipitation at Shenzhen Station changed insignificantly, and the total amount of precipitation showed a weakly increasing trend. Although the intensity of precipitation decreased, the number of persistently dry days decreased, the number of persistently wet days increased (though not significantly), and the frequencies of light and heavy rainfall increased. These results indicated that extreme wet events were more frequent and that the risk of heavy rainfall and flooding increased from 1953 to 2020.
2. The MR composite score shows that the models with the best to worst ability to simulate extreme precipitation indices in Shenzhen are BCC-CSM2-MR > PI-ESM1-2-LR > GOALS-g3 > PSL-CM6A-L > MCC-CM2-SR5 > CanESM5. The series of extreme precipitation index changes under the four scenarios indicate that future precipitation will tend to be unstable. Except for the R95p index, which shows a significant decrease in the future, other extreme precipitation indexes will generally increase. R99p, Rx1day, Rx5day, R20, and R25 will be the most important factors leading to flood events. In the future, extreme weather events will increase, and the risk of precipitation in Shenzhen will also increase.
3. The causes of flooding in Shenzhen are multifaceted, complex, and comprehensive. The weather process of short-duration heavy precipitation is the direct meteorological factor triggering flooding in Shenzhen, and the drainage capacity is the key factor for the occurrence of flooding. Specific resilience strategies for integrated urban flood risk management include strengthening the construction of new smart cities, promoting smart water utilities, and sponge city construction. In addition, risk control is the main measure when the flood standard is $\leq$ once in 50 years. When the flood standard is > once in 50 years, the main consideration is to transfer the flood risk by market-based means such as catastrophic insurance.

Author Contributions: Conceptualization, W.S. and J.L. (Jiahong Liu); methodology, W.S. and X.S.; software, X.S. and J.L. (Jie Lu); validation, J.L. (Jie Lu).; formal analysis, W.S., J.L. (Jiahong Liu), and Z.Y.; investigation, X.S., J.L. (Jie Lu), C.M., C.L., and J.L. (Jiahui Lu); resources, W.S. and J.L. (Jiahong Liu); data curation, X.S. and J.L. (Jie Lu); writing—original draft preparation, W.S., X.S., and J.L. (Jie Lu); writing—review and editing, W.S., X.S., J.L. (Jie Lu), and J.L. (Jiahong Liu); visualization, W.S., X.S. and J.L. (Jie Lu); supervision, J.L. (Jiahong Liu), Z.Y., and C.M.; project administration, W.S. and J.L. (Jiahong Liu). All authors have read and agreed to the published version of the manuscript.

Funding: This research received no external funding.

Institutional Review Board Statement: Not applicable.

Informed Consent Statement: Not applicable.

Data Availability Statement: All data used in this study are available upon request.

Acknowledgments: The researchers would like to extend their thanks to the Chinese National Key Research and Development Program (2018YFC1508203) and Chinese National Natural Science Foundation (Nos. 51979285, 51739011 and 51879274). We also want to express the thanks to the Climate Risk and Resilience in China (CRR) project supported by GIZ (Deutsche Gesellschaft für Internationale Zusammenarbeit) and SDC (Swiss Agency for Development and Cooperation). We also thank the anonymous reviewers and the Editor for their insightful comments and helpful suggestions to improve this manuscript. We also thank the help of World Climate Research Programme's Working Group on Coupled Modelling, which is responsible for CMIP.

Conflicts of Interest: The authors declare no conflict of interest.

References

1. Xiong, Y.; Ta, Z.; Gan, M.; Yang, M.L.; Yu, Y. Evaluation of cmip5 climate models using historical surface air temperatures in central Asia. *Atmosphere* **2021**, *12*, 308. [CrossRef]
2. Intergovernmental Panel on Climate Change (IPCC). *Climate Change 2013: The Physical Science Basis. Contribution of Working Group I to the Fifth Assessment Report of the Intergovernmental Panel on Climate Change*; Cambridge University Press: Cambridge, UK; New York, NY, USA, 2013; p. 1535.
3. IPCC. *Climate Change 2014: Synthesis Report*; Cambridge University Press: Cambridge, UK; New York, NY, USA, 2014.
4. Piras, M.; Mascaro, G.; Deidda, R.; Vivoni, E.R. Impacts of climate change on precipitation and discharge extremes through the use of statistical downscaling approaches in a Mediterranean basin. *Sci. Total Environ.* **2016**, *543*, 952–964. [CrossRef]
5. Aslam, A.Q.; Ahmad, S.R.; Ahmad, I.; Hussain, Y.; Hussain, M.S. Vulnerability and impact assessment of extreme climatic event: A case study of southern Punjab, Pakistan. *Sci. Total Environ.* **2017**, *580*, 468–481. [CrossRef] [PubMed]
6. Maxwell, S.L.; Butt, N.; Maron, M.; McAlpine, C.A.; Chapman, S.; Ullmann, A.; Segan, D.B.; Watson, J.E. Conservation implications of ecological responses to extreme weather and climate events. *Divers. Distrib.* **2019**, *25*, 613–625. [CrossRef]
7. Lancia, M.; Zheng, C.; He, X.; Lerner, D.N.; Andrews, C.; Tian, Y. Hydrogeological constraints and opportunities for "Sponge City" development: Shenzhen, Southern China. *J. Hydrol. Reg. Stud.* **2020**, *28*, 100679. [CrossRef]
8. Nyaupane, N.; Thakur, B.; Kalra, A.; Ahmad, S. Evaluating future flood scenarios using CMIP5 climate projections. *Water* **2018**, *10*, 1866. [CrossRef]
9. Lehtonen, I.; Jylhä, K. Tendency towards a more extreme precipitation climate in the Coupled Model Intercomparison Project Phase 5 models. *Atmos. Sci. Lett.* **2019**, *20*, e895. [CrossRef]
10. Alfieri, L.; Feyen, L.; Di Baldassarre, G. Increasing flood risk under climate change: A pan-European assessment of the benefits of four adaptation strategies. *Clim. Chang.* **2016**, *136*, 507–521. [CrossRef]
11. Wang, L.; Chen, S.; Zhu, W.; Ren, H.; Zhang, L.; Zhu, L. Spatiotemporal variations of extreme precipitation and its potential driving factors in China's North-South Transition Zone during 1960–2017. *Atmos. Res.* **2021**, *252*, 105429. [CrossRef]
12. Tegegne, G.; Melesse, A.M.; Alamirew, T. Projected changes in extreme precipitation indices from CORDEX simulations over Ethiopia, East Africa. *Atmos. Res.* **2020**, *247*, 105156. [CrossRef]
13. Liang, K.; Bai, P.; Li, J.; Liu, C. Variability of temperature extremes in the Yellow River basin during 1961–2011. *Quat. Int.* **2014**, *336*, 52–64. [CrossRef]
14. Sun, W.; Mu, X.; Song, X.; Wu, D.; Cheng, A.; Qiu, B. Changes in extreme temperature and precipitation events in the Loess Plateau (China) during 1960–2013 under global warming. *Atmos. Res.* **2016**, *168*, 33–48. [CrossRef]
15. Zolina, O.; Kapala, A.; Simmer, C.; Gulev, S.K. Analysis of extreme precipitation over Europe from different reanalyses: A comparative assessment. *Glob. Planet. Chang.* **2004**, *44*, 129–161. [CrossRef]
16. Liu, K.; Nie, G.; Zhang, S. Study on the Spatiotemporal Evolution of Temperature and Precipitation in China from 1951 to 2018. *Adv. Earth Sci.* **2020**, *35*, 1113–1126. [CrossRef]
17. Eyring, V.; Bony, S.; Meehl, G.A.; Senior, C.A.; Stevens, B.; Stouffer, R.J.; Taylor, K.E. Overview of the Coupled Model Intercomparison Project Phase 6 (CMIP6) experimental design and organization. *Geosci. Model Dev.* **2016**, *9*, 1937–1958. [CrossRef]

18. O'Neill, B.C.; Tebaldi, C.; Vuuren, D.P.V.; Eyring, V.; Friedlingstein, P.; Hurtt, G.; Knutti, R.; Kriegler, E.; Lamarque, J.; Lowe, J.; et al. The scenario model intercomparison project (ScenarioMIP) for CMIP6. *Geosci. Model Dev.* **2016**, *9*, 3461–3482. [CrossRef]
19. Ukkola, A.M.; de Kauwe, M.G.; Roderick, M.L.; Abramowitz, G.; Pitman, A.J. Robust future changes in meteorological drought in CMIP6 projections despite uncertainty in precipitation. *Geophys. Res. Lett.* **2020**, *47*, e2020GL087820. [CrossRef]
20. Ma, F.; Yuan, X.; Jiao, Y.; Ji, P. Unprecedented Europe heat in June–July 2019: Risk in the historical and future context. *Geophys. Res. Lett.* **2020**, *47*, e2020GL087809. [CrossRef]
21. Bracegirdle, T.J.; Krinner, G.; Tonelli, M.; Haumann, F.A.; Naughten, K.A.; Rackow, T.; Roach, L.A.; Wainer, I. Twenty first century changes in Antarctic and Southern Ocean surface climate in CMIP6. *Atmos. Sci. Lett.* **2020**, *21*, e984. [CrossRef]
22. Akinsanola, A.A.; Kooperman, G.J.; Pendergrass, A.G.; Hannah, W.M.; Reed, K.A. Seasonal representation of extreme precipitation indices over the United States in CMIP6 present-day simulations. *Environ. Res. Lett.* **2020**, *15*, 094003. [CrossRef]
23. Chen, H.; Sun, J.; Lin, W.; Xu, H. Comparison of CMIP6 and CMIP5 models in simulating climate extremes. *Sci. Bull.* **2020**, *65*, 1415–1418. [CrossRef]
24. Palmer, T.N.; Doblas-Reyes, F.J.; Hagedorn, R.; Weisheimer, A. Probabilistic prediction of climate using multi-model ensembles: From basics to applications. *Trans. R. Soc. B Biol. Sci.* **2005**, *360*, 1991–1998. [CrossRef] [PubMed]
25. Thomson, M.C.; Doblas-Reyes, F.J.; Mason, S.J.; Hagedorn, R.; Connor, S.J.; Phindela, T.; Morse, A.P.; Palmer, T.N. Malaria early warnings based on seasonal climate forecasts from multi-model ensembles. *Nature* **2006**, *439*, 576–579. [CrossRef] [PubMed]
26. Liu, Z.; Cai, Y.; Wang, S.; Lan, F.; Wu, X. Small and medium-scale river flood controls in highly urbanized areas: A whole region perspective. *Water* **2020**, *12*, 182. [CrossRef]
27. Linham, M.M.; Green, C.H.; Nicholls, R.J. *Costs of Adaptation to the Effects of Climate in the World's Largest Port Cities*; A VOID: London, UK, 2010; p. 225.
28. Lasage, R.; Veldkamp, T.I.E.; de Moel, H.; Van, T.C.; Phi, H.L.; Vellinga, P.; Aerts, J.C.J.H. Assessment of the effectiveness of flood adaptation strategies for HCMC. *Nat. Hazards Earth Syst. Sci.* **2014**, *14*, 1441–1457. [CrossRef]
29. Li, Y.; Guo, F.; Mao, K.; Chen, F. Response strategy for drought and flood in sponge city construction risk under the background of climate change. In *IOP Conference Series: Earth and Environmental Science*; IOP Publishing: Bristol, UK, 2019; Volume 252, p. 042015. [CrossRef]
30. Shenzhen Climate Bulletin. 2008. Available online: http://weather.sz.gov.cn/ (accessed on 20 April 2021).
31. Wu, W.; Mu, H.; Liang, Z.; Liu, X. Projected changes in extreme temperature and precipitation events in Shanghai based on CMIP5 simulations. *Clim. Environ. Res.* **2016**, *21*, 269–281. [CrossRef]
32. Holling, C.S. Resilience and Stability of Ecological Systems. *Annu. Rev. Ecol. Syst.* **1973**, *4*, 1–23. [CrossRef]
33. Alberti, M.; Marzluff, J.M.; Shulenberger, E.; Bradley, G.; Ryan, C.; Zumbrunnen, C. Integrating humans into ecology: Opportunities and challenges for studying urban ecosystems. *BioScience* **2003**, *53*, 1169–1179. [CrossRef]
34. Alliance, R. *Urban Resilience Research Prospectus. Australia*; CSIRO: Canberra, Australia, 2007; Volume 2. Available online: http://81.47.175.201/ET2050_library/docs/scenarios/urban_resilence.pdf (accessed on 20 April 2021).
35. Berkes, F.; Colding, J.; Folke, C. (Eds.) *Navigating Social-Ecological Systems: Building Resilience for Complexity and Change*; Cambridge University Press: Cambridge, UK, 2008.
36. Jabareen, Y. Planning the resilient city: Concepts and strategies for coping with climate change and environmental risk. *Cities* **2013**, *31*, 220–229. [CrossRef]
37. Wardekker, J.A.; de Jong, A.; Knoop, J.M.; van der Sluijs, J.P. Operationalising a resilience approach to adapting an urban delta to uncertain climate changes. *Technol. Forecast. Soc. Chang.* **2010**, *77*, 987–998. [CrossRef]
38. Meyer, V.; Priest, S.; Kuhlicke, C. Economic evaluation of structural and non-structural flood risk management measures: Examples from the Mulde River. *Nat. Hazards* **2011**, *62*, 301–324. [CrossRef]
39. Wang, H.; Mei, C.; Liu, J.; Shao, W. A new strategy for integrated urban water management in China: Sponge city. *Sci. China Ser. E Technol. Sci.* **2018**, *61*, 317–329. [CrossRef]
40. Liang, X. Integrated Economic and Financial Analysis of China's Sponge City Program for Water-resilient Urban Development. *Sustainability* **2018**, *10*, 669. [CrossRef]
41. Xia, J.; Zhang, Y.; Xiong, L.; He, S.; Wang, L.; Yu, Z. Opportunities and challenges of the Sponge City construction related to urban water issues in China. *Sci. China Earth Sci.* **2017**, *60*, 652–658. [CrossRef]
42. Yuanyuan, W.; Ping, L.; Wenze, S.; Xinchun, Y. A New Framework on Regional Smart Water. *Procedia Comput. Sci.* **2017**, *107*, 122–128. [CrossRef]
43. Zhang, Y.; Luo, W.; Yu, F. Construction of Chinese Smart Water Conservancy Platform Based on the Blockchain: Technology Integration and Innovation Application. *Sustainability* **2020**, *12*, 8306. [CrossRef]
44. Li, T. Study on the application of smart water affairs based on the concept of sponge city. *Water Supply Drain* **2017**, *53*, 129–135. [CrossRef]
45. Jiang, Y.; Luo, Y.; Xu, X. Flood insurance in China: Recommendations based on a comparative analysis of flood insurance in developed countries. *Environ. Earth Sci.* **2019**, *78*, 1–11. [CrossRef]
46. Taolue, F. Emergency Management of Foreign Cities from the Perspective of Flood Fighting. Available online: https://3g.163.com/news/article/FIQ74UR70519D828.html (accessed on 18 April 2021).

47. Cheng, X. *Strengthening Flood and Drought Disaster Management. The Common Trend of Global Water Control Strategy Adjustment. China Water Conservancy Technology Information Center. Summary of Urban and Rural Drinking Water Source Safety and Development*; China Water Conservancy Technology Information Center: Beijing, China, 2009; pp. 58–61.
48. Zhou, A. The difficulties and countermeasures of flood insurance in my country. *Prod. Res.* **2010**, *03*, 171–172. [CrossRef]
49. Xu, D.; Ouyang, Z.; Wu, T.; Han, B. Dynamic Trends of Urban Flooding Mitigation Services in Shenzhen, China. *Sustainability* **2020**, *12*, 4799. [CrossRef]
50. Zhang, L.; Chen, X.; Xin, X. Overview and review of CMIP6 Scenario Model Comparison Program (ScenarioMIP). *Clim. Chang. Res.* **2019**, *15*, 519–525. [CrossRef]
51. Yin, H.; Sun, Y. Characteristics of extreme temperature and precipitation in China in 2017 based on ETCCDI indices. *Adv. Clim. Chang. Res.* **2018**, *9*, 218–226. [CrossRef]
52. Taylor, K.E. Summarizing multiple aspects of model performance in a single diagram. *J. Geophys. Res. Atmos.* **2001**, *106*, 7183–7192. [CrossRef]
53. Tebaldi, C.; Hayhoe, K.; Arblaster, J.M.; Meehl, G.A. Going to the extremes: An intercomparison of model-simulated historical and future changes in extreme events. *Clim. Chang.* **2006**, *79*, 185–211. [CrossRef]
54. Li, L. *Spatial and Temporal Evolution Patterns of Extreme Temperature, Precipitation and Drought Events and Their Multi-Model Predictions*; Northwest Agriculture and Forestry University: Shaanxi, China, 2019.
55. Li, Y.; Yan, D.; Peng, H.; Xiao, S. Evaluation of precipitation in CMIP6 over the Yangtze River Basin. *Atmos. Res.* **2020**, *253*, 105406. [CrossRef]
56. Chen, X. Comparison and analysis of the causes of rainstorms and waterlogging in Beijing "7.21" and Shenzhen "6.13". *Water Resour. Dev. Res.* **2013**, *13*, 39–43. [CrossRef]
57. Huang, J.; Cao, W.; Wang, H.; Wang, Z. Affect path to flood protective coping behaviors using SEM based on a survey in Shenzhen, China. *Int. J. Environ. Res. Public Health* **2020**, *17*, 940. [CrossRef]
58. Wu, Y.; Li, H. Meteorological disasters and risk assessment in Shenzhen since 2000. *Guangdong Meteorol.* **2009**, *31*, 43–45. [CrossRef]
59. Schuenemann, K.C.; Cassano, J.J. Changes in synoptic weather patterns and Greenland precipitation in the 20th and 21st centuries: Evaluation of late 20th century simulations from IPCC models. *J. Geophys. Res.* **2009**, *114*, 1–20. [CrossRef]
60. Lu, X.; Rao, X.; Dong, W. Model evaluation and uncertainties in projected changes of drought over northern China based on CMIP5 models. *Int. J. Climatol.* **2021**, *41*, E3085–E3100. [CrossRef]
61. Seo, K.H.; Ok, J. Assessing future changes in the East Asian summer monsoon using CMIP3 models: Results from the best model ensemble. *J. Clim.* **2013**, *26*, 1807–1817. [CrossRef]
62. Zhao, T.; Chen, L.; Ma, Z. Simulation of historical and projected climate change in arid and semiarid areas by CMIP5 models. *Chin. Sci. Bull.* **2014**, *59*, 412–429. [CrossRef]
63. Schuetze, T.; Chelleri, L. Integrating decentralized rainwater management in urban planning and design: Flood resilient and sustainable water management using the example of coastal cities in the Netherlands and Taiwan. *Water* **2013**, *5*, 593–616. [CrossRef]
64. Liao, G. *Research on the Construction of a New Type of Smart City in Shenzhen*; Shenzhen University: Shenzhen, China, 2018.
65. Tian, Y.; Yang, M.; Jiang, Y. Research on Urban Smart Water Resources Emergency Management. *Appl. Mech. Mater.* **2013**, *409*, 75–78. [CrossRef]
66. Antzoulatos, G.; Mourtzios, C.; Stournara, P.; Kouloglou, I.-O.; Papadimitriou, N.; Spyrou, D.; Mentes, A.; Nikolaidis, E.; Karakostas, A.; Kourtesis, D.; et al. Making urban water smart: The SMART-WATER solution. *Water Sci. Technol.* **2020**, *82*, 2691–2710. [CrossRef]
67. Shahanas, K.M.; Sivakumar, P.B. Framework for a smart water management system in the context of smart city initiatives in India. *Procedia Comput. Sci.* **2016**, *92*, 142–147. [CrossRef]
68. Zhu, X.; Yin, Z.; Liu, Y.; Feng, B.; Wang, Y. Study on Framework Design of Smart Water Management System in Shenzhen. In *IOP Conference Series: Earth and Environmental Science*; IOP Publishing: Bristol, UK, 2019; Volume 330, p. 032008. [CrossRef]
69. Chan, F.K.S.; Griffiths, J.A.; Higgitt, D.; Xu, S.; Zhu, F.; Tang, Y.T.; Xu, Y.; Thorne, C.R. "Sponge City" in China—A breakthrough of planning and flood risk management in the urban context. *Land Use Policy* **2018**, *76*, 772–778. [CrossRef]
70. Shao, W.; Liu, J.; Yang, Z.; Yang, Z.; Yu, Y.; Li, W. Carbon reduction effects of Sponge City construction: A case study of the city of Xiamen. *Energy Procedia* **2018**, *152*, 1145–1151. [CrossRef]
71. Li, Y.J.; Zhang, C.; Leng, X.Y. Exploration and expectation of smart sponge city. *South North Water Transf. Water Sci. Technol.* **2016**, *14*, 161–164. [CrossRef]
72. Liu, J.; Gong, X.; Ren, X.; Liu, C.; Wu, L.; Wu, Y. Sponge city construction and innovation in Shenzhen. *J. Shenzhen Univ. Sci. Eng.* **2020**, *37*, 334–346. [CrossRef]
73. Zhang, W.; Che, W. Connotation and multi-angle analysis of sponge city construction. *Water Resour. Prot.* **2016**, *32*, 19–26.
74. Cheng, X.; Wu, H. *Flood Risk Scenario Analysis Method and Practice—Taking Taihu Lake Basin as an Example*; China Water Resources and Hydropower Press: Beijing, China, 2019.
75. Ma, R.-Y.; Du, Y.; Li, K. Study on flood control risk of flood control engineering system based on the clustering of measured data. *Clust. Comput.* **2019**, *22*, S6541–S6549. [CrossRef]

76. Ge, Y.; Liu, J.; Li, F. Research on assessment method for vulnerability transfer of insured enterprise: A case study of Changsha Region. *J. Nat. Disasters* **2008**, *17*, 81–85.
77. Scott, I.N. *The National Flood Insurance Program: Background, Issues and Reauthorization*; Nova Science Publishers, Inc.: Hauppauge, NY, USA, 2021.
78. Surminski, S. The role of insurance in reducing direct risk: The case of flood insurance. *Int. Rev. Environ. Resour. Econ.* **2014**, *7*, 241–278. [CrossRef]
79. Landry, C.; Turner, D. Risk Perceptions and Flood Insurance: Insights from Homeowners on the Georgia Coast. *Sustainability* **2020**, *12*, 10372. [CrossRef]
80. Atreya, A.; Czajkowski, J. Graduated flood risks and property prices in Galveston County. *Real Estate Econ.* **2019**, *47*, 807–844. [CrossRef]

atmosphere

MDPI

Article

Effects of Densification on Urban Microclimate—A Case Study for the City of Vienna

Wolfgang Loibl [1,*], Milena Vuckovic [2], Ghazal Etminan [1], Matthias Ratheiser [3], Simon Tschannett [3] and Doris Österreicher [4]

1 AIT Austrian Institute of Technology, Giefinggasse 6, 1210 Vienna, Austria; ghazal.etminan@ait.ac.at
2 VRVis Zentrum für Virtual Reality und Visualisierung Forschungs-GmbH, Donau-City-Straße 11, 1220 Vienna, Austria; vuckovic@vrvis.at
3 Weatherpark GmbH, Gardegasse 3, 1070 Vienna, Austria; matthias.ratheiser@weatherpark.com (M.R.); simon.tschannett@weatherpark.com (S.T.)
4 Department of Landscape, Spatial and Infrastructure Sciences, Institute of Spatial Planning, Environmental Planning and Land Rearrangement, University of Natural Resources and Life Sciences, 1190 Vienna, Austria; doris.oesterreicher@boku.ac.at
* Correspondence: wolfgang.loibl@ait.ac.at; Tel.: +43-505509-4587

Abstract: Climate adaptation, mitigation, and protecting strategies are becoming even more important as climate change is intensifying. The impacts of climate change are especially tangible in dense urban areas due to the inherent characteristics of urban structure and materiality. To assess impacts of densification on urban climate and potential adaptation strategies a densely populated Viennese district was modeled as a typical sample area for the city of Vienna. The case study analyzed the large-scale densification potential and its potential effects on microclimate, air flow, comfort, and energy demand by developing 3D models of the area showing the base case and densification scenarios. Three methods were deployed to assess the impact of urban densification: Micro-climate analysis (1) explored urban heat island phenomena, wind pattern analysis (2) investigated ventilation and wind comfort at street level, and energy and indoor climate comfort analysis (3) compared construction types and greening scenarios and analyzed their impact on the energy demand and indoor temperatures. Densification has negative impacts on urban microclimates because of reducing wind speeds and thus weakening ventilation of street canyons, as well as accelerating heat island effects and associated impact on the buildings. However, densification also has daytime cooling effects because of larger shaded areas. On buildings, densification may have negative effects especially in the new upper, sun-exposed floors. Construction material has less impact than glazing area and rooftop greening. Regarding adaptation to climate change, the impacts of street greening, green facades, and green roofs were simulated: The 24-h average mean radiant temperature (MRT) at street level can be reduced by up to 15 K during daytime. At night there is only a slight reduction by a few tenths of 1 K MRT. Green facades have a similar effect on MRT reduction, while green roofs show only a slight reduction by a few tenths of 1 K MRT on street level. The results show that if appropriate measures were applied, negative effects of densification could be reduced, and positive effects could be achieved.

Keywords: urban microclimate; urban fabric; urban densification; microclimate simulations; urban heat island effect; climate adaption measures; thermal comfort; building refurbishment; building simulation

Citation: Loibl, W.; Vuckovic, M.; Etminan, G.; Ratheiser, M.; Tschannett, S.; Österreicher, D. Effects of Densification on Urban Microclimate—A Case Study for the City of Vienna. *Atmosphere* **2021**, *12*, 511. https://doi.org/10.3390/atmos12040511

Academic Editors: Hideki Takebayashi and Jihui Yuan

Received: 28 March 2021
Accepted: 13 April 2021
Published: 17 April 2021

1. Introduction

Although adaptation to climate change has always been highlighted as important alongside the mitigation of climate change [1], more emphasis has been laid in the past on climate protection. The Paris Agreement in November 2016 [2] has finally stressed protection and adaptation as equally important, strengthening the efforts of many countries that have already developed national adaptation strategies and motivating others to follow. Urban climate and climate change are issues of growing importance with respect to urban neighborhoods as well as entire cities.

Urban land cover is estimated to substantially increase by around 1,527,000 km^2 until 2030 [3] and the urban population is growing worldwide at a steady rate with a rise from 55% in 2018 to a projected 68% of people living in cities across the globe in 2050 [4]. On the other hand, the number of people per household is decreasing, a fact that could be mostly attributed to socio-demographic trends that include aging population, declining fertility rates, single parent households, and single households. At the same time, the floor space of individual homes is increasing, especially in developed countries [5]. Globally, over the last 25 years in over 40% of urban centers the rate of demographic growth has been greater than the one of spatial expansion [6].

The city of Vienna is amongst those growing cities, with an estimated rise of almost 300,000 inhabitants until 2048 [7]. The administrative boundaries of the city have conversely remained untouched since 1954 and this will change in the future. To accommodate the rising population moving to the city, new residential buildings and associated infrastructure must be provided. Since the city does not want to expand excessively and is planning to keep its green belt (with forest, nature reserve, and agricultural areas) along its borders, the density of the already built-up area needs to be increased. But this shall be carried out by increasing the building height in a moderate way and to a lesser extent through sealing of additional surface area. However, much of the quality of life in the city is due to its large quantity of green spaces, with 31% of the overall urban area consisting of publicly available green spaces [8].

As the urban population steadily grows, anthropogenic heat flux is similarly increasing and subsequently speeding up the urban heat island effect. Thus, climate dynamics must be considered in combination with urban changes, requiring adaptive activities to better cope with urban dynamics under future climate change. This is of importance as temperatures are increasingly on the rise during the summertime. For Vienna, a moderate increase of up to 25 summer days (days with maximum air temperature exceeding 25 °C) is expected for the period 2021–2050 and 20 to 50 days per year for the period until 2100 [9]. While (horizontal) urban growth has been considered with respect to urban microclimate and climate change [10,11], urban densification has been less explored to be an issue affecting urban microclimate, specifically with respect to heat island effects and air flow. Overall densification measures may have a substantial impact on both microclimate and energy demand in cities; thus, energy-sensitive densification strategies must be integrated in long-term urban regeneration policies [12]. At the same time, adaptation and mitigation actions need to be carefully balanced in densely populated areas [13].

In a research project within the funding framework of the Austrian Climate Research Program (ACRP 10th call, grant number KR17AC0K13790), the project "CLUDEX—Climate Change and urban densification impact exploration" has carried out a series of interdisciplinary investigations into the effects of urban densification measures. This paper builds on the results and recommendations of the final activity report [14]. The project investigated dependencies between urban densification and urban climate under current climate and global warming conditions by means of wind, microclimate, and building simulations based on a case study in a densely built-up urban district of the city of Vienna. The study builds on the hypothesis that specific urban and building densification measures can positively or negatively impact the urban microclimate and energy demand, as well as comfort for its citizens. Subsequently, the research questions focused on how densification affects the urban microclimate and which mitigation measures on an urban and building planning level could be most effective. The key findings and conclusions are subsequently summarized in this paper.

2. Background

As climate change is progressively affecting our rural and urban habitats, the need for adaptation and mitigation measures is intensifying. In densely populated and built-up urban areas, the impact of climate change is accelerated as natural surfaces are largely replaced by sealed cover and building sites. This has the destructive effect that less green area is available and that the built infrastructure multiplies the amount of thermal mass and consequently heat storing materials in the third dimension. An increasing population, however, needs additional living space and associated infrastructure; thus, densification shall be conducted without additional land use. It is therefore key to specifically identify climate-sensitive urban systems and provide recommendations for appropriate measures.

Mahtta et al. identified in their analysis on 478 cities across the world with one million or more inhabitants five distinct urban growth and densification typologies: Budding outward, stabilized, outward, upward and outward, and mature upward [15]. Densification measures can also be combined with building refurbishment and can subsequently significantly contribute to an upgrade of certain neighborhoods. Attic extensions, if carried out to high efficiency standards, can increase the building quality and property value, and can improve the living comfort of the inhabitants and reduce the overall energy demand for heating and cooling [16].

When assessing urban growth and densification measures, especially also related to horizontal expansion where additional land might be sealed, the effects of greening should be considered. Blue-green infrastructure, defined as an interconnected network of natural and designed landscape components, is considered an essential element in mitigating the effects of climate change [17] and in particular in reducing the consequences that lead to increased flooding [18]. Green infrastructure can be implemented in a variety of measures. Categorizing them in terms of approach can help to accelerate implementation. Actions can be practical, educational, or participatory, and can be seen as positive in terms of their landscape for the micro-climate, health, and aesthetic value [19]. The urban microclimate can be improved if suitable mitigation measures are considered. Increasing vegetation, de-sealing surfaces and roof greening are cited as suitable actions within the "Urban Heat Island Strategy Plan" of the city of Vienna [20].

The retrofitting of roofs and facades with plants can thus contribute to ease the urban heat island effect [21]. Especially vulnerable communities, with less potential for alternative cooling options, can benefit from green roofs [22]. For facades it is, however, important to note that the plants should not hinder solar access during the winter months. Thus, green facades that lose their foliage in winter are preferable [23]. A study by Chun et al. has also shown that green urban spaces reduce temperatures during the summer and increase them during winter months, thus concluding that seasonal effects must be considered when implementing green infrastructure [24].

However as renewable energy increasingly needs to be incorporated into the building shell if ambitious climate targets are to be met, green roofs are also competing for space with the application of photovoltaic or solar thermal systems. Combining the benefits of green roofs with the generation of renewable energy is thus of utmost importance. Nevertheless, the number of solar green roofs is still low, with only a few applications in 2018 and virtually none a few years before [25]. Novel applications such as the "PV-Rooftop-Garden" provide suitable solutions to combine the positive effects of green roofs and PVs [26].

In buildings, summertime overheating becomes, also in temperate climates, increasingly problematic as consecutive days with high temperatures are on the rise. Several studies have already found that Urban Heat Islands (UHI) influence cooling energy consumption [27,28]. By reducing the internal room temperature through extensive roof greening, less energy is also required for active cooling systems [29]. Future energy scenarios show that highly insulated buildings, such as net zero energy dwellings, can have an increasing overheating risk [30]. A similar effect has also been assessed for passive houses [31]. Oikonomou et al. came to the conclusion that the thermal quality of a dwelling can have a greater impact on indoor temperatures than the UHI [32]. The current high insulation standards can thus have an adverse effect related to overheating and rapid cooling by lower night temperatures, not being easily achieved [33]. Nevertheless, there are a series of passive architectural measures, such as external solar shading, high thermal mass, and night ventilation that can also be implemented in order to avoid summertime overheating in well-insulated buildings [34]. Specific strategies can mitigate the effects of the high-quality building shell, even in passive houses and super-insulated buildings [35–37]. The type of construction similarly affects the energy demand but also the internal room temperature. A heavy-weight construction shows through the higher inertia a slower thermal response than a light-weight construction [38]. For Vienna, overheating in buildings is regulated within the respective building codes and evidence must be provided that overheating is kept within specific temperature boundaries [39].

The densification of the urban fabric, however, does not only affect the buildings as such but also the overall urban morphology of the city. Changing the height and shape of a larger number of buildings within a district can change the aerodynamic properties of the urban fabric over time. It can also increase or decrease ventilation and accelerate heat trapping in the city. The goal is therefore to improve the urban microclimate and at the same time increase thermal comfort in and around buildings and optimize energy efficiency. Addressing and evaluating the climate impact of the vertical extension of buildings is rather complex, especially if a variety of scenarios based on different policies must be explored.

On a building level, precise information about physical and functional characteristics can be carried out with a variety of simulations tools and organized in a building information model (BIM) [40]. Moving from the single-building dimension up to the urban or at least neighborhood scale requires a somewhat similar but to some extent different approach which is generally defined as city information modeling (CIM) and which is meant to enable street block and finally city-wide detailed knowledge of all relevant objects. To this extent, semantic 3D city models are relatively new, capable of dealing with the third dimension, and are conceived to help in overcoming the intrinsic difficulty of creating, collecting, and homogenizing large amounts of both spatial and non-spatial heterogeneous data [41]. The term semantic 3D city model does not only refer to geometry but also to semantics (e.g., building usage, construction date, and materials) and topology (e.g., adjacency to other buildings, shared walls). In this context CityGML (City Geography Markup Language) is applied—an open XML-based format for the storage and exchange of virtual 3D city models, accepted as international data model standard [42]. The benefits tied to a spatial-semantically coherent urban model [43] are various, as well as the possibilities to exploit such a model for applications ranging from urban planning to augmented reality to utility network management to energy simulation tools. An extensive list of further applications can be found in Biljecki et al. [44].

There is however a clear research gap in bringing the urban level (microclimate, comfort) together with the building level (internal comfort, energy) in relation to vertical densification. This is one of the key aspects this study aims at addressing in order to provide recommendations for future densification strategies in cities.

3. Methods

In order to assess impacts of densification on urban climate and potential adaptation strategies a densely populated Viennese district was modeled as a typical sample area for the city of Vienna. The case study focused on Meidling, the 12th district of Vienna, and analyzed the large-scale densification potential and its potential effects on microclimate, air flow, comfort, and energy demand. For the analysis, three methods as outlined below were deployed to assess the impact of urban densification. Micro-climate analysis (1) assessed the effect related to urban heat island phenomena. Wind pattern analysis (2) investigated wind speed and wind comfort at street level. The energy and internal comfort analysis (3) compared different construction types and greening scenarios and analyzed their impact on the energy demand and internal temperatures of buildings. In addition, adaptation alternatives were discussed with public stakeholders, and the results were fed into urban planning guidelines. More specifically, stakeholders from local authorities were involved to define adaptation measures to better cope with local heat increase, as well as to the effects of urban densification.

The densification scenarios were conducted for the entire central Meidling as a study area based on the building height zoning maxima as outlined in the current land use and zoning plan and subsequent zoning height regulations for the city of Vienna. Spatially explicit scenarios were carried out through 3D city models, considering building footprints and building height. The initial 3D city model was generated using the built-in generative algorithms within the Rhinoceros 3D plug-in environment Grasshopper [45]. The base date was derived from geodata, provided by the city of Vienna, showing building footprints with building height information, from street level to building eave. Information regarding the maximum allowed building height based on the Vienna zoning plan [45] was integrated into the geodata base through a spatial relation of the height class annotation of the development plan—the height zoning—to the building footprints of the geodata base using ArcGIS software. Once the initial geometric input was prepared, 3D geometry representing both base case and densification scenarios was generated. The process provided the automated upward extrusion of building footprints transferring the buildings' height information into 3D volumes (Figures 1 and 2). The difference between current and possible maximum building heights allowed us to estimate the additional number of possible floors (with and without attic extension) for each building. Based on the building footprint size and the number of additional floors, the additional potential gross floorspace was estimated and summarized for the entire area.

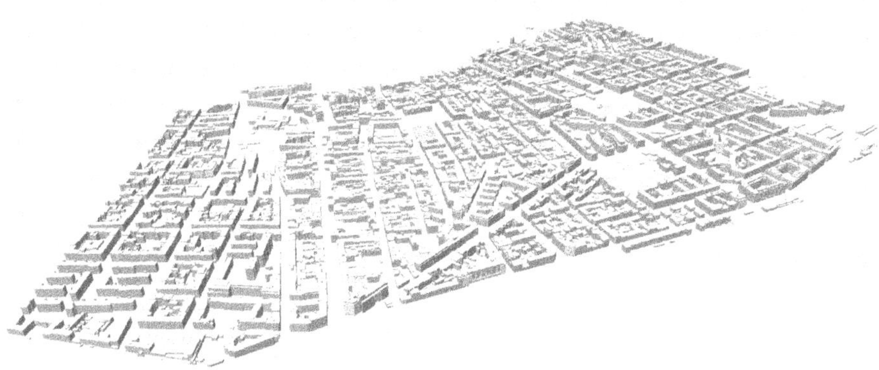

Figure 1. 3D model of the study area: Base case depicting the current volumes developed from the Land Use and Zoning Plan of the City of Vienna [46].

The overall floorspace extension potential was quantified for the residential and mixed-use buildings by taking current building height, maximum building height, and building footprint area from the building footprint geodata base. Hypothetical height extensions were only considered for buildings if the footprint exceeded 100 m^2 in order to cater for a realistic size for an additional flat. Differences between the current and the potential maximum elevation were only considered when the height difference was higher than 1.5 m to serve as additional floor space, at least as attic floor space. Backyard buildings were only included for height extension if their current height was above 4 m, otherwise they were considered as workshop buildings and backyard sheds, not feasible for residential floorspace extension.

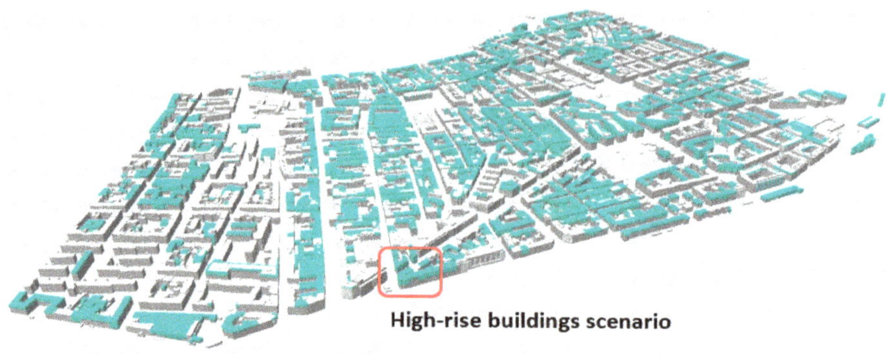

Figure 2. 3D model of the study area: Densification scenario including the volumes for maximum allowable building height (marked in blue) and the location of a hypothetical high-rise buildings cluster.

Compared to the actual building height distribution in Meidling, the height zoning regulation allowed for a substantial extension of many buildings where the height is currently well below the height zoning limits. The current gross floor area (GFS) in the study area was around 2.8 million m^2 with the GFS extension potential by means of regular floors amounting to 467.000 m^2 and including attic conversions up to 701.000 m^2. This resulted in a maximum GFS extension potential of 16%. With the inclusion of attic conversions, the growth potential could add up to 25% of the GFS. Thus, urban densification in central Meidling would allow for an additional 6.600 to 9.900 flats by considering an average flat size of around 94 m^2 GFS [47]. See Figure 2 for the general densification scenario including the maximum allowable building height. In addition, an area for a hypothetical high-rise building cluster with two buildings of 80 and 100 m, respectively, was defined to study the effects of increased heights on the urban microclimate as also indicated in Figure 2.

The base case and densification scenarios provide the basis for the analysis related to microclimate, wind, energy, and indoor climate as outlined below.

3.1. Microclimate Analysis

Microclimate simulations were conducted applying a set of plug-ins from the family of Ladybug tools such as the Grasshopper toolset within Rhinoceros 3D [48]. The Ladybug plug-in components inherit the physical principles and functionalities of its underlying simulation engines.

The Ladybug component Honeybee enabled us to simulate microclimatic effects in urban environments under different climate framework conditions to investigate urban heat island phenomena and related adaptation measures, by modeling the small-scale interactions between individual buildings, surfaces, and plants. The Ladybug components allowed the computation of complex interactions of the built environment and local climate by considering a unique position of the study domain and by solving intersections between solar vectors and physical obstructions of the built environment such as buildings or trees [49]. The simulations were conducted in three subsets as shown in Figure 3 to consider both the extent of the study area as well as more detailed simulations. One subset was subsequently carried out for the entire study area along the Meidlinger Hauptstraße (Figure 3a,b), a second one was conducted on a smaller sample area of some building blocks in the center of the Meidlinger Hauptstraße (Figure 3c,d), and a third one for a theoretical high-rise building cluster at the edge of the study area (Figure 3e,f). For a future climate scenario, the input weather file was modified using extreme heat day characteristics from

Austrian climate simulations, downscaled for Vienna to a 1×1 km grid for the year 2041, which constitutes the most extreme year until 2050 [50].

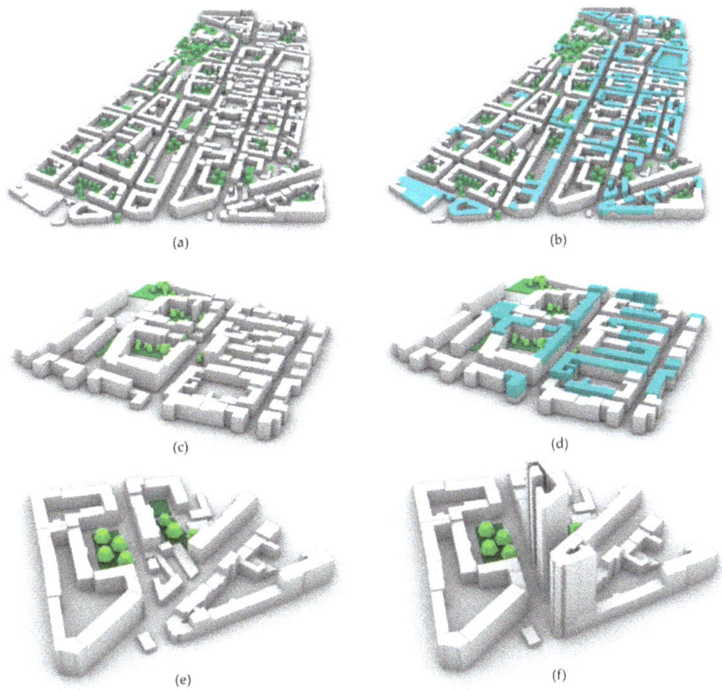

Figure 3. 3D model for the microclimate simulations for three subsets of the study area: Large domain for simulating large-scale general densification impacts for base case (**a**) and densification scenarios (**b**); sub-domain for simulating impacts of adaptation measures for base case (**c**) and densification scenarios (**d**); location for the simulation of the impact of a high-rise cluster for the base case (**e**) and densification scenarios (**f**).

3.2. Wind Analysis

To conduct wind field simulations for the main wind directions and speeds, the frequencies of wind directions and windspeed classes were explored based on the monitoring results for a 30-year climate period (1989–2018) and for the subset of heat days (>30 °C Tmax) during this period. The wind simulations were carried out using a computational fluid dynamics (CFD) model applying the software STAR-CCM+ [51] to calculate full 3D flow fields. Wind fields were modeled for the study area considering the current building height and the densification scenario by applying the main wind directions and wind speeds measured during heat waves. In addition, local wind fields were modeled for the hypothetical high-rise building cluster in the south of the study area. The goal was to combine the advantages of CFD wind simulations with a methodology for providing fresh air supply. While the classic urban ventilation models covering large areas used mostly parameterized land cover categories as surface roughness to represent buildings and terrain (e.g., [52]), with a CFD model, every single building of an urban quarter could be explicitly modeled. This enabled the generation of results that quantified the impact of large building structures or street canyons on the city's ventilation.

For the 30-year wind pattern analysis (1989–2018), data from a close weather station was applied. The analysis showed that the prevailing wind direction during heat days was southeast with rather low to moderate wind speeds, mostly between 3 and 5 m/s.

Wind from all other directions during hot days showed wind speeds between 1 and 3 m/s (Figure 4).

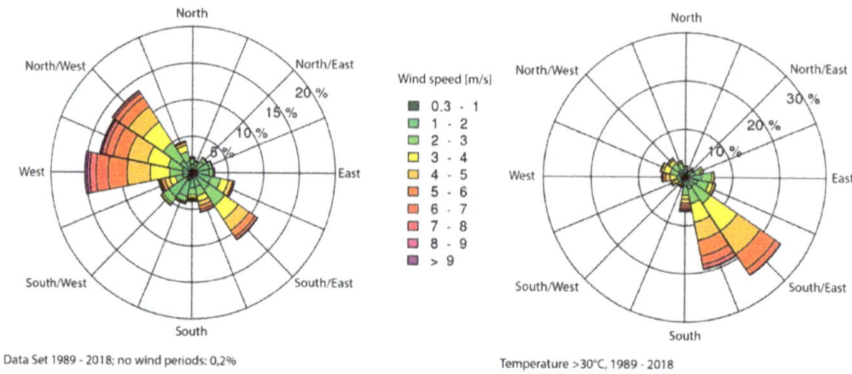

Figure 4. Windrose diagrams: Average wind speeds (color) and directions (bars) observed at the inner city weather station in Vienna for all weather situations (**left**) and for hours with temperatures > 30 °C (**right**).

Based on these datasets, uncomfortable wind speeds (exceeding 12 km/h) were analyzed. The map below displays the wind characteristics in central Meidling, combining all 16 wind directions related to uncomfortable conditions (Figure 5). Street areas where the number of hours per year with uncomfortable conditions exceeded 1.200 h (yellow, orange, and red) were considered to have a low wind comfort. Streets marked in green indicated high wind comfort. In the study area, most streets showed high wind comfort, with only several crossings and long, straight roads in line with the main wind directions, as well as the areas bordering the railway tracks in the south and the western, more open areas having lower wind comforts.

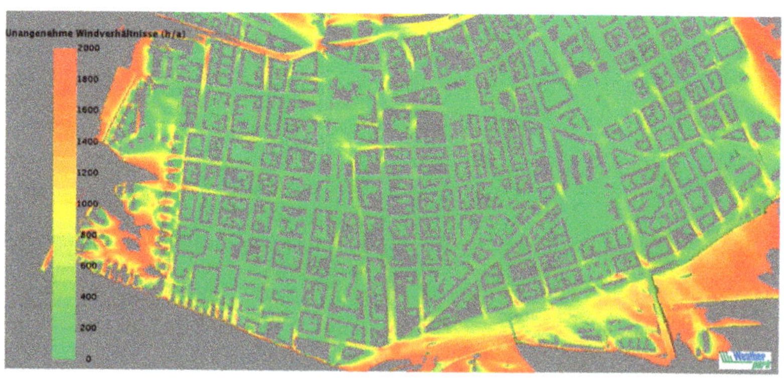

Figure 5. Number of hours per year (in the map's legend abbreviated as "h/a" (hours per annum)) with uncomfortable wind speeds.

3.3. *Indoor Climate Analysis*

Densification measures may also affect indoor climate and thus the overall energy demand of buildings for heating and cooling to provide indoor comfort temperatures. In order to assess effects of building height increase (of existing buildings) two exemplary buildings within the study area which showed a high potential for vertical extension and represent typical building types were selected for a detailed analysis.

The height of two buildings was (theoretically) extended through modeling to the maximum allowable building height based on the height zoning plan. For this theoretical extension, constructive aspects (such as potentials of load bearing elements) or building design improvement aspects (like facade structuring or inclusion of balconies) were discounted in this context. The buildings were subsequently simply extended based on the footprint of the underlying, existing buildings. The facade of the additional floors was assumed to match with the existing building (window to wall ratio). Only for the additional glazing case was the window area doubled in the last floor. Subsequently, the building could integrate 3 or 4 additional levels for flat roof extension and 2 or 3 levels and an attic extension for the slanted roof extension. In Figure 6, the additional floors of the extended building parts are shown in green and the attic conversions are shown in red.

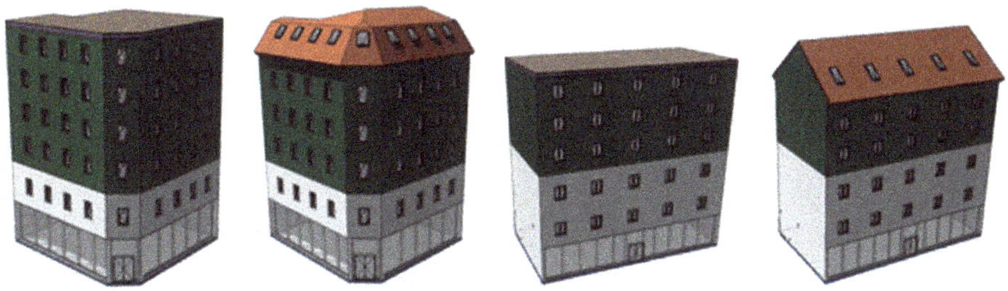

Figure 6. Theoretical vertical extensions of the case study buildings with the floor extensions shown in green with flat and slanted roofs [53,54].

The extensions of both building were modeled and analyzed with different construction types (lightweight and heavy-weight construction) and with greening of the roof and facade. The modeling was conducted in AutoCAD and SketchUp. Energy demand analysis was carried out with ArchiPhysik [55] and indoor climate comfort analysis with Thesim3d [56]. The simulation was compiled within two master thesis works within the framework of the project as documented in [53,54].

To compare the effects of different construction types, various aspects of the building shell and structure were altered as outlined in Table 1.

Table 1. Overview: buildings extensions with different construction types [53].

A. Roof Type	B. Glazing	C. Light-Weight Construction	D. Heavy-Weight Construction
Flat roof	Equal to existing building	Wall with cellulose insulation and inverted roof	Solid wood wall with intermediate insulation and inverted roof
		Ventilated wall with wooden paneling and inverted roof	Concrete wall with external insulation and inverted roof
	Twice that of existing building in last floor	Wall with cellulose insulation and inverted roof	Solid wood wall with intermediate insulation and inverted roof
		Ventilated wall with wooden paneling and inverted roof	Concrete wall with external insulation and inverted roof
Sloping roof	Equal to existing building	Wall with cellulose insulation and ventilated rafter roof	Solid wood wall with intermediate insulation and rafter roof
		Ventilated wall with wooden paneling and ventilated rafter roof	Concrete wall and roof with external insulation
	Twice that of existing building in last floor	Wall with cellulose insulation and ventilated rafter roof	Solid wood wall with intermediate insulation and rafter roof
		Ventilated wall with wooden paneling and ventilated rafter roof	Concrete wall and roof with external insulation

Roofs (column A) were modeled as flat roof and sloping roof. The glazing areas (column B) of the new highest floors are simulated in 2 versions: in version 1 the glazing area of the highest floor is identical to the glazing area of the floors below and in version 2 the glazing area of the last floor is twice as large as the glazing area of the lower floors. The overall structure of the buildings was compared light-weight- (column C) and heavy-weight-construction types (column D).

The build-up of the building shell followed the minimum requirements as defined in the Viennese Building Code, which in turn refers to the standards of the OIB Guideline 6 (OIB, 2019) in terms of energy efficiency in buildings. The U-values for the different construction types varied based on the actual wall or roof layers and overall build-up (Table 2). The different construction types were defined to meet but not exceed the requirements of the guideline. This was to ensure that only economically feasible construction types were applied for the different scenarios.

Table 2. Overview: Buildings extensions with different construction types [53].

Construction Type Wall	U-Value [W/m²K]	Construction Type Roof	U-Value [W/m²K]
Wall with cellulose insulation	0.190	Inverted roof	0.196
Ventilated wall with wooden paneling	0.162	Ventilated rafter roof	0.104
Solid wood wall with intermediate insulation	0.194	Rafter roof	0.115
Concrete wall with external insulation	0.152	Roof with external insulation	0.129

In addition to the different construction types, different greening scenarios for the roofs and facades were analyzed (Table 3). Differentiations were made in terms of the extent of the greening of the roof and the roof types were modeled with both a flat roof construction and sloping roof construction (column A). The construction of the roof was altered (column B), the flat roof with five different variants ranging from a traditional roof with no green layer to an extensive green roof and three options for the intensive green roofs with varying vegetation options between 10 and 90 cm. The U-values (column C) decreased with the thickness of the substrate as the additional green layer acted like additional insulation. For the sloping roof, a distinction was only made between no greening and a minor extensive greening with 15 cm. The limitation of the sloping roof lay in the actual angle, as only low height was feasible since substrate and vegetation layers could slide off with thicker construction heights. Since two different roof types were compared, the U-values differed, as the traditional roof was ventilated and thus featured a slightly better U-value.

Table 3. Overview: Buildings extensions with different green roof types [54].

A. Roof Type	B. Greening of the Roof	U-Value [W/m²K]
Flat Roof	No greening (no substrate layer)	0.186
	Extensive greening (10 cm)	0.181
	Intensive greening (20 cm)	0.179
	Intensive greening (45 cm)	0.174
	Intensive greening (90 cm)	0.159
Sloping roof	No greening (no substrate layer)	0.111
	Extensive greening (15 cm)	0.161

4. Results

In this section the key findings were compiled based on the three methodologies described above: The (1) microclimate analysis, the (2) wind analysis, and the (3) energy and indoor climate comfort analysis. Each analysis addressed the same area as previously outlined, covering larger or smaller sample areas depending on the actual specific research question. The different scales of the simulations allowed for a detailed assessment of the overall effects of the densification within the specified areas.

4.1. Microclimate Analysis: Key Results

The microclimate analysis was carried by modeling the mean radiant temperature (MRT) since the ambient air temperature alone did not show significant differences between shaded and sunny places. The MRT is regarded as the weighted sum of all long- and short-wave radiant fluxes (direct, reflected, and diffuse), to which a human body is exposed. For an outdoor space, the MRT thus depends on the temperature of the sky, ground, vegetation, and surrounding buildings [57]. The MRT was assessed as the daily average for 24 h, for the 12 daytime as well as the 12 nighttime hours using a reference period between 10 and 12 August 2014 to display representative boundary conditions for a heat wave in the Vienna. Within this analysis, the effect of the densification (increase in building height) was analyzed compared to the base case as well as the impacts of several greening scenarios on building and street level as outlined in the following Figures.

In a first assessment, the entire study area as displayed in Figure 3a,b was applied. A 24-h analysis was carried out for both the base case as well as the densification scenario (Figure 7).

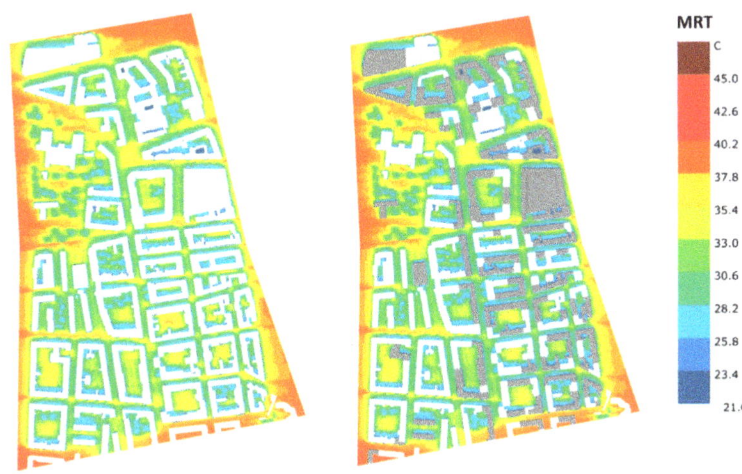

Figure 7. Twenty-four-hour mean radiant temperature (MRT) at street level for large-scale general densification impacts for base case (**left**) and densification scenario (**right**).

In order to assess the impact in more detail, a smaller sample area was extracted as previously depicted in Figure 3c,d. For this area a 12-h timeframe was chosen in addition to the 24-h to differentiate more clearly between daytime and nighttime temperatures (Figures 8 and 9).

The results show that general densification led to an obvious increase of shaded areas, which lead to in increased cooling over a period of time as the average exchange of radiative energy is reduced due to the incoming radiative flux being impeded by the deepened urban street canyons. This can be seen at street level between the buildings but also in the inner courtyards of the various building blocks. When buildings are extended on both sides of the street, the shading effect could be smaller as the upper floors are still exposed to incoming short-wave radiation. This effect could be more clearly seen in the smaller sample area, where the 24-h MRT showed a reduction of up to 4 K (Figure 8).

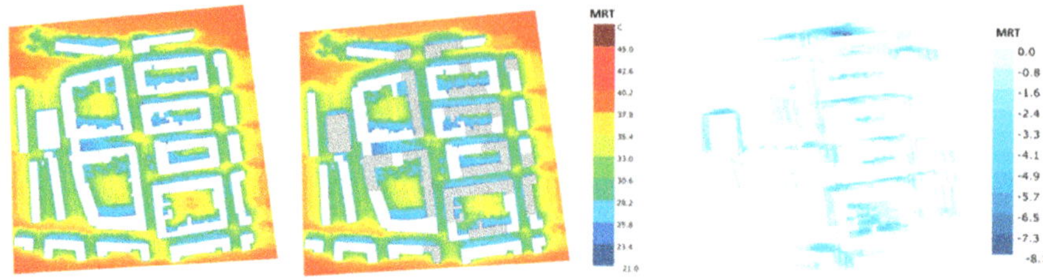

Figure 8. Twenty-four-hour MRT at street level for smaller sample area for base case (**left**) and densification scenario (**middle**) and 24-h MRT differences at street level between base case and densification scenario.

Whilst the 24-h analysis provided a meaningful indication of the overall shading effect due to the densification, the differentiation between daytime and nighttime temperatures delivered a clearer picture on the effects over time (Figure 9 (left) versus Figure 9 (right)) During the day, the cooling effect predominated due to the increased shading and subsequent reduced solar radiation. The long-wave radiation of the stored heat within the buildings did not change much at street levels due to the shadows of the higher buildings, thus protecting street level and lower floors from incoming short-wave radiation. The nighttime temperatures of the densification scenario were, however, higher compared to the based case due to the reduced sky view factor and subsequent lower nighttime cooling. The extended upper floors increased the total thermal mass of the building; however, at street level the downward air flow was limited inside the street canyons and courtyards.

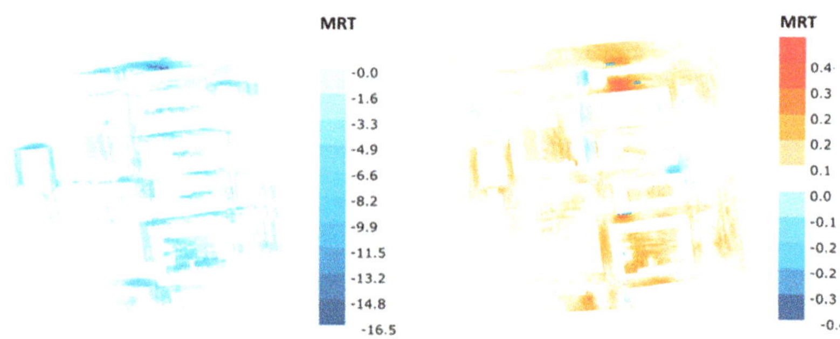

Figure 9. Twelve-hour MRT differences at street level between base case and densification scenario for smaller sample area for daytime (**left**) and nighttime (**right**).

In the third sample area, the impact of a high-rise building cluster as displayed in Figure 3e,f) was assessed. High-rise buildings lead to larger shaded areas over a longer daytime period in the vicinity of the buildings and subsequently to lower heat loads due to reduced solar radiation during the day. The results show that close to the high-rise building, there is a significant drop of MRT at street level from 38 to 23 °C over a period of 24 h (Figure 10, top left and right). The 12-h MRT analysis highlighted that the daytime temperatures could reach comfortable levels due to the heavily shaded areas, whilst the nighttime MRT did not change significantly and ranged only up to an increase of 0.5 K (Figure 10 bottom left and right).

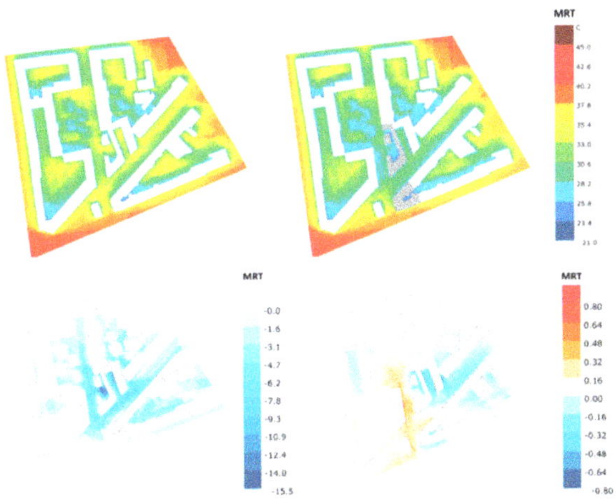

Figure 10. Twenty-four-hour MRT at street level for high-rise cluster for base case (top left) and densification scenario (top right) and 12-h MRT differences at street level between base case and densification scenario for smaller sample area for daytime (**left**) and nighttime (**right**).

In addition to the densification scenarios, the effect of several greening options on street and building level was explored within the small-scale sample area as depicted in Figure 3c,d. For this comparative analysis, only the scenario with the already increased building height was selected. In Figure 11, the scenarios for additional green roofs, green facades, and trees at street level are displayed in 3D models.

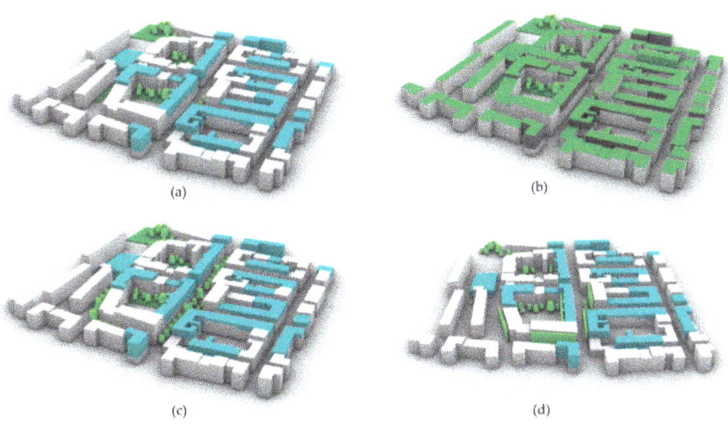

Figure 11. The 3D model of the smaller sample area for the microclimate simulations: for the initial densification scenario (**a**), for the densification scenario with green roofs (**b**), with additional trees at street level (**c**), and with green facades in a south-western street block (**d**).

The results of the 24-h street-level MRT distribution for the greening of the streets with additional trees are shown in Figure 12b: A positive cooling effect was observed due to the increased shading of the trees with an overall 24-h MRT reduction of up to 7 K. Trees can shield building facades from the incoming solar radiation, leading to lower energy demand for cooling in buildings and shielding road surfaces, mitigating incoming solar radiation to the surface and allowing additional evapotranspiration.

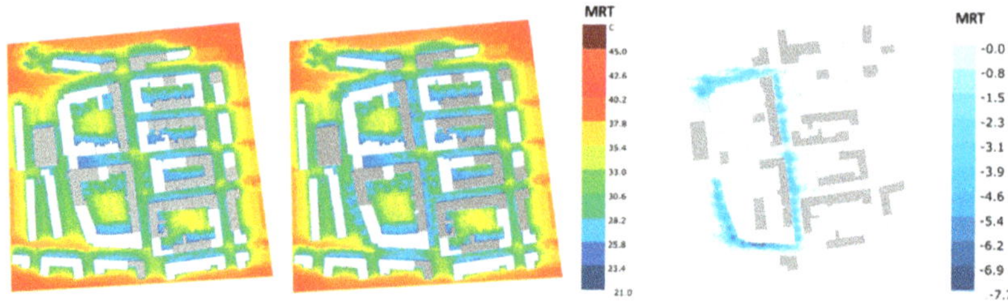

Figure 12. Twenty-four-hour MRT at street level for the smaller sample area with densification (**left**), densification and added trees at street level (**middle**), and the differences at street level between densification scenario without and with trees (**right**).

Whilst planting of additional green on street level showed clear benefits, the green roofs (see Figure 11c) did not show much positive effect on street level temperatures with an overall reduction of the 24-h MRT of around 0.5 K (Figure 13—left). The green roofs mostly contribute to lowering the air temperature on roof-top level rather than on street-level. Green facades (see Figure 11d) lead to lower temperatures in urban street canyons by shielding buildings from incoming solar radiation. Subsequently, heat exposure s reduced and cooling by evapotranspiration is increased, leading to a 24-h MRT reduction of up to 5 K (Figure 13—right).

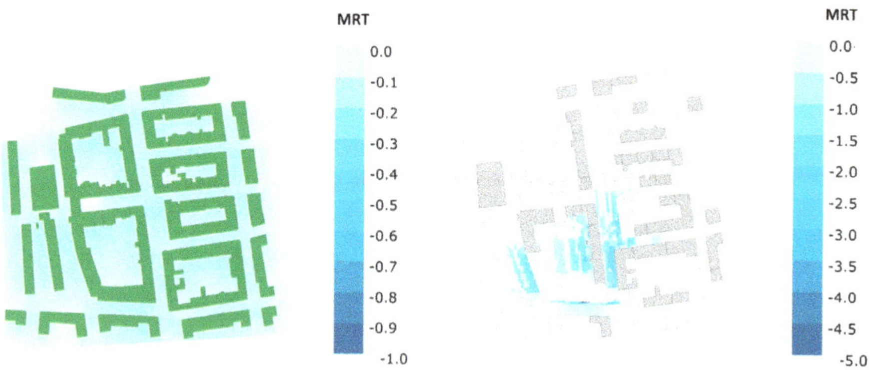

Figure 13. Twenty-four-hour MRT at street level for the smaller sample area with densification with additional green roofs (**left**) and green facades (**right**).

4.2. Wind Analysis: Key Results

The wind analysis highlighted the most significant effects the densification scenarios had on the wind speed and directions and subsequently the wind-comfort within the area. Wind simulations were carried out for the entire central Meidling area as shown in the previous section in Figure 5. The following Figures show examples of the key results. In Figure 14, which depicts one of the larger crossings at Meidlinger Hauptstraße, it can be seen that whilst the densification increased the wind speed at the crossing itself, the wind speed at the neighboring streets was decreased.

The length and direction of the green arrows in the Figure depict the changes in wind speed and direction; the yellow marker highlights the streets with the most significant changes.

Another street crossing was examined with wind direction from south-east (heat-wave scenario), as shown in Figure 15. The effect of increased building height could be observed in the right image compared with the left image. The wind was stronger in the base case, as it was not blocked by higher buildings. During heat waves, wind speeds can substantially decline by up to 50% in the densification scenario, which results in lower wind comfort as the ventilation in the street canyons decreases. Only on specific crossings, wind speeds could increase by up to 25%. During a typical west wind scenario, however, the conditions did not significantly change.

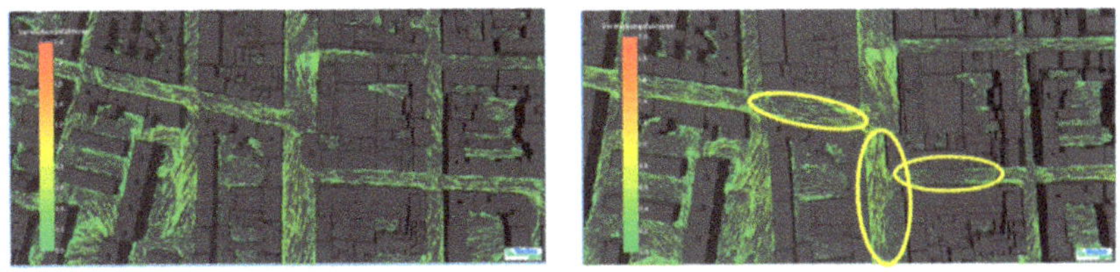

Figure 14. Changes in wind speed and direction for wind direction west for base case (**left**) and densification scenario (**right**) with streets with significant changes marked in yellow.

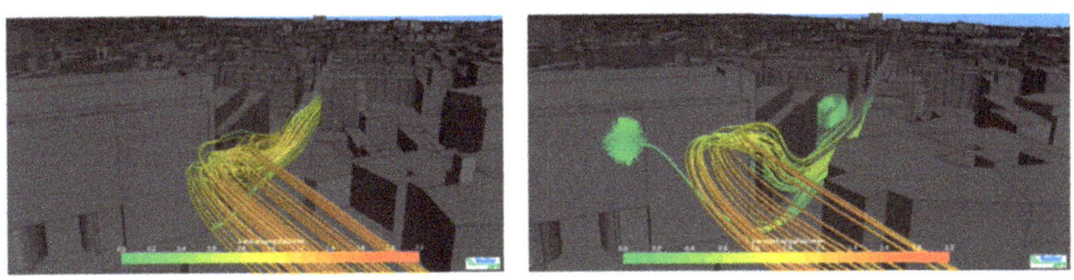

Figure 15. Changes in redirecting wind flow reducing wind speed for wind direction south-east for base case (**left**) and densification scenario (**right**).

A specific test case was carried out for the high-rise building densification scenario at the southern end of the development as depicted in Figure 3e,f with a heat-wave simulation (south-east wind). Figure 16 shows that at the very street level on the corners and along the facades of the high-rise buildings, turbulences occur, leading to uncomfortable situations for pedestrians.

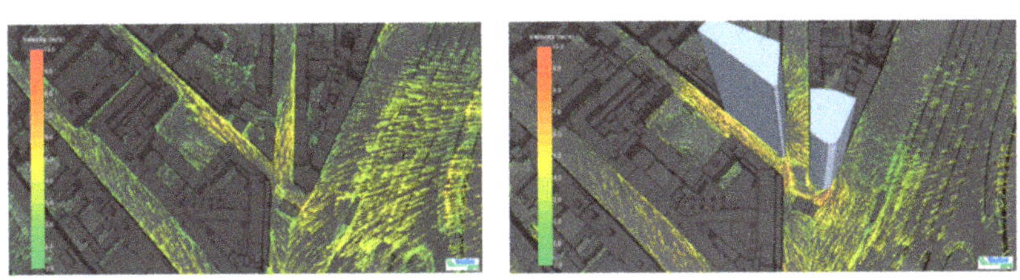

Figure 16. Changes in wind speed and direction for south-east wind direction for base case (**left**) and the high-rise buildings densification scenario (**right**).

Whilst south-easterly winds showed decreasing windspeed when high-rise buildings were established, westerly winds showed the opposite effect on wind comfort, as depicted in Figure 17. Hours of low wind comfort increased significantly with westerly winds, leading to uncomfortable wind situations in the streets near the potential high-rise buildings.

Overall, the high-rise building scenario showed that the impact on wind speed, direction, and pedestrian comfort could be significant in the vicinity of high-rise buildings; however, it depends very much on shape and location of the buildings to be erected.

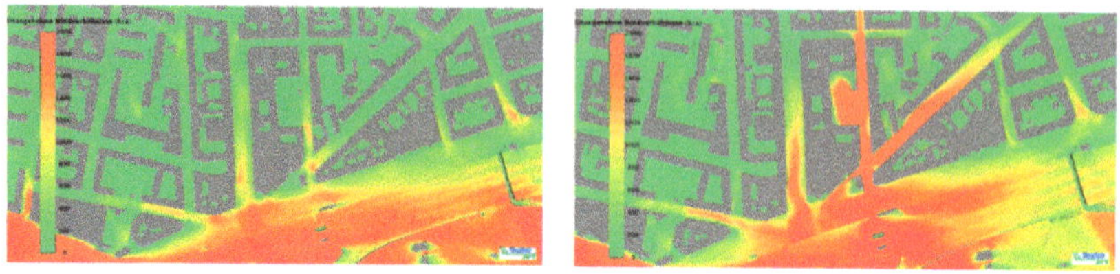

Figure 17. Changes in wind comfort for westerly wind direction for base case (**left**) and a high-rise buildings densification scenario (**right**).

4.3. Energy and Indoor Climate Comfort Analysis: Key Results

The energy and indoor climate comfort analysis was carried out to highlight the effect of the different construction types as outlined in Tables 1 and 2. For the indoor temperature simulations, the framework conditions were defined for a potential worst-case scenario: A top floor room with the potential highest solar gains was selected as the reference case, simulating the effects during one of the hottest days of the year with external daytime temperatures exceeding 30 °C. In order to create realistic circumstances, the room was modeled with natural nighttime ventilation with open windows between 11 p.m. and 8 a.m. If modeled without night-ventilation, the temperatures largely exceeded comfort limits well above 30 °C. Therefore, a scenario with ventilation needed to be considered in order to adequately compare effects of the different construction types. The framework conditions for the heating energy demand followed the current building regulations as noted in the previous section with weather data for the city of Vienna and an assessment over an entire year.

Figure 18 shows that the heating energy demand varied negligibly between the different variants (see the wall types in Table 4 below). The heavy-weight construction had only a slightly positive effect both in terms of heating energy demand as well as indoor room temperature. In the variants with the sloping roof, the indoor temperatures were generally 2–3° K higher compared to the flat roof construction due to the inclination of the roof and windows and the subsequently increased solar radiation. However, even with the sloping roof, the heavy-weight construction reached only around 1 K difference to the light-weight construction in terms of indoor room temperature.

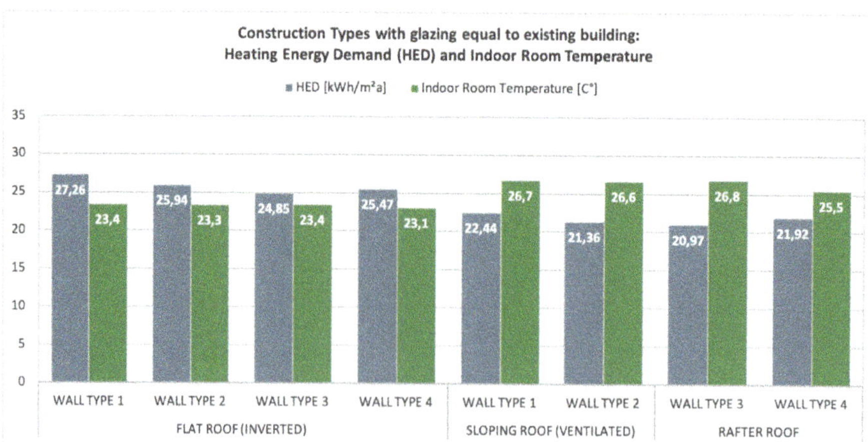

Figure 18. Heating energy demand and indoor room temperature for construction types with glazing equal to existing building (for wall type description see Table 4) [53].

Table 4. Wall type numbering and description for Figures 18 and 19 [53].

Wall Type Number	Wall Type Description
Wall type 1	Wall with cellulose insulation
Wall type 2	Ventilated wall with wooden paneling
Wall type 3	Solid wood wall with intermediate insulation
Wall type 4	Concrete wall with external insulation

The same variants were further analyzed with an increased glazing area. Figure 19 shows the results with twice the glazing area in the top floors compared to the existing building (see wall type descriptions in Table 4). Whilst the heating energy demand barely changed compared to the previous simulation with less glazing, the indoor temperature was, especially in the variant with the sloping roof and the subsequently angled glazing, considerably higher, although the effect of heavy-weight versus light-weight construction was only minimal. This showed that the transparent elements could have a far greater effect on indoor temperatures than different construction types, if the building quality of the building envelope (U-values) was already generally higher than noted in the building regulations. It is also evident that without further measures (e.g., adequate shading of the glazed areas; increased ventilation), the indoor temperatures by far exceeded comfortable indoor thermal conditions. Thus, the increased solar gains had to be specifically considered when planning with high glazing areas, which is especially relevant for the planning of attic extensions.

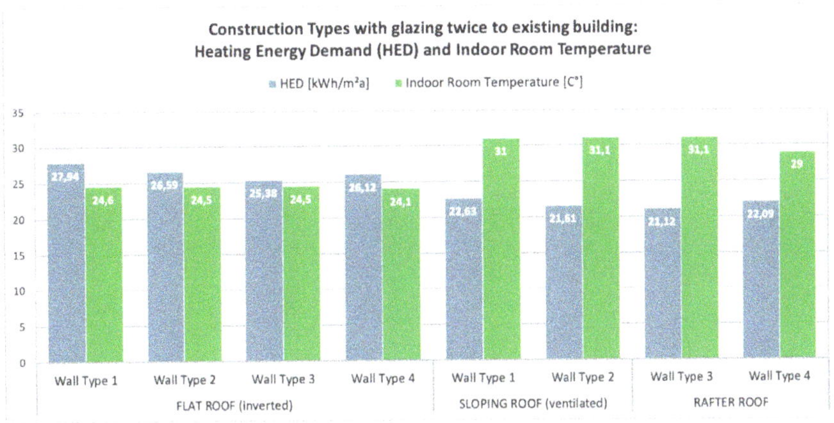

Figure 19. Heating energy demand and indoor room temperature for construction types with glazing twice to existing building (for wall type description see Table 4) [53].

In another assessment, the effects of green roofs and green facades were analyzed. The same construction types and framework conditions as outlined above were applied. In a first step, differentiations were made in terms of the extent of the greening of the roof (Table 3). In a first instance, roof types were modeled with both a flat roof construction and sloping roof construction. Secondly, the construction of the roof was altered. For the greening variants, the heating energy demand as well as the indoor room temperature were analyzed. Equal to the assessment for the construction types, for the calculation of the indoor room temperature, the same framework conditions were applied.

Figure 20 shows that there was only a negligible difference between the various greening options related to the heating energy demand. Similarly, the indoor room temperature in Figure 21 varied only marginally and was only slightly higher in the sloped roof option. This was, however, mainly due to the inclination of the roof and the increased solar radiation compared to the flat roof options. It showed that if the overall construction was already of high quality and subsequently well insulated, the added layer of the green roof generally had little effect on the heating energy demand and indoor temperatures.

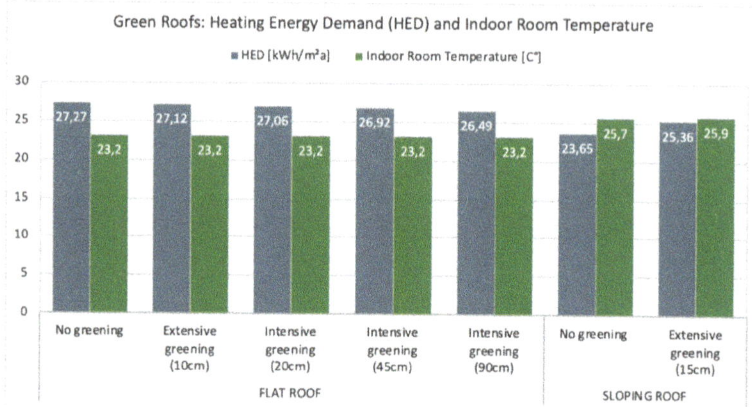

Figure 20. Heating energy demand and indoor room temperature for green roof construction types [54].

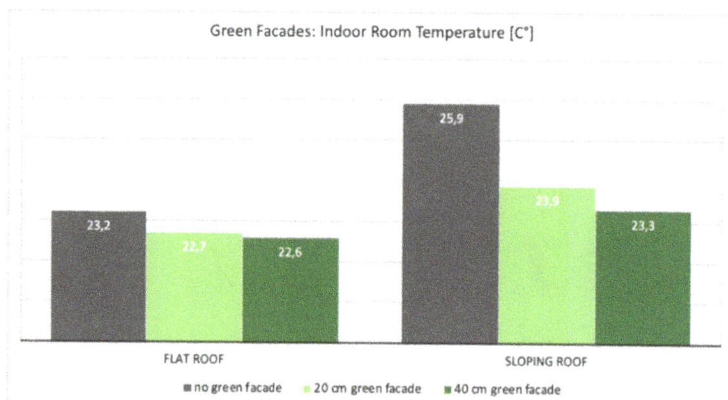

Figure 21. Indoor temperature for green roof facade construction types [54].

A further attempt was made to simulate green facades. However, due to the limited availability of adequate monitored data in relation to the physical properties and effects of the greenery, the analysis was limited to the shading effect of the plants. Thus, the actual effect of the evapotranspiration of the plants and the change in the microclimate was not considered. The green facades were modeled with an external layer that protruded either 20 or 40 cm from the actual facade. The greening was thus modeled with a shading effect that influenced the solar radiation reaching the windows. In the flat roof variant, the reduction in the maximum indoor temperature between the un-greened facade and a green area with a 20-cm overhang was only 0.5 K and thus almost negligible. With a 40-cm thick green area, the maximum temperature reduction is 0.6 K. In the variant with the sloping roof the partial shading of the windows with the facade greening created a slightly stronger effect. With 20 cm facade greening, the temperature was reduced by 2 K, and with 40 cm cantilevered greening, the difference increased to 2.6 K. However, it should be noted that the same shading effect could be achieved with any other material.

5. Discussion

Based on the hypothesis that specific urban and building densification measures could have a positive or negative impact on the urban microclimate, the building energy demand as well as the external (street level) and internal (building) comfort level, this paper focused on the quantification of these impacts. As the main aim was to assess which measures related to urban and building planning could be most effective, the key findings can be summarized as follows.

The results of the microclimate simulations show that large-scale densification led to an increase in shaded areas and enhanced local cooling at street level and inner courtyards during the day as the mean radiation temperature dropped between 3 and 10 K MRT (mean radiant temperature), especially in west-east-oriented streets. At night, the mean radiation temperature only increased slightly up to 0.5 K MRT. A reduction in wind speed by 50% (from approximately 6 m to 3 m/s) led to an increase in the mean radiation temperature of 1 to 3 K MRT during the day and an increase of 0.1 to 0.3 K MRT at night. High-rise buildings led to larger shaded areas over a longer period and due to limited solar radiation, which resulted in mean radiation temperature drops by up to 15 K MRT. The effect during the night was almost negligible with a simulated increase of up to 0.5 K MRT.

Greening the urban landscape on street or building level is widely seen as an appropriate adaptation measure to mitigate the urban heat island effect. Street greening as an adaptation measure contributes to a significant reduction in the radiation temperature. Shaded zones on street level can be reduced by up to 15 K MRT during daytime. At night there is only a slight reduction in the radiation temperature by a few tenths of 1 K MRT. Green facades contributed to a significant reduction in the external radiation temperature. During the day, the radiation temperature in the shaded zones was reduced by up to 15 K MRT, whilst during nighttime there was only a slight reduction in the radiation temperature by a few tenths of 1 K MRT.

Green roofs caused a slight reduction in the external radiation temperature by a few 1/10 K MRT at street level, both during day and night. On roof level, green roofs could reduce the radiation temperature by 0.5 K MRT (without the application of intensive greening such as large trees). The effect of green roofs was negligible if the building was already well insulated. Any effect was thus mainly related to the height of the substrate layer, which acted like an added insulating layer on top of the building. Green facades can influence the indoor temperatures due to their shading properties. Partially shading the windows (e.g., through floor-based facade greening) had the greatest effect in reducing the indoor temperature as it acted like an external shading system. However, it must be noted that due to missing data related to the physical properties of plants (e.g., evapotranspiration), the potential change in microclimate was not considered in the assessment of the building energy demand or internal conditions.

The effects of heavy-weight or light-weight rooftop extensions on the energy demand and thermal comfort within the building were overall minimal compared to the impact of the glazing area. High thermal storage capacities could not adequately compensate for a lack of shading measures for large glazing areas or a lack of appropriate ventilation measures. For window areas that are not shaded by external shading systems, the size of the window had a greater effect than the thermal mass. Night ventilation as a passive measure was highly relevant in light-weight and heavy-weight constructions to keep the indoor temperatures within the comfort limits. This also confirmed other studies related to passive measures to avoid summertime overheating [33,34].

Based on these results and stakeholder consultations with relevant experts from the city administration and planning departments in Vienna, several key recommendations were derived:

(1) Large scale densification with a moderate increase in building height should consider the shading of south and west oriented facades through facade greening and trees, the shading of rooftop extensions, and proper orientation of buildings and streets to reduce solar irradiance and improve natural ventilation. Vertical densification

should be linked to specific requirements related to external shading and ventilation measures (e.g., cross and night ventilation);

(2) For high-rise densification, the building design and orientation should be planned in consideration of main wind-directions and reflections of glass facades. These aspects must be evaluated by means of mandatory individual microclimate and wind-comfort assessments;

(3) Urban greening should focus on the street levels in order to increase evaporation and humidity, as well as decrease heat exposure through shading. Trees should be planted along the northern and eastern street sides, exposed to the sun during the hours with highest sun radiation. Soil sealing should be reduced including in private yards and large public places to increase evaporation. Greening on street level should be mandatory on both the plot area (courtyard areas) and on public areas (private and public sector);

(4) Building greening on facades should be considered in conjunction with the use of renewable energy systems, since an increase in indoor comfort in summer is achieved primarily through shading during hot periods. Facade greening should also be primarily be implemented at street level, since the evaporation and shading effect can reduce the radiation temperature and the perceived temperature at street level. Rooftop greening has hardly an influence on microclimate at street level. (Extensive) green roofs act primarily as thermal insulation for the buildings; however, the better the insulation of the roof underneath, the smaller the effect of the green roof on the energy demand and indoor temperatures of the building.

6. Conclusions

Densification in cities is undoubtedly a valid strategy to avoid extensive urban sprawl into surrounding suburbia. Cities growing in population are, however, increasingly confronted with the conflicting targets of catering enough living and associated infrastructure space for their inhabitants and at the same time providing a high quality of life with comfortable external and indoor environments. The increased sealing of surfaces, the inherent density of tight urban spaces, and the high thermal mass influence the urban microclimate in relation to the ventilation of the streetscape, the heat island effect and the subsequent effect on the buildings. Densification, however, also provides a chance to implement sustainable climate adaptation and mitigation strategies if appropriate measures are selected. The results from this paper highlight that the effects of densification are diverse and manifold and can both have positive as well as negative implications on the urban microclimate, energy demand, as well as external and internal comfort.

Therefore, one of the key conclusions is that it is absolutely essential to simulate and assess the specific densification measures at an early planning stage in order to exploit the potential benefits such as increased comfort temperatures at street level, whilst at the same time limiting the negative effects, such as, e.g., uncomfortable wind conditions. Providing simulations at an early urban planning level for land-use and zoning plans could support an adequate urban development, which takes the impact on the urban microclimate into account. Improving the respective tools to combine the assessments within one environment, especially one that considers both the effects of microclimate on an external, urban scale as well as on the internal, building scale will be a prerequisite in order to foster a truly integrated approach.

Author Contributions: Conceptualization, W.L.; methodology, analysis, and visualizations, W.L., M.V., M.R., S.T., and D.Ö.; writing—original draft preparation, W.L. and D.Ö.; writing—review and editing, W.L. and D.Ö.; project administration and funding acquisition, G.E. All authors have read and agreed to the published version of the manuscript.

Funding: This research was funded within the project "CLUDEX—Climate Change and urban densification impact exploration—heat exposure reduction through roof scape adaptation—a large scale case study" under the 10th Call of the Austrian Climate Research Program (ACRP) by the Austrian Climate and Energy Fund, Grant number KR17AC0K13790, Project Number B769976.

Institutional Review Board Statement: Not applicable.

Informed Consent Statement: Not applicable.

Data Availability Statement: The study contains model results and do not contain monitoring or personal data. Spatial results are array data depicted as maps included as Figures to the article. Results of simulations of thermal Indoor conditions are again arrays of gridded data. They are averaged and depicted in tables and bar charts—numbers are included in the charts. Array data can be provided through the authors on request.

Conflicts of Interest: The authors declare no conflict of interest.

References

1. IPCC. *Climate Change Synthesis Report. Contribution of Working Groups I, II and III to the Fifth Assessment Report of the Intergovernmental Panel on Climate Change*; Core Writing Team, Pachauri, R.K., Meyer, L.A., Eds.; IPCC: Geneva, Switzerland, 2014.
2. United Nations Climate Change. The Paris Agreements, 4th November 2016. Available online: https://unfccc.int/files/essential_background/convention/application/pdf/english_paris_agreement.pdf (accessed on 16 June 2020).
3. Seto, K.C.; Fragkias, M.; Guneralp, B.; Reilly, M.K. A Meta-Analysis of Global Urban Land Expansion. *PLoS ONE* **2011**, *6*, 9. [CrossRef] [PubMed]
4. United Nations, Department of Economic and Social Affairs, News. 68% of the World Population Projected to Live in Urban Areas by 2050, Says UN. Available online: https://www.un.org/development/desa/en/news/population/2018-revision-of-world-urbanization-prospects.html (accessed on 22 July 2020).
5. Bradbury, M.; Peterson, M.N.; Liu, J. Long-term dynamics of household size and their environmental implications. *Popul. Environ.* **2014**, *36*, 73–84. [CrossRef]
6. Carneiro Freire, S.; Corban, C.; Ehrlich, D.; Florczyk, A.; Kemper, T.; Maffenini, L.; Melchiorri, M.; Pesaresi, M.; Schiavina, M.; Tommasi, P. *Atlas of the Human Planet*; Publications Office of the European Union: Luxembourg, 2019; ISBN 978-92-76-14132-7. [CrossRef]
7. Stadt Wien, Magistratsabteilung 23–Wirtschaft, Arbeit und Statistik, Wien in Zahlen, Bevölkerungsprognose. 2018. Available online: https://www.wien.gv.at/statistik/pdf/bev-prog-2018.pdf (accessed on 16 June 2020).
8. Stadt Wien, Wiener Umweltschutzabteilung (MA22). Öffentlich Zugängliche Grünflächen–Analyse. Stadt Wien. 2020. Available online: https://www.wien.gv.at/umweltschutz/umweltgut/oeffentlich.html (accessed on 10 May 2020).
9. Zuvela-Aloise, M. *Future of Climatic Urban Heat Stress Impacts—Adaption and Mitigation of the Climate Change Impact on Urban Heat Stress Based on Model Runs Derived with an Urban Climate Model*; Endbericht zum Projekt FOCUS I, Klima- und Energiefonds: Wien, Austria, 2013. Available online: https://www.klimafonds.gv.at/wp-content/uploads/sites/6/03032015FOCUSZuvela-AloiseEBACRP2B060373.pdf (accessed on 10 May 2020).
10. Fokaides, P.A.; Kylili, A.; Nicolaou, L.; Ioannou, B. The effect of soil sealing on the urban heat island phenomenon. *Indoor Built Environ.* **2016**, *25*, 1136–1147. [CrossRef]
11. Kottmeier, C.; Biegert, C.; Corsmeier, U. Effects of urban land use on surface temperature in Berlin: Case study. *J. Urban Plan. Dev.* **2007**, *133*, 128–137. [CrossRef]
12. Conticelli, E.; Proli, S.; Tondelli, S. Integrating energy efficiency and urban densification policies: Two Italian case studies. *Energy Build.* **2017**, *155*, 308–323. [CrossRef]
13. McEvoy, D.; Lindley, S.; Handley, J. Adaptation and mitigation in urban areas: Synergies and conflicts. *Proc. Inst. Civil Eng. Munic. Eng.* **2006**, *159*, 185–191. [CrossRef]
14. Loibl, W.; Vuckovics, M.; Etminan, G.; Österreicher, D.; Ratheiser, M.; Tschannett, S. *Climate Change and Urban Densification Impact Exploration—Heat Exposure Reduction Through Roof Scape Adaptation—A Large Scale Case Study*; Project Number: B769976, ACRP Final Report–Activity Report; AIT Austrian Institute of Technology: Vienna, Austria, 2020.
15. Mahtta, R.; Mahendra, A.; Seto, K.C. Building up or spreading out? Typologies of urban growth across 478 cities of 1 million+. *Environ. Res. Lett.* **2019**, *14*, 13. [CrossRef]
16. Jaksch, S.; Franke, A.; Österreicher, D.; Treberspurg, M. A systematic approach to sustainable urban densification using prefabricated timber-based attic extension modules. In *Energy Procedia, Proceedings of the SBE16 Tallinn and Helsinki Conference; Build Green and Renovate Deep, Tallinn, Estonia, 5–7 October 2016*; Elsevier Ltd.: Amsterdam, The Netherlands, 2016; ISBN 1876-6102. [CrossRef]
17. Ghofrani, Z.; Sposito, V.; Faggian, R. A Comprehensive Review of Blue-Green Infrastructure Concepts. *Int. J. Environ. Sustain.* **2017**, *6*, 15–36. [CrossRef]
18. Alves, A.; Vojinovic, Z.; Kapelan, Z.; Sanchez, A.; Gersonius, B. Exploring trade-offs among the multiple benefits of green-blue-grey infrastructure for urban flood mitigation. *Sci. Total Environ.* **2020**, *703*, 14. [CrossRef]
19. Belcakova, I.; Swiader, M.; Bartyna-Zielinska, M. The Green Infrastructure in Cities as a Tool for Climate Change Adaptation and Mitigation: Slovakian and Polish Experiences. *Atmosphere* **2019**, *10*, 552. [CrossRef]
20. Brandenburg, C.; Damyanovic, D.; Reinwald, F.; Allex, B.; Gantner, B.; Czachs, C. *Urban. Heat Islands Strategieplan*; Magistrat der Stadt Wien; Wiener Umweltschutzabteilung (MA 21): Vienna, Austria, 2015.

21. Virk, G.; Jansz, A.; Mavrogianni, A.; Mylona, A.; Stocker, J.; Davies, M. The effectiveness of retrofitted green and cool roofs at reducing overheating in a naturally ventilated office in London: Direct and indirect effects in current and future climates. *Indoor Built Environ.* **2014**, *23*, 504–520. [CrossRef]
22. Sharma, A.; Woodruff, S.; Budhathoki, M.; Hamlet, A.F.; Chen, F.; Fernando, H.J.S. Role of green roofs in reducing heat stress in vulnerable urban communities-a multidisciplinary approach. *Environ. Res. Lett.* **2018**, *13*, 13. [CrossRef]
23. Treberspurg, M.; Streicher, W.; Reim, T.H. *Sommertauglichkeit im Gebäudebestand. Wohnbauforschung des Bundes*; F1494; Smart CityWien: Vienna, Austria, 2011.
24. Chun, B.; Guldmann, J.M. Impact of greening on the urban heat island: Seasonal variations and mitigation strategies. *Comput. Environ. Urban. Syst.* **2018**, *71*, 165–176. [CrossRef]
25. Ciriminna, R.; Meneguzzo, F.; Pecoraino, M.; Pagliaro, M. Solar Green Roofs: A Unified Outlook 20 Years On. *Energy Technol.* **2019**, *7*, 7. [CrossRef]
26. Sattler, S.; Zluwa, I.; Österreicher, D. The "PV Rooftop Garden": Providing Recreational Green Roofs and Renewable Energy as a Multifunctional System within One Surface Area. *Appl. Sci.* **2020**, *10*, 1791. [CrossRef]
27. Skelhorn, C.P.; Levermore, G.; Lindley, S.J. Impacts on cooling energy consumption due to the UHI and vegetation changes in Manchester, UK. *Energy Build.* **2016**, *122*, 150–159. [CrossRef]
28. Duan, S.P.; Luo, Z.W.; Yang, X.Y.; Li, Y.G. The impact of building operations on urban heat/cool islands under urban densification: A comparison between naturally-ventilated and air-conditioned buildings. *Appl. Energy* **2019**, *235*, 129–138. [CrossRef]
29. Bevilacqua, P.; Mazzeo, D.; Arcuri, N. Thermal inertia assessment of an experimental extensive green roof in summer conditions. *Build. Environ.* **2018**, *131*, 264–276. [CrossRef]
30. Gupta, R.; Gregg, M. Assessing energy use and overheating risk in net zero energy dwellings in UK. *Energy Build.* **2018**, *158*, 897–905. [CrossRef]
31. McLeod, R.S.; Hopfe, C.J.; Kwan, A. An investigation into future performance and overheating risks in Passivhaus dwellings. *Build. Environ.* **2013**, *70*, 189–209. [CrossRef]
32. Oikonomou, E.; Davies, M.; Mavrogianni, A.; Biddulph, P.; Wilkinson, P.; Kolokotroni, M. Modelling the relative importance of the urban heat island and the thermal quality of dwellings for overheating in London. *Build. Environ.* **2012**, *57*, 223–238. [CrossRef]
33. Handler, S.; Korjenic, A.; Bednar, T. The influence of external thermal insulation composite systems on the summer performance of buildings. *Bauphysik* **2011**, *33*, 225–233. [CrossRef]
34. Österreicher, D.; Sattler, S. Maintaining Comfortable Summertime Indoor Temperatures by Means of Passive Design Measures to Mitigate the Urban Heat Island Effect—A Sensitivity Analysis for Residential Buildings in the City of Vienna. *Urban. Sci.* **2018**, *2*, 66. [CrossRef]
35. Mlakar, J.; Strancar, J. Overheating in residential passive house: Solution strategies revealed and confirmed through data analysis and simulations. *Energy Build.* **2011**, *43*, 1443–1451. [CrossRef]
36. Stazi, F.; Tomassoni, E.; Di Perna, C. Super-insulated wooden envelopes in Mediterranean climate: Summer overheating, thermal comfort optimization, environmental impact on an Italian case study. *Energy Build.* **2017**, *138*, 716–732. [CrossRef]
37. Sameni, S.M.T.; Gaterell, M.; Montazami, A.; Ahmed, A. Overheating investigation in UK social housing flats built to the Passivhaus standard. *Building Environ.* **2015**, *92*, 222–235. [CrossRef]
38. Adekunle, T.O.; Nikolopoulou, M. Thermal comfort, summertime temperatures and overheating in prefabricated timber housing. *Building Environ.* **2016**, *103*, 21–35. [CrossRef]
39. Austrian Standards. ÖNORM B 8110-3. *Wärmeschutz im Hochbau—Teil 3: Vermeidung Sommerlicher Überwärmung*; Austrian Standards: Vienna, Austria, 2020.
40. Kim, H.; Anderson, K. Energy Modeling System Using Building Information Modeling Open Standards. *J. Comput. Civ. Eng.* **2013**, *27*, 203–211. [CrossRef]
41. Perez, D.; Robinson, D. Urban Energy Flow Modelling: A Data-Aware Approach. In *Digital Urban Modeling and Simulation. Communications in Computer and Information Science*; Arisona, S.M., Aschwanden, G., Halatsch, J., Wonka, P., Eds.; Springer: Berlin/Heidelberg, Germany, 2012; Volume 242.
42. Groger, G.; Plumer, L. CityGML—Interoperable semantic 3D city models. *ISPRS J. Photogramm. Remote Sens.* **2012**, *71*, 12–33. [CrossRef]
43. Agugiaro, G. First steps towards an integrated CityGML-based 3D model of Vienna. In Proceedings of the XXIII ISPRS Congress, Commission, Prague, Czech Republic, 12–19 July 2016. [CrossRef]
44. Biljecki, F.; Ledoux, H.; Stoter, J. Does a finer level of detail of a 3D city model bring an improvement for estimating shadows? In *Advances in 3D Geoinformation*; Springer International Publishing: Cham, Switzerland, 2016.
45. Rhinoceros, G. Available online: https://www.rhino3d.com (accessed on 24 July 2020).
46. City of Vienna, MA21. Land Use and Zoning Plan. 2020. Available online: https://www.wien.gv.at/flaechenwidmung/public/ (accessed on 24 July 2020).

47. Loibl, W.; Etminan, G.; Österreicher, D.; Ratheiser, M.; Stollnberger, R.; Tschannett, S.; Tötzer, T.; Vuckovic, M.; Walal, K. Urban densification and urban climate change–assessing interaction through densification scenarios and climate simulations. In Proceedings of the 24th International Conference on Urban Planning and Regional Development in the Information Society, GeoMultimedia, Karlsruhe, Germany, 2–4 April 2019. Available online: https://conference.corp.at/index.php?id=38&L=0 (accessed on 19 August 2020).
48. Food4Rhino. Available online: https://www.food4rhino.com/app/ladybug-tools (accessed on 24 July 2020).
49. Vuckovic, M.; Loibl, W.; Tötzer, T.; Stollnberger, R. Potential of Urban Densification to Mitigate the Effects of Heat Island in Vienna, Austria. *Environments* **2019**, *6*, 82. Available online: https://doi.org/10.3390/environments6070082/ (accessed on 19 August 2020). [CrossRef]
50. Loibl, W.; Züger, H.; Köstl, M. Reclip:century-Entwicklung eines Basisdatensatzes regionaler Klimaszenarien. Proceedings of 11. Österreichischen Klimatag "Klima im Wandel, Auswirkungen und Strategien" 2010. Available online: http://www.austroclim.at/fileadmin/user_upload/ppt_11.Klimatag/Tagungsband_upload.pdf (accessed on 19 August 2020).
51. STAR-CCM+. 2020. Available online: https://www.plm.automation.siemens.com/global/en/products/simcenter/STAR-CCM.html (accessed on 19 August 2020).
52. Yuan, C.; Rena, C.; Ng, E. GIS-based surface roughness evaluation in the urban planning system to improve the wind environment–A study in Wuhan, China. *Urban. Clim.* **2014**, *10*, 585–593. Available online: https://doi.org/10.1016/j.uclim.2014.06.005 (accessed on 19 August 2020). [CrossRef]
53. Zwirschitz, S. Energetische Analyse von Nachverdichtungsvarianten im Urbanen Wohnbau am Beispiel eines dicht Bebauten Wiener Stadtviertels. Master's Thesis, University of Natural Resources and Life Sciences, Vienna, Austria, 2020.
54. Vansco, D. Auswirkung der Gebäudebegrünung auf den Innenraumkomfort im Urbanen Wohnbau am Beispiel eines dicht bebauten Wiener Stadtviertels. Master's Thesis, University of Natural Resources and Life Sciences, Vienna, Austria, 2020.
55. ArchiPhysik. 2020. Available online: https://archiphysik.at/ (accessed on 24 July 2020).
56. Thesim3d. 2020. Available online: http://www.thesim.at (accessed on 24 July 2020).
57. Lindberg, F.; Holmer, B.; Thorsson, S.; Rayner, D. Characteristics of the mean radiant temperature in high latitude cities—implications for sensitive climate planning applications. *Int. J. Biometeorol.* **2014**, *58*, 613–627. [CrossRef] [PubMed]

atmosphere

Article

Impact of Urbanization on Sunshine Duration from 1987 to 2016 in Hangzhou City, China

Kai Jin [1,2,*], Peng Qin [1], Chunxia Liu [1], Quanli Zong [1] and Shaoxia Wang [1,*]

1 Qingdao Engineering Research Center for Rural Environment, College of Resources and Environment, Qingdao Agricultural University, Qingdao 266109, China; qinpeng@qau.edu.cn (P.Q.); 201901150@qau.edu.cn (C.L.); zongql@qau.edu.cn (Q.Z.)

2 State Key Laboratory of Soil Erosion and Dryland Farming on the Loess Plateau, Institute of Water and Soil Conservation, Northwest A&F University, Yangling 712100, China

* Correspondence: jinkai@qau.edu.cn (K.J.); wang-sx@qau.edu.cn (S.W.); Tel.: +86-150-6682-4968 (K.J.); +86-136-1642-9118 (S.W.)

Abstract: Worldwide solar dimming from the 1960s to the 1980s has been widely recognized, but the occurrence of solar brightening since the late 1980s is still under debate—particularly in China. This study aims to properly examine the biases of urbanization in the observed sunshine duration series from 1987 to 2016 and explore the related driving factors based on five meteorological stations around Hangzhou City, China. The results inferred a weak and insignificant decreasing trend in annual mean sunshine duration (-0.09 h/d decade^{-1}) from 1987 to 2016 in the Hangzhou region, indicating a solar dimming tendency. However, large differences in sunshine duration changes between rural, suburban, and urban stations were observed on the annual, seasonal, and monthly scales, which can be attributed to the varied urbanization effects. Using rural stations as a baseline, we found evident urbanization effects on the annual mean sunshine duration series at urban and suburban stations—particularly in the period of 2002–2016. The effects of urbanization on the annual mean sunshine duration trends during 1987–2016 were estimated to be -0.16 and -0.35 h/d decade^{-1} at suburban and urban stations, respectively. For urban stations, the strongest urbanization effect was observed in summer (-0.46 h/d decade^{-1}) on the seasonal scale and in June (-0.63 h/d decade^{-1}) on the monthly scale. The notable negative impact of urbanization on local solar radiation changes was closely related to the changes in anthropogenic pollutions, which largely reduced the estimations of solar radiation trends in the Hangzhou region. This result highlights the necessity to carefully consider urbanization impacts when analyzing the trend in regional solar radiation and designing cities for sustainable development.

Keywords: sunshine duration; solar radiation; change trend; urbanization effect

Citation: Jin, K.; Qin, P.; Liu, C.; Zong, Q.; Wang, S. Impact of Urbanization on Sunshine Duration from 1987 to 2016 in Hangzhou City, China. *Atmosphere* **2021**, *12*, 211. https://doi.org/10.3390/atmos12020211

Academic Editor: Hideki Takebayashi and Jihui Yuan

Received: 11 January 2021
Accepted: 31 January 2021
Published: 4 February 2021

1. Introduction

Anthropogenic interference with climate and the hydrological cycle occurs primarily through modification of radiative fluxes in the climate system [1]. Heavily populated cities consume significant amounts of fossil energy sources [2,3]. Consequently, a large number of pollutants are emitted into the atmosphere, which increases the concentration of aerosols in urbanized and industrialized areas [4,5]. Aerosols can attenuate incoming surface solar radiation (*SSR*) through direct (scattering and solar radiation absorption) and indirect (increasing cloud reflectivity and lifetime) radiative forcings [5,6]. Natural factors—including cloud cover variability and radiatively active gases—can also influence the change in *SSR* [7–9]. The reduction of solar energy may affect the sustainable development of cities that focus on the use of clean energy. Therefore, determining the long-term changes in *SSR* and the dominant driving forces is critically important for understanding regional climatic changes and sustainable development of cities [10,11].

Based on *SSR* data and that of its proxies—such as sunshine duration (*SSD*) and diurnal surface air temperature range (*DTR*)—previous studies inferred worldwide solar dimming from the 1960s to the 1980s [12–14]. In contrast, solar brightening since the late 1980s occurred in many regions around the world, including Northwest Italy [11] and South America [15]. However, these observations may contain certain biases from urbanization effects, as many meteorological stations are located within or close to cities (i.e., urban or suburban stations). For example, based on the 172 pairs of urban and nearby rural stations in China, Wang et al. found that the declining rate of sunshine duration in rural areas is around two-thirds of that in urban areas in the dimming phase [12]. Compared with remote stations away from urban areas (i.e., rural stations), urbanization effects—such as increasing atmospheric aerosol—on solar radiation are more pronounced at urban stations [16]. Based on the large Tel Aviv pyranometer network in Israel, Stanhill and Cohen found a maximum urban dimming effect of 7% that was significantly negatively related to the number of vehicles on the roads [17]. This finding highlights the need to further investigate urbanization effects on the long-term changes in solar radiation.

In general, the impacts of urbanization on climate change can be estimated by comparing the difference between the climatic time-series of urban and rural stations [18–20]. This method referred to as urban minus rural (UMR), has been frequently applied in previous urban warming research, with relatively effective and reliable results [21,22]. The UMR method has also been applied to estimate the effects of urbanization on solar radiation changes. For example, based on 105 urban-rural station pairs across the world, Wang et al. found that the impact of urbanization on mean solar radiation at urban stations varied from -30 to 30 W m^{-2} during 1961–1990 [10]. The UMR method assumes that the *SSD* trends of rural stations are rarely affected by urbanization, which highlights the importance of accurately identifying and classifying rural stations. The population is one of the most frequently used indexes to reflect the degree of urbanization for station classification [4,12]. While population information is spatially generalized and outdated [17], the urban fraction (*UF*)—the proportion of urban built-up areas surrounding stations—is a more current and accurate representation of urbanization level [22]. Different indexes used to classify stations may lead to errors in the results, as discussed in existing studies [18,23,24].

China has experienced a dramatic environmental change in response to rapid urbanization, particularly in developed regions [23]. Many studies have reported pronounced urbanization effects on solar radiation in China. For example, using the UMR method, Wang et al. showed that urbanization largely contributed to solar dimming during 1960–1989 in urban areas of China [12]. Based on temperature data at 549 stations, Qian reported that urbanization-related land use/cover change caused a *DTR* change of -0.051 °C decade^{-1} during 1979–2008 in East China, inferring a negative impact on solar radiation [13]. Song et al. found that urbanization caused an *SSD* trend of -0.065 h/d decade^{-1} during 1961–2014 in East China [14]. In contrast, Wang et al. inferred a small urbanization effect on the solar radiation trend over China during 1961–1990 [10]. The above-mentioned divergences are likely associated with the representativeness of rural stations and the quality of the datasets used [25]. Selecting rural stations with less representative regional climate conditions may underestimate the urbanization effect [24]. Inhomogeneities in the climatic time-series, mainly relating to station relocation, are rarely considered in previous results, which may lead to large uncertainties when estimating urbanization effects on solar radiation [7,23]. Urbanization effects on solar radiation and the uncertainties in detections contributed to the debate on the occurrence of solar brightening since the late 1980s in China [4,6,12].

Previous studies generally attributed urbanization effects on solar radiation to atmospheric pollutions [4,12,26,27]. However, the impact of urbanization on other climatic factors could also affect the variation in solar radiation, which was rarely considered. Notably, cloud cover, which can block the sun, is generally higher in cities than in rural areas due to the urban rain island effect [28]. Therefore, it is necessary to consider the role of the cloud in *SSD* variations in urban areas. Moreover, urbanization effects may vary seasonally and monthly owing to fluctuations in climatic and atmospheric conditions [29], highlighting the need for this research at different temporal scales.

This study thus aims to properly examine how the observed *SSD* changes are influenced by urbanization from 1987 to 2016 in Hangzhou City, China, at different temporal scales. The findings will answer whether the impact of urbanization will affect the estimation of solar brightening trends and what are the main driving factors. This study would improve our understanding of regional environmental and climatic changes caused by human activities.

2. Materials and Methods

2.1. Study Area

Hangzhou City is located on the East coast of China (Figure 1); it is the capital of Zhejiang province and a central city of the Yangtze River Delta. Hangzhou has a subtropical monsoon climate, with a mean annual precipitation of 1421.7 mm and a mean annual temperature of 16.2 °C [16,30]. The city has experienced rapid urbanization and increased energy consumption in response to economic and population growth since the 1980s. In 2019, Hangzhou's population size and urbanization level reached 10.36 million and 78.5%, respectively. Urbanization has led to severe environmental issues, including elevated concentrations of anthropogenic aerosols in the Hangzhou region [31].

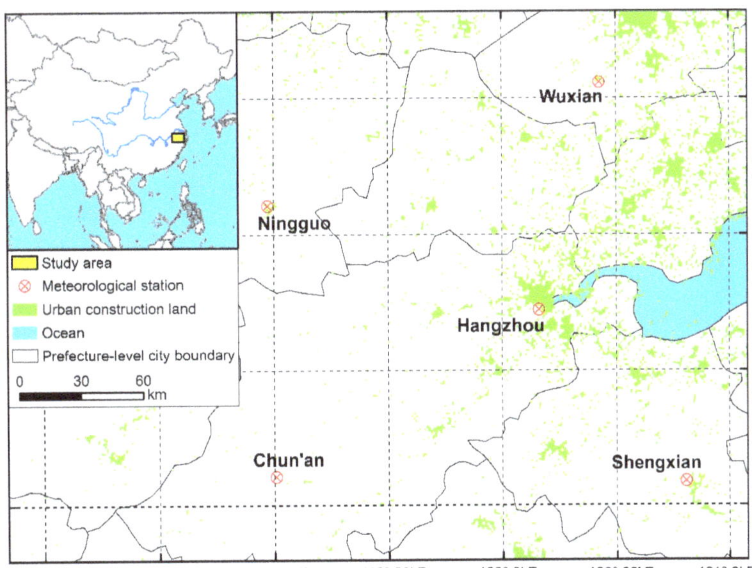

Figure 1. Locations of the five meteorological stations near Hangzhou City, China.

2.2. Data and Preprocesses

The daily *SSD* dataset obtained from the National Meteorological Station of China was provided by the China Meteorological Data Service Center (http://data.cma.cn/en). The unit of daily *SSD* is h/d. quality control procedures, such as consistency checks and extremum checks, were applied to the *SSD* dataset. The *SSD* dataset has been used in many studies on climate variability in China [12].

To avoid the impact of inhomogeneities in the *SSD* time-series on analysis results caused by station relocation, we only selected meteorological stations that have not been relocated. We selected stations with records covering the period 1987–2016, which is a period of solar brightening [12], dramatic warming [20], and rapid urbanization in China [32]. A total of five stations located near Hangzhou City were selected for this study (Figure 1). The altitudes of the five stations ranged from 17.5 to 171.4 m, and their mean annual *SSD*s ranged from 4.5 to 5.1 h/d (Table 1). Missing values at each station accounted for less than 0.05% of the total records, which were estimated using the average *SSD*s of adjacent days. Monthly, seasonal, and annual mean *SSD*s were then calculated based on the daily *SSD* data for each station over the period 1987–2016. Seasons were defined in terms of the international standard of season division, in which spring includes March–May, summer includes June–August, autumn includes September–November, and winter includes December–February [14].

Table 1. Basic information for the selected five meteorological stations near Hangzhou City, China.

Station Name	Latitude (degree)	Longitude (degree)	Altitude (m)	Distance [1] (km)	*SSD* [2] (h/d)	UF_{1990} [3] (%)	UF_{2015} [4] (%)	Category
Chun'an	29.62	119.02	171.4	130	5.0	0.6	1.3	Rural
Hangzhou	30.23	120.17	41.7	-	4.5	29.2	55.2	Urban
Ningguo	30.62	118.98	89.4	120	4.6	5.8	11.0	Suburban
Shengxian	29.60	120.82	104.3	94	4.7	9.1	16.2	Suburban
Wuxian	31.07	120.43	17.5	97	5.1	3.2	7.8	Rural

[1] distance from a given station to Hangzhou station; [2] mean annual sunshine duration from 1987 to 2016; [3] urban fraction in 1990, which is equal to the proportion of urban built-up areas in the 7 km circular buffer surrounding each meteorological station in 1990. [4] same as [3], but for 2015.

Land cover maps of China in 1990 and 2015 with a resolution of 1 km were provided by the Data Center for Resources and Environmental Sciences, Chinese Academy of Sciences (RESDC) (http://www.resdc.cn). The land-use dataset was acquired by the digital interpretation method using high-resolution remotely sensed images: Landsat 8 OLI and GF-2 remote sensing images [33]. The land cover map contained six land-use types: cultivated land, forest, grassland, water bodies, built-up areas, and unused land. In this study, built-up areas were used to calculate the *UF* surrounding each meteorological station.

In addition, this study adopted total cloud cover (*TCC*), total population (*TP*), and the total number of motor vehicles (*TMV*) to explore the changes in the atmospheric environment in the urban district of Hangzhou City. *TCC* data with a resolution of 0.25 degrees were obtained from the fifth-generation ECMWF reanalysis data (ERA5), which were provided by the Copernicus Climate Data Store (https://cds.climate.copernicus.eu/cdsapp#!/home). This parameter is the proportion of a grid box covered by cloud and varies from 0 to 100%. City data, including *TP* and *TMV*, were from Statistical Yearbook issued by the Hangzhou Statistic Bureau (http://tjj.hangzhou.gov.cn/).

2.3. Classification of Meteorological Stations

Based on the study by Song et al. [14], we calculated the *UF* in the 7 km circular buffer surrounding each station. In general, a larger *UF* implies a higher urbanization level. However, the thresholds of *UF* for station classification are inconsistent in previous studies. For example, Liao et al. classified stations with a $UF < 15\%$ within circular buffers as rural stations [34]; while Wang and Ge classified stations with a $UF > 12\%$ within circular buffers as urban stations [35], Song et al. categorized stations with a $UF > 20\%$ as urban stations [14]. In order to reduce the impact of urbanization on the *SSD* series of rural stations, we adopted a stricter threshold of *UF* for rural stations ($UF_{2015} < 10\%$) compared to previous studies mentioned above. Then, the non-rural stations were classified into two categories based on the *UF* in 2015: a station with low urbanization impact ($10\% < UF < 20\%$, suburban station), and a station with a high urbanization impact ($UF > 20\%$, urban station). Finally, Hangzhou station was classified as an urban station, Ningguo and Shengxian stations were classified as suburban stations, and Wuxian and Chun'an stations were classified as rural stations (Table 1). The average distance from Hangzhou station to the remaining four stations is approximately 110 km. Based on previous studies [32,36], the background climate change was expected to be nearly homogenous for these stations. However, the changes in *UF* from 1990 to 2015 were very different among the five stations, implying different levels in urbanization effects on local climate changes.

2.4. Estimation of Urbanization Effects on SSD Trends

We used *SSD* as a proxy for solar radiation, as solar radiation and *SSD* are strongly linearly correlated [25]. We assumed that the trends in *SSD* at rural stations were free from the effects of urbanization [12]. The *SSD* reference series was calculated by averaging the *SSD* series of the two selected rural stations to reflect the impact of background climate change on *SSD* changes. Based on the UMR method [18], the *SSD* series of urban and suburban stations were compared with the reference series to quantify the impacts of urbanization on observed *SSD* changes. The urbanization effect was calculated using the following equation:

$$\Delta SSD = SSD_{\mathrm{u}} - SSD_{\mathrm{r}}, \quad (1)$$

where ΔSSD indicates the difference in the sunshine duration series between urban (suburban) and rural stations in units of h/d; SSD_{u} indicates the sunshine duration series of urban or suburban stations (h/d), and SSD_{r} indicates the reference series of sunshine duration (h/d).

2.5. Statistical Analysis

Linear trends in annual, seasonal, and monthly mean *SSD* and ΔSSD from 1987 to 2016 were examined using ordinary least-squares regression [3]. The increasing (decreasing) trend in ΔSSD indicates an increase (decrease) in *SSD* in urbanized areas compared to that of the rural areas. In addition, we used ordinary least-squares regression to explore linear trends in annual *TCC*, *TP*, and *TMV* from 2002 to 2016. The significance of the trends in variables was determined using a t-test at significance levels of 0.10, 0.05, and 0.01 [6]. Relationships between *TP* and *SSD*, between *TMV* and *SSD*, and between *TCC* and *SSD* were explored using correlation analyses.

3. Results

3.1. Temporal Changes in SSD

The annual mean *SSD* series of the three station categories showed similar interannual variations from 1987 to 2016, but their trends were notably different (Figure 2). The annual mean *SSD* series of rural stations showed a weak and increasing trend during 1987–2016, but that of suburban and urban stations showed decreasing trends. Only the urban station (Hangzhou station) showed a significant change in the annual mean *SSD* ($p < 0.01$) (Table 2). The annual mean *SSD* trends for rural, suburban, and urban stations were 0.05, −0.12, and −0.31 h/d decade^{-1}, respectively.

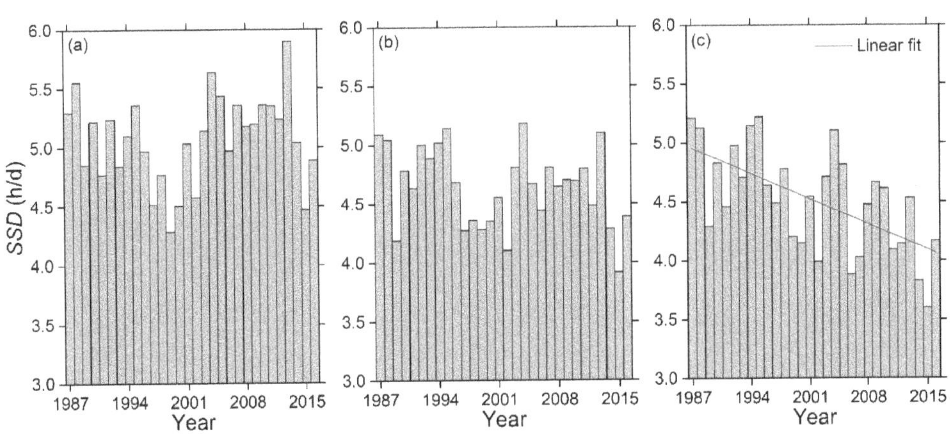

Figure 2. Interannual variation in annual mean sunshine duration (*SSD*) at (**a**) rural, (**b**) suburban, and (**c**) urban stations from 1987 to 2016.

Table 2. Change trends of annual and seasonal mean sunshine duration (*SSD*) at rural, urban, and suburban stations during 1987–2016.

	Trend in *SSD* (h/d decade^{-1})			
	Rural Station	**Suburban Station**	**Urban Station**	**All Stations**
Annual	0.05	−0.12	−0.31 **	−0.09
Spring	0.41 **	0.25 *	0.07	0.28 *
Summer	−0.02	−0.28	−0.48 *	−0.21
Autumn	−0.25	−0.35 *	−0.59 **	−0.35 *
Winter	0.11	0.01	−0.13	0.02

Note: All stations indicate the five selected stations in the Hangzhou region, including two rural stations, two suburban stations, and one urban station. * and ** indicate that the trends in *SSD* are significant at the 0.05 and 0.01 significance levels, respectively.

We used the average *SSD* series of the five selected stations to estimate the average trends in *SSD* for all stations. On average, the annual mean *SSD* trends for all stations was −0.09 h/d decade^{-1} during 1987–2016, while the seasonal mean *SSD* trends for all stations ranged from −0.35 h/d decade^{-1} in autumn to 0.28 h/d decade^{-1} in spring. Moreover, *SSD*s in the three station categories showed extremely different trends between seasons (Table 2). The seasonal mean *SSD* trends at the rural station occurred in the following descending order: spring (0.41 h/d decade^{-1}) > winter (0.11 h/d decade^{-1}) > summer (−0.02 h/d decade^{-1}) > autumn (−0.25 h/d decade^{-1}). The order of the seasonal mean *SSD* trends at suburban and urban stations was the same as those of the rural stations. The seasonal mean *SSD* trends of urban (suburban) stations ranged from −0.59 (−0.35) to 0.07 (0.25) h/d decade^{-1}. For each season, the rural station showed the largest *SSD* trend among the three station categories, while urban stations showed the lowest *SSD* trend.

For rural stations, the monthly mean *SSD* of six months (January, March, April, May, July, and August) increased from 1987 to 2016, with the largest increasing trend in March (0.63 h/d decade^{-1}, $p < 0.01$). November exhibited the strongest decreasing trend in *SSD* at −0.51 h/d decade^{-1} ($p < 0.05$) (Figure 3). In contrast, for urban stations, only March and April showed increasing trends in *SSD*, but the trends were weak and insignificant. *SSD*s in June, October, and November exhibited strong and significant decreasing trends ($p < 0.05$ or 0.01). The monthly mean *SSD* trends at the urban (suburban) station ranged from −0.98 (−0.58) h/d decade^{-1} in June to 0.25 (0.50) h/d decade^{-1} in March. Overall, for each month, the *SSD* trend was the largest in the rural station and the lowest in the urban station. The difference in *SSD* trends between the three station categories highlights the varying degrees of urbanization effects.

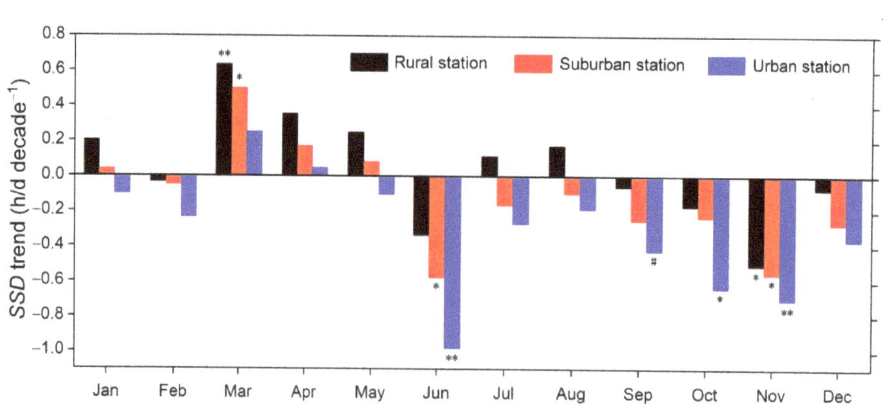

Figure 3. Change trends of monthly mean sunshine duration (*SSD*) at urban, suburban, and rural Scheme 1987. to 2016. #, *, and ** indicate that the trends are significant at the 0.10, 0.05, and 0.01 significance levels, respectively.

3.2. Urbanization Effects on SSD Change

On average, the difference in annual mean *SSD* from 1987 to 2016 between rural and suburban (urban) stations was −0.42 (−0.55) h/d (Table 3), indicating less *SSD* in urban areas than that in rural areas due to urbanization effects (e.g., anthropogenic pollution). This phenomenon was also observed at a seasonal scale, particularly in autumn. Additionally, the effect of urbanization on *SSD* was more severe since the beginning of the 21st century (Figure 4). On average, the urbanization effect on the annual mean *SSD* from 2002 to 2016 at suburban (urban) stations was −0.58 (−0.87) h/d, which was more severe than that during 1987–2001 (Table 3). This divergence was also evident for the four seasons—particularly for summer. The average effect of urbanization on summer mean *SSD* from 2002 to 2016 at the urban station was −1.09 h/d, which was more severe than that during 1987–2001 (−0.18 h/d).

Table 3. Average effects of urbanization on the annual and seasonal mean sunshine duration (*SSD*) series at suburban and urban stations over different periods.

Period	Suburban Station (h/d)					Urban Station (h/d)				
	Annual	Spring	Summer	Autumn	Winter	Annual	Spring	Summer	Autumn	Winter
1987–2001	−0.27	−0.16	−0.16	−0.54	−0.23	−0.24	−0.03	−0.18	−0.51	−0.23
2002–2016	−0.58	−0.45	−0.66	−0.77	−0.42	−0.87	−0.63	−1.09	−1.11	−0.67
1987–2016	−0.42	−0.30	−0.41	−0.66	−0.32	−0.55	−0.33	−0.63	−0.81	−0.45

Overall, the annual mean ΔSSD trends at suburban and urban stations were estimated to be −0.16 and −0.35 h/d decade^{-1}, respectively, from 1987 to 2016 ($p < 0.01$) (Figure 4a). Moreover, ΔSSD trends showed large divergence on the seasonal and monthly scales (Figures 4 and 5). ΔSSD trend was stronger in summer than that in spring, autumn, and winter, with the rates of −0.26 and −0.46 h/d decade^{-1} at suburban and urban stations, respectively. At the urban station, the most rapid decline in ΔSSD was observed in June (−0.63 h/d decade^{-1}), and the slowest decline in ΔSSD was observed in November (−0.19 h/d decade^{-1}) (Figure 5). At suburban stations, the most rapid decline in ΔSSD was observed in July (−0.28 h/d decade^{-1}), and the slowest decline in ΔSSD was observed in February (−0.01 h/d decade^{-1}) (Figure 5). The declining rates of ΔSSD indicate that urbanization exhibited an increasingly strong impact on *SSD* in urban areas, particularly in summer.

3.3. Environmental Variations Associated with SSD in Hangzhou City

Because data associated with the atmospheric environment are not available from the earlier period, we adopted *TP* and *TMV* to reflect the change in atmospheric aerosols and pollutions resulting from human activities during 2002–2016. *TP* and *TMV* exhibited a dramatic increase during 2002–2016 in the urban district of Hangzhou City, which may result in an increase in atmospheric pollution (Figure 6a,b). Moreover, *TCC* exhibited an increasing trend of 3.2% decade^{-1} during 2002–2016 in the urban district of Hangzhou City ($p < 0.10$) (Figure 6c). Correlation analyses suggested that *SSD* change was significantly and negatively correlated with the changes in *TCC* ($r = -0.682$, $p < 0.01$), *TP* ($r = -0.5292$, $p < 0.05$), and *TMV* ($r = -0.455$, $p < 0.10$) in Hangzhou City (Figure 6d–f). These results suggested that the increased cloud cover and anthropogenic pollutions could lead to a decrease in *SSD* in Hangzhou City.

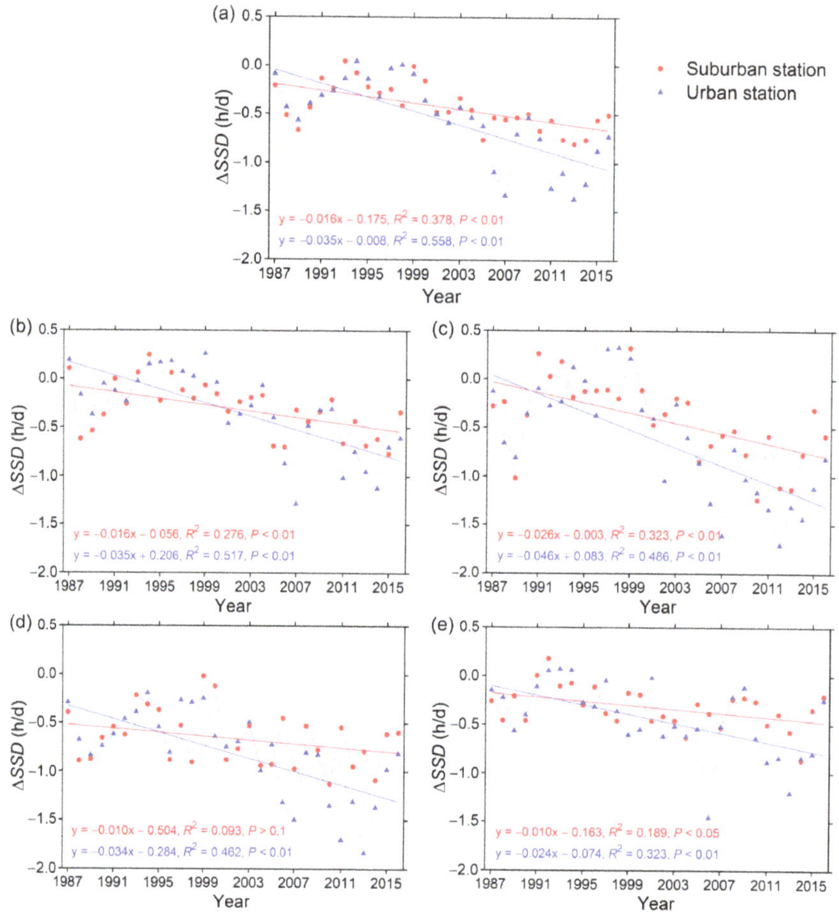

Figure 4. Urbanization effects on (**a**) annual, (**b**) spring, (**c**) summer, (**d**) autumn, and (**e**) winter variations of sunshine duration (ΔSSD) at suburban and urban stations from 1987 to 2016.

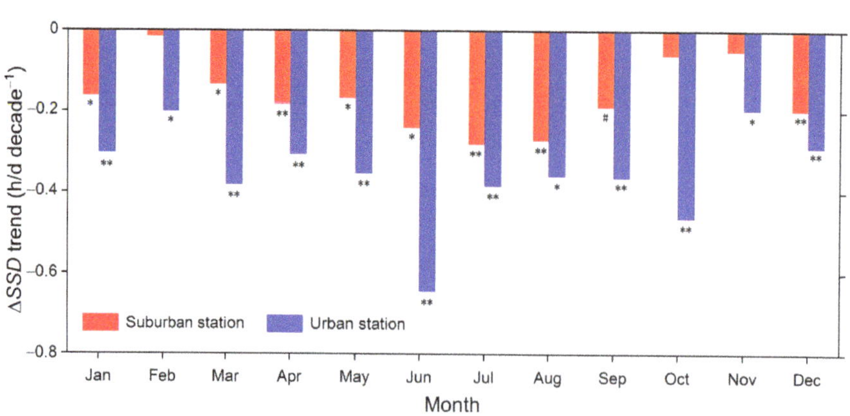

Figure 5. Urbanization effects on the trends of monthly mean sunshine duration (ΔSSD) at suburban and urban stations from 1987 to 2016. #, *, and ** indicate that the trends are significant at the 0.10, 0.05, and 0.01 significance levels, respectively.

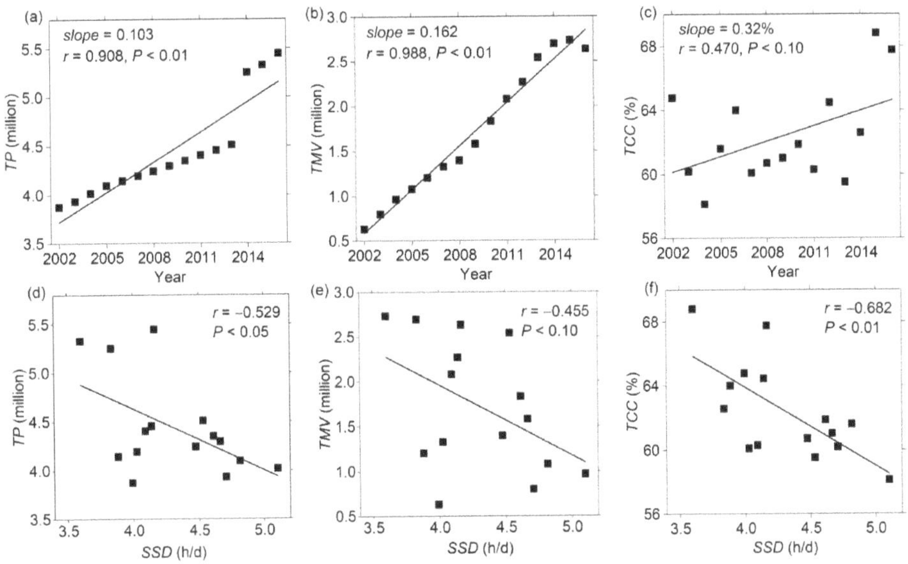

Figure 6. Variations in (**a**) total population (*TP*), (**b**) the total number of motor vehicles (*TMV*), and (**c**) total cloud cover (*TCC*) from 2002 to 2016 in the urban district of Hangzhou City and relationships (**d**) between *TP* and sunshine duration (*SSD*), (**e**) between *TMV* and *SSD*, and (**f**) between *TCC* and *SSD*.

4. Discussion

In this study, we compared the *SSD* time-series between two rural stations, two suburban stations, and an urban station (i.e., Hangzhou station) in the Hangzhou region to explore the urbanization effects on *SSD* at different temporal scales. We observed a weak and insignificant increase in the annual mean *SSD* at the rural stations from 1987 to 2016, with an average trend of 0.05 h/d decade^{-1}. This was consistent with the findings of Tang et al., who inferred stable solar radiation across China since 1990 [6]. However, solar radiation variability generally shows large spatial heterogeneity across China [37]. For example, Haerbin underwent a dimming tendency from 1991 to 2010, but Lhasa underwent a brightening tendency [8]. The brightening tendency since the late 1980s was also observed in Germany [9], Italy [11], South America [15], Japan [25], and New Zealand [38]. In contrast, we observed a rapid and significant dimming tendency at Hangzhou Station from 1987 to 2016, with an annual mean *SSD* trend of -0.31 h/d decade^{-1}. Suburban stations also exhibited a certain decrease in annual mean *SSD*. These results highlight the significant divergence of solar radiation changes—even for a relatively small-scale region.

Changes in solar radiation are generally attributed to the role of aerosols, clouds, and radiatively active gases [6,8,9]. Our results suggested that the increased cloud cover and anthropogenic pollutions could lead to a decrease in *SSD* in Hangzhou station. The same results were found in other regions, including South China [39], the Mediterranean [40], and Poland [41]. However, further analyses exhibited that the increasing trends in *TCC* for rural and suburban stations were very close to that of urban stations during 2002–2016 in the Hangzhou region (Figures 6c and 7). This indicates a consistent change in cloud cover among the three station categories. Therefore, the difference in *SSD* trends among the three station categories is mainly caused by the different levels of urbanization effect related to human activities. Alpert et al. also found greater declines in solar radiation in populated urban areas compared to rural areas, which was attributed to the rapid increase in aerosols due to industry activities [26].

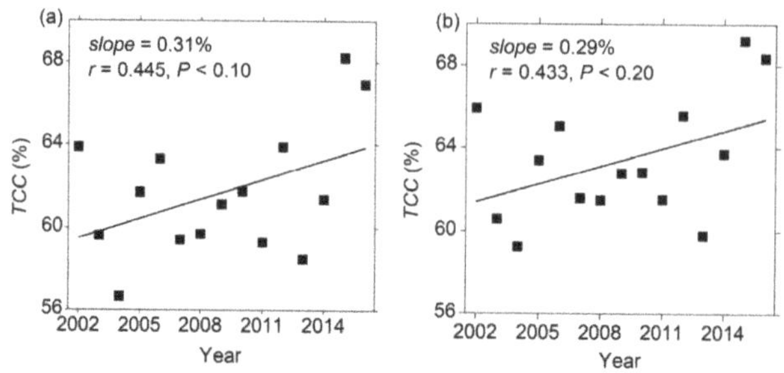

Figure 7. Variations in total cloud cover (*TCC*) from 2002 to 2016 at (**a**) rural and (**b**) suburban stations.

Based on qualitative analysis, Li et al. demonstrated that *SSD* was strongly correlated with visibility under clear-sky conditions in the low-latitude belt of South China [39]. In this study, we found that urbanization effects on the *SSD* series increased from 1987 to 2016, with significant *SSD* trends of −0.16 and −0.35 h/d decade^{-1} at suburban and urban stations, respectively (Figure 4). Our results are consistent with those of Qian [13] and Wang et al. [42], who observed a rapidly decreasing trend in *DTR* during the brightening phase in China in response to the urbanization effect. However, Wang et al. demonstrated that the brightening tendency since the late 1980s was closely related to the decrease in aerosol concentration following the implementation of policies to control pollution levels in China [12]. This suggests that the negative effect of urbanization on *SSD* has been reducing in urban areas. The different results may be caused by the difference in study regions and period. Hangzhou is one of the most prosperous cities in China and has experienced rapid economic development, urbanization, population increase, and energy consumption since the late 1980s [3]. As a result, large amounts of pollutants, such as $PM_{2.5}$ and PM_{10}, have been emitted into the atmosphere in the Hangzhou region [31]. Chang et al. suggested that past emission policies were unable to adequately control pollution levels in China [43]. Figure 6d,e showed that *SSD* change was significantly and negatively correlated with the changes in *TP* and *TMV* in Hangzhou City, indicating an important role of anthropogenic pollutions in the decreasing *SSD* trend.

Moreover, urbanization negatively impacted the seasonal *SSD* variability at suburban and urban stations at varying levels (Figure 4). This may be related to the significant seasonal variation in aerosols due to human activities [29,44]. In North China, pollution concentrations are generally higher in winter than in summer due to increased coal combustion for domestic heating [45]. As a result, the urbanization effect on the *SSD* trend in winter was found to be higher than that in summer in North China [27]. In contrast, we found that urbanization had a stronger effect on the *SSD* trends in summer (−0.46 h/d decade^{-1}) and a weaker effect in winter (−0.24 h/d decade^{-1}) at the Hangzhou station in South China. Song et al. also noticed greater urbanization effect on summer *SSD* compared to other seasons [14]. This divergent result can be explained by a number of factors. For example, coal is rarely used for domestic heating in South China due to its warmer climate, and anthropogenic pollution may be more in summer than in winter due to many energy consumptions for cooling [46]. Therefore, the work on energy conservation and emission-reduction are imperative in Hangzhou City, particularly during summer. Moreover, the specific reasons for the considerable differences in the urbanization effects on seasonal *SSD* change need to be further investigated to shed more light on this disparity.

Since urbanized areas only occupy a small portion of the global land area, the impact of urbanization on solar radiation change is considered a local phenomenon [17,26]. In China, the number of meteorological stations with long-term records—especially for solar radiation observations—are very few and are generally located near cities [6,37]. If urban or suburban stations are used to estimate regionally averaged trends in solar radiation, the dimming trend from the 1960s to the 1980s would be overestimated, but the brightening trend since the 1980s would be underestimated [6,12]. For example, based on the five stations selected in this study, the average trend of annual mean *SSD* in the Hangzhou region was estimated to be -0.09 h/d decade^{-1} from 1987 to 2016 (Table 2), which is lower than the average *SSD* trend at rural stations (0.05 h/d decade^{-1}).

The strong and significant impacts of urbanization on *SSD* variations at the Hangzhou station—particularly in summer—reflect a reduction in solar energy received by the urban land surface. This urbanization effect may impact the ecosystem, environment, and land surface energy balance of urban areas. These indirect effects, together with the complicated mechanisms of urbanization effects on the *SSD* change, highlight the need for further consideration in future studies.

5. Conclusions

In this study, we investigated the trends in *SSD* recorded at rural, suburban, and urban meteorological stations around Hangzhou City in China. The impacts of urbanization on *SSD* trends at suburban and urban stations and the related driving factors were explored.

Based on the annual mean *SSDs* at the five selected stations, a solar dimming trend (-0.09 h/d decade^{-1}) was observed in the Hangzhou region from 1987 to 2016, which was opposite to previous studies in China. However, *SSD* variability showed large differences between rural, suburban, and urban stations from 1987 to 2016 at different temporal scales. Using rural stations as a baseline, we found evident urbanization effects on the annual mean *SSD* series at urban and suburban stations—particularly in the period of 2002–2016. Over the three decades, the impacts of urbanization on the annual mean *SSD* trends at suburban and urban stations were estimated to be -0.16 and -0.35 h/d decade^{-1}, respectively. Therefore, the solar dimming trend since the 1980s in the Hangzhou region is largely attributed to the effects of urbanization. Moreover, urbanization impacts on the *SSD* trends showed a large divergence between seasons (months). At Hangzhou station, the strongest urbanization effect was observed in summer (-0.46 h/d decade^{-1}) on the seasonal scale and in June (-0.63 h/d decade^{-1}) on the monthly scale. The notable negative impacts of urbanization on local *SSD* were closely related to the changes in the atmospheric environment due to human activities. The urbanization effect on SSD may significantly impact the ecosystem, environment, and land surface energy balance of urban areas. Therefore, urbanization effects must be considered in future solar radiation research. Moreover, we suggested that the effort on energy conservation and emission-reduction need to be enhanced in Hangzhou City, particularly during summer; optimizing urban wind paths can be taken into account when conducting urban planning to improve urban heat environment and air quality.

Author Contributions: All authors contributed meaningfully to this study. K.J. and S.W. conceived the research topic. K.J., P.Q. designed the methodology, data acquisition, and analysis. P.Q., C.L., and Q.Z. provided methodology support, continuous follow-up of the research process, and K.J. drafted the manuscript. S.W. reviewed the manuscript, provided suggestions on the entire writing process, and polished the manuscript. All authors have read and agreed to the published version of the manuscript.

Funding: This research was funded by the Foundation of High-level Talents of Qingdao Agricultural University, grant number 665/1120041, 663/1120104.

Institutional Review Board Statement: Not applicable.

Informed Consent Statement: Not applicable.

Data Availability Statement: Data are contained within the article.

Acknowledgments: The authors would like to thank the China Meteorological Data Service Center for providing *SSD* data.

Conflicts of Interest: The authors declare no conflict of interest.

References

1. Zhao, T.B.; Li, C.X.; Zuo, Z.Y. Contributions of anthropogenic and external natural forcings to climate changes over China based on CMIP5 model simulations. *Sci. China Earth Sci.* **2016**, *59*, 503–517. [CrossRef]
2. Grimm, N.B.; Faeth, S.H.; Golubiewski, N.E.; Redman, C.L.; Wu, J.; Bai, X.; Briggs, J.M. Global Change and the Ecology of Cities. *Science* **2008**, *319*, 756–760. [CrossRef]
3. Jin, K.; Wang, F.; Wang, S. Assessing the spatiotemporal variation in anthropogenic heat and its impact on the surface thermal environment over global land areas. *Sustain. Cities Soc.* **2020**, *63*, 102488. [CrossRef]
4. Fu, C.B.; Dan, L.; Chen, Y.L.; Tang, J.X. Trends of the sunshine duration and diffuse radiation percentage on sunny days in urban agglomerations of China during 1960–2005. *J. Environ. Sci.* **2015**, *34*, 206–211. [CrossRef] [PubMed]
5. McNeill, V.F. Atmospheric Aerosols: Clouds, Chemistry, and Climate. *Annu. Rev. Chem. Biomol. Eng.* **2017**, *8*, 427–444. [CrossRef]
6. Tang, W.J.; Yang, K.; Qin, J.; Cheng, C.C.K.; He, J. Solar radiation trend across China in recent decades: A revisit with quality-controlled data. *Atmos. Chem. Phys.* **2011**, *11*, 393–406. [CrossRef]
7. Sanchez-Lorenzo, A.; Calbó, J.; Wild, M. Global and diffuse solar radiation in Spain: Building a homogeneous dataset and assessing their trends. *Glob. Planet. Chang.* **2013**, *100*, 343–352. [CrossRef]
8. Liu, J.; Linderholm, H.; Chen, D.; Zhou, X.; Flerchinger, G.; Yu, Q.; Du, J.; Wu, D.; Shen, Y.; Yang, Z. Changes in the relationship between solar radiation and sunshine duration in large cities of China. *Energy* **2015**, *82*, 589–600. [CrossRef]
9. Wild, M. Decadal changes in radiative fluxes at land and ocean surfaces and their relevance for global warming. *WIREs Clim. Chang.* **2016**, *7*, 91–107. [CrossRef]
10. Wang, K.; Ma, Q.; Wang, X.Y.; Wild, M. Urban impacts on mean and trend of surface incident solar radiation. *Geophys. Res. Lett.* **2014**, *41*, 4664–4668. [CrossRef]
11. Manara, V.; Beltrano, M.C.; Brunetti, M.; Maugeri, M.; Sanchez-Lorenzo, A.; Simolo, C.; Sorrenti, S. Sunshine duration variability and trends in Italy from homogenized instrumental time series (1936–2013). *J. Geophys. Res. Atmos.* **2015**, *120*, 3622–3641. [CrossRef]
12. Wang, Y.; Wild, M.; Sanchez-Lorenzo, A.; Manara, V. Urbanization effect on trends in sunshine duration in China. *Ann. Geophys.* **2017**, *35*, 839–851. [CrossRef]
13. Qian, C. Impact of land use/land cover change on changes in surface solar radiation in eastern China since the reform and opening up. *Theor. Appl. Climatol.* **2016**, *123*, 131–139. [CrossRef]
14. Song, Z.Y.; Chen, L.T.; Wang, Y.J.; Liu, X.P.; Lin, L.J.; Luo, M. Effects of urbanization on the decrease in sunshine duration over eastern China. *Urban Clim.* **2019**, *28*, 100471. [CrossRef]
15. Raichijk, C. Observed trends in sunshine duration over South America. *Int. J. Climatol.* **2012**, *32*, 669–680. [CrossRef]
16. Li, H.; Meier, F.; Lee, X.; Chakraborty, T.; Liu, J.; Schaap, M.; Sodoudi, S. Interaction between urban heat island and urban pollution island during summer in Berlin. *Sci. Total Environ.* **2018**, *636*, 818–828. [CrossRef]
17. Stanhill, G.; Cohen, S. Is solar dimming global or urban? Evidence from measurements in Israel between 1954 and 2007. *J. Geophys. Res.* **2009**, *114*. [CrossRef]
18. Ren, G.; Li, J.; Ren, Y.; Chu, Z.; Zhang, A.; Zhou, Y.; Zhang, L.; Zhang, Y.; Bian, T. An integrated procedure to determine a reference station network for evaluating and adjusting urban bias in surface air temperature data. *J. Appl. Meteorol. Climatol.* **2015**, *54*, 1248–1266. [CrossRef]
19. Xu, Y.L. Characteristics of Precipitation in Hangzhou City from 1951 to 2014. *J. Yangtze River Sci. Res. Inst.* **2018**, *35*, 25–29. (In Chinese) [CrossRef]
20. Chao, L.; Huang, B.; Yang, Y.; Jones, P.; Cheng, J.; Yang, Y.; Li, Q. A new evaluation of the role of urbanization to warming at various spatial scales: Evidence from the Guangdong-Hong Kong-Macau region, China. *Geophys. Res. Lett.* **2020**, *47*, e2020GL089152. [CrossRef]
21. Bian, T.; Ren, G.; Yue, Y. Effect of Urbanization on Land-Surface Temperature at an Urban Climate Station in North China. *Bound. Layer Meteorol.* **2017**, *165*, 553–567. [CrossRef]
22. Tysa, S.K.; Ren, G.; Qin, Y.; Zhang, P.; Ren, Y.; Jia, W.; Wen, K. Urbanization effect in regional temperature series based on a remote sensing classification scheme of stations. *J. Geophys. Res. Atmos.* **2019**, *124*, 10646–10661. [CrossRef]
23. Wang, J.; Yan, Z.W. Urbanization-related warming in local temperature records: A review. *Atmos. Ocean. Sci. Lett.* **2016**, *9*, 129–138. [CrossRef]
24. Jin, K.; Wang, F.; Yu, Q.; Gou, J.; Liu, H. Varied degrees of urbanization effects on observed surface air temperature trends in China. *Clim. Res.* **2018**, *76*, 131–143. [CrossRef]
25. Stanhill, G.; Cohen, S. Solar radiation changes in Japan during the 20th century: Evidence from sunshine duration measurements. *J. Meteorol. Soc. Jpn.* **2008**, *86*, 1503–1512. [CrossRef]
26. Alpert, P.; Kishcha, P.; Kaufman, Y.J.; Schwarzbard, R. Global dimming or local dimming?: Effect of urbanization on sunlight availability. *Geophys. Res. Lett.* **2005**, *32*, L17802. [CrossRef]
27. Wang, Y.; Yang, Y.; Zhao, N.; Liu, C.; Wang, Q. The magnitude of the effect of air pollution on sunshine hours in China. *J. Geophys. Res. Atmos.* **2012**, *117*, D00V14. [CrossRef]
28. Liang, P.; Ding, Y. The Long-term Variation of Extreme Heavy Precipitation and Its Link to Urbanization Effects in Shanghai during 1916–2014. *Adv. Atmos. Sci.* **2017**, *34*, 321–334. [CrossRef]

29. Robaa, S.M. A study of solar radiation climate at Cairo urban area, Egypt and its environs. *Int. J. Climatol.* **2006**, *26*, 1913–1928. [CrossRef]
30. Zhang, M.; Song, K.; Da, L. The Diversity Distribution Pattern of Ruderal Community under the Rapid Urbanization in Hangzhou, East China. *Diversity* **2020**, *12*, 116. [CrossRef]
31. Feng, R. Investigating wintertime air pollution in Hangzhou, China. *Air Qual. Atmos. Health* **2020**, *13*, 321–328. [CrossRef]
32. Jin, K.; Wang, F.; Zong, Q.; Qin, P.; Liu, C. An Updated Estimate of the Urban Heat Island Effect on Observed Local Warming Trends in Mainland China's 45 Urban Stations. *J. Meteorol. Soc. Jpn.* **2020**, *98*, 787–799. [CrossRef]
33. Ning, J.; Liu, J.; Kuang, W.; Xu, X.; Zhang, S.; Yan, C.; Li, R.; Wu, S.; Hu, Y.; Du, G.; et al. Spatio-temporal patterns and characteristics of land-use change in China during 2010-2015. *J. Geogr. Sci.* **2018**, *28*, 547–562. [CrossRef]
34. Liao, W.L.; Wang, D.G.; Liu, X.P.; Wang, G.L.; Zhang, J.B. Estimated influence of urbanization on surface warming in Eastern China using time-varying land use data. *Int. J. Climatol.* **2017**, *37*, 3197–3208. [CrossRef]
35. Wang, F.; Ge, Q.S. Estimation of urbanization bias in observed surface temperature change in China from 1980 to 2009 using satellite land-use data. *Chin. Sci. Bull.* **2012**, *57*, 1708–1715. [CrossRef]
36. Wang, F.; Ge, Q.S.; Wang, S.W.; Li, Q.X.; Jones, P.D. A new estimation of urbanization's contribution to the warming trend in China. *J. Clim.* **2015**, *28*, 8923–8938. [CrossRef]
37. Wang, Y.W.; Yang, Y.H. China's dimming and brightening: Evidence, causes and hydrological implications. *Ann. Geophys.* **2014**, *32*, 41–55. [CrossRef]
38. Liley, J.B. New Zealand dimming and brightening. *J. Geophys. Res.* **2009**, *114*, D00D10. [CrossRef]
39. Li, W.; Hou, M.; Xin, J. Low-cloud and sunshine duration in the low-latitude belt of South China for the period 1961–2005. *Theor. Appl. Climatol.* **2011**, *104*, 473–478. [CrossRef]
40. Founda, D.; Kalimeris, A.; Pierros, F. Multi annual variability and climatic signal analysis of sunshine duration at a large urban area of Mediterranean (Athens). *Urban Clim.* **2014**, *10*, 815–830. [CrossRef]
41. Bartoszek, K.; Matuszko, D.; Soroka, J. Relationships between cloudiness, aerosol optical thickness, and sunshine duration in Poland. *Atmos. Res.* **2020**, *245*, 105097. [CrossRef]
42. Wang, K.; Ye, H.; Chen, F.; Xiong, Y.; Wang, C. Urbanization effect on the diurnal temperature range: Different roles under solar dimming and brightening. *J. Clim.* **2012**, *25*, 1022–1027. [CrossRef]
43. Chang, W.; Zhan, J.; Zhang, Y.; Li, Z.; Xing, J.; Li, J. Emission-driven changes in anthropogenic aerosol concentrations in China during 1970–2010 and its implications for PM2.5 control policy. *Atmos. Res.* **2018**, *212*, 106–119. [CrossRef]
44. Xia, X. Spatiotemporal changes in sunshine duration and cloud amount as well as their relationship in China during 1954–2005. *J. Geophys. Res.* **2010**, *115*, D00K06. [CrossRef]
45. Jiang, L.; Bai, L. Spatio-temporal characteristics of urban air pollutions and their causal relationships: Evidence from Beijing and its neighboring cities. *Sci. Rep.* **2018**, *8*, 1279. [CrossRef] [PubMed]
46. Li, R.; Wang, Z.; Cui, L.; Fu, H.; Zhang, L.; Kong, L.; Chen, W.; Chen, J. Air pollution characteristics in China during 2015–2016: Spatiotemporal variations and key meteorological factors. *Sci. Total Environ.* **2019**, *648*, 902–915. [CrossRef] [PubMed]

atmosphere

MDPI

Article

Experimental Determination of Pedestrian Thermal Comfort on Water-Retaining Pavement for UHI Adaptation Strategy

Yasuhiro Shimazaki [1,2,*], Masashige Aoki [3], Jumpei Nitta [2,3], Hodaka Okajima [3] and Atsumasa Yoshida [4]

1 Department of Architecture and Civil Engineering, Toyohashi University of Technology, Toyohashi 441-8580, Japan
2 Department of Human Information Engineering, Okayama Prefectural University, Soja 719-1197, Japan; jumpei_nitta@taiseirotec.co.jp
3 Taisei Rotec Corporation, Konosu 365-0027, Japan; masashige_aoki@taiseirotec.co.jp (M.A.); hodaka_okajima@taiseirotec.co.jp (H.O.)
4 Department of Mechanical Engineering, Osaka Prefecture University, Sakai 599-8531, Japan; ayoshida@me.osakafu-u.ac.jp
* Correspondence: shimazaki@ace.tut.ac.jp; Tel.: +81-532-44-6838

Abstract: Artificial impervious surfaces are one of the most significant factors contributing to urban heat islands (UHIs). Adapting to UHIs is a challenge in achieving thermal comfort. We conducted a quantitative and subjective evaluation of a closely paved novel water-retaining pavement (WR) and a conventional dense-asphalt pavement (AS). We investigated the thermal states of humans based on the human energy balance known as "human thermal load" as an indicator for the assessment, and the original human thermal load method was improved for assessing nonuniform environments such as pavements. We looked for individual thermal perceptions simultaneously. The experiment was conducted in typical summer weather. The surface temperature of the WR was found to be significantly lower, by 9.5 °C, while the air temperature and humidity above both pavements were not significantly different. Thus, air conditions did not directly affect the sensible and latent heat loss. The reflected solar radiation was significantly larger, and the infrared radiation was significantly smaller on the WR than on the AS due to the lower surface temperature from the water evaporation and higher reflectance. Further, the human thermal load at a pedestrian level of 1.5 m was found to be significantly different: 237 W/m^2 for AS and 215 W/m^2 for WR. In a subjective evaluation, the perceptions of WR tend to be distributed in smaller human thermal load, thereby resulting in a cooler and comfortable sensation. Therefore, we demonstrated that when compared to AS, WR significantly improves thermal comfort.

Keywords: human thermal load; surface material; evaporation; watering; subjective experiment

Citation: Shimazaki, Y.; Aoki, M.; Nitta, J.; Okajima, H.; Yoshida, A. Experimental Determination of Pedestrian Thermal Comfort on Water-Retaining Pavement for UHI Adaptation Strategy. *Atmosphere* **2021**, *12*, 127. https://doi.org/10.3390/atmos12020127

Received: 14 December 2020
Accepted: 18 January 2021
Published: 20 January 2021

Publisher's Note: MDPI stays neutral with regard to jurisdictional claims in published maps and institutional affiliations.

1. Introduction

Climate change is a fundamental problem of our time. Like other countries or regions, Japanese cities have been setting new records for the highest temperatures almost every year because of the climate warming trend and urban heat islands (UHIs). UHIs lead to high energy consumption and the deterioration of the quality of life in densely populated areas. This scenario is linked to the outdoor thermal comfort condition because the level of comfort is an essential factor for promoting outdoor activity, especially in urban areas.

It is well known that different types of ground-surface covers affect the climate of the built environment. The thermo-physical properties of surfaces in cities (e.g., asphalt) result in modified surface radiation and heat balance. In addition, pavements account for 20–40% of the surface area of a typical city, and therefore, they have the potential to mitigate UHIs [1]. Air temperatures depend on the absorption of solar radiation. Conventional absorptive pavements become hotter because of their higher absorption or lower albedo, and the hotter pavement warms the surrounding air. In contrast, natural ground contains water, which suppresses the rise in its temperature. Artificial impervious surfaces are considered one of the most significant causes of UHIs. The hot pavements aggravate UHIs by warming the surrounding air.

Temperature-lowering functional "cool pavements," which use novel materials and design modifications, can provide cooling compared to conventional pavements. Cool pavements can be categorized into two categories: reflective and evaporative [2]. Cool-pavement technologies have been developed mostly by enhancing the surface reflectivity, owing to the cost and ease of installation [3]. Because the solar heat absorbed by the pavement results in a rise in surface temperature, the pavement temperature could be decreased by decreasing surface absorptance (by increasing the reflectance). An approximately 20 °C difference in surface temperature was reported between black-colored and white-colored pavements during summer experiments [4]. Increasing reflectance by 0.1 could be analytically estimated to reduce the maximum pavement surface temperature by about 3.3 °C [5]. Researchers in Japan have been diligently working on water-retaining pavements. Because the energy is taken during the evaporation process, evaporative pavements contribute to cooling of the pavement surfaces. Temperature reductions of up to about 20 °C compared to conventional pavements have been reported [6–9]. The Japanese have a traditional "Uchimizu" watering custom, in which people sprinkle leftover water, such as used bath water, on the streets on hot summer days to create a cooling effect from evaporation. Similarly, water-retaining-pavement technology is believed to be environmentally friendly because it uses frequent natural rainfall effectively. In order to evaluate the effectiveness and the applicability of evaporative pavements, verification experiments were conducted in real outdoor environments.

Thermal-comfort studies in Japan are extensive, because achieving thermal comfort is a challenge due to the hot and humid summers in the country. These studies are mostly based on the measurement of physical environmental quantities. Because the sun represents the main heat source outdoors and influences the outdoor thermal environment directly or indirectly, assessing radiative components is important. Heat mitigation in pavement studies tends to focus on changes in surface temperature. This is only one of the factors that influences human thermal comfort. Few studies have focused on the effects of human aspects on functional evaporative pavements and how humans respond to these pavements. Although outdoor thermal indices such as the physiological equivalent temperature (PET) [10] and the universal thermal climate index (UTCI) [11] have become increasingly common, existing thermal indices developed for indoor conditions, such as the standard effective temperature (SET*) and the predicted mean vote (PMV), have been applied to outdoor conditions directly or with modifications [12]. Thermal comfort can be predicted using two different methods: the rational model and the adaptive model [13–15]. The rational model is based on human energy balance, and the adaptive model is based on thermal adaptation. To predict the thermal comfort conditions accurately, it is desirable to consider both aspects simultaneously. It is necessary to understand the subjective performance through an experimental approach.

Pedestrian spaces in cities are public spaces, and they play an important role in the physical and mental well-being of pedestrians. Although a variety of UHI mitigation and adaptation evaluation measures are already present, understanding how humans interact with their environment is fundamentally important for developing an evaluation tool. An experimental assessment method for pedestrian comfort is proposed here to provide a comfortable urban space using a functional evaporative pavement. The thermal environment on a water-retaining pavement and on a conventional asphalt pavement was investigated with Japanese participants.

2. Experiments

2.1. Material

Test pavements were constructed at Okayama Prefectural University in Okayama. The campus is located at 34°41′32″ N, 133°46′54.5″ E in Japan's western region, along the

Seto Inland Seacoast. Okayama is known as the "land of sunshine" because of its minimal rain and mild climate throughout the year. Previously, evaporative cooling duration for the WR was reported to be a few days to one week [2]. Based on the data from the Japan Meteorological Agency, the average rainfall in Okayama during summer is 171 mm and 17 days for June, 160 mm and 17 days for July, 87 mm and 13 days for August, and 134 mm and 17 days for September; for practical applications, the amount of rainfall is considered sufficient for watering. The location of the test was an open space. The pavement materials used in the experiment were WR and AS. The WR pavement was made of water-retentive blocks composed of glass (10%) and ceramic (90%), and thus, it contained fine voids to hold water (water retention capacity $\geq$ 15 vol%). The general dimensions of the water-retentive block were: length $\times$ width = 100 mm $\times$ 200 mm, and thickness = 60 mm. Both pavements had an area of 7.0 m $\times$ 7.0 m, as shown in Figure 1.

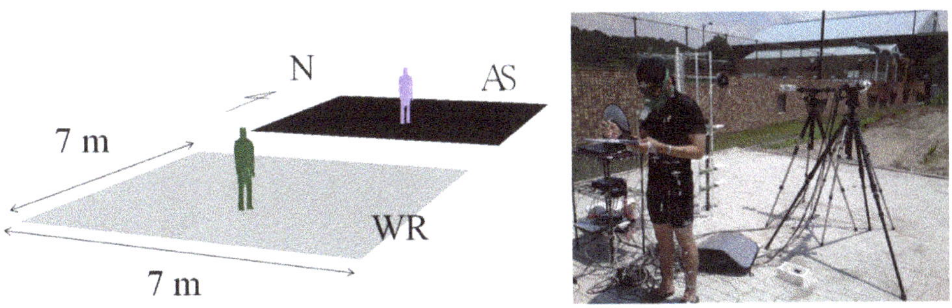

Figure 1. The experimental landscape.

2.2. *Methods*

2.2.1. Experimental Setup

We measured the physiological and psychological reactions of the human body to different environmental conditions due to the pavements. Further, the surrounding weather factors were also measured. The experiment was conducted between 10:00 a.m.–16:00 p.m. JST from 22 August to 26 August 2016.

A total of 14 healthy male Japanese university students participated in the study for both pavements. Only one participant stayed on each pavement at a time, and the experiment was conducted on both pavements simultaneously. The height, weight, and age of volunteers were 172.7 $\pm$ 4.5 cm, 63.4 $\pm$ 6.6 kg, and 22.4 $\pm$ 1.3 years, respectively. The body mass index (BMI) of the participants was 21.1 $\pm$ 1.7, which was categorized in the healthy weight range. Informed consent was obtained from all volunteers, and the research was conducted in accordance with the guidelines and approval of the research ethics committee of the institute.

Each trial lasted for 30 min, during which the participants maintained a standing-still posture. To ensure consistent clothing insulation, each participant dressed in the same tight-fit garments and footwear; the properties of these garments were determined before the trial (whole clothing insulation I = 0.35 clo). Each participant drank 200 mL of water 30 min before the experiment to ensure suitable hydration.

2.2.2. Climatic Observation

Physical environmental quantities including air temperature, humidity, solar and infrared radiation, and wind speed were measured. Air temperature and humidity were measured at heights of 0.35 m, 0.5 m, and 1.5 m from the surface using a capacitive thermo-hygrometer recorder (TR-73U, T&D Corporation, Nagano, Japan). The solar radiation and infrared radiation in upward and downward directions were measured at a height of 1.5 m using a net radiometer (MR-60, EKO, Instruments, Tokyo, Japan). The wind speed was measured at a height of 1.5 m using a hot-wire anemometer (Climomaster Model-6531, Kanomax Japan Inc., Tokyo, Japan). A measurement height of 1.5 m was selected to indicate the pedestrian height [12,16]. To understand the effect of the height on the environment, near-surface values at 0.35 and 0.5 m were measured for temperature and humidity. Every environmental quantity was measured and recorded at 1 min intervals using a data-logger (LR-8400, Hioki E.E. Corp., Nagano, Japan).

2.2.3. Pavement Conditions

The surface temperatures at five points on both pavements were measured using J-type thermocouples. A thermocouple was placed at the center of the pavement, and four thermocouples were placed 0.5 m away from the center of the pavement in the north, east, south, and west directions. Infrared images were periodically captured during the experiment using thermography (InfReC H2640, Nippon Avionics Co., Yokohama, Japan). The emissivity was set to 1.0 to obtain infrared images for both pavements.

The degree of reflectance (albedo) is an important factor in the formation of urban climate [17]. Because the experimental pavements were relatively large and ensured uniformity, the values of albedo ρ were obtained simply as the ratio of the amount of global solar radiation $S\downarrow$ and the reflected solar radiation $S\uparrow$ measured by the net radiometer [18].

$$\rho = \frac{S\uparrow}{S\downarrow} \tag{1}$$

A water hose was used to sprinkle water until the WR blocks could not retain water any more to ensure uniformity in the conditions before each experiment.

2.2.4. Human Thermal States on Pavements

To design a better environment for humans by improving ground coverings, we investigated and quantified the thermal states of humans on pavements. We previously developed a measure for evaluating human thermal environment outdoors known as "human thermal load" [19]; this measure is based on human energy balance. When the human body and the surrounding environment are in a state of thermal equilibrium, the thermal condition of the human body can be expressed by the heat balance as:

$$M + R_{net} = W + C + E \tag{2}$$

where M, R_{net}, W, C, and E denote the metabolic rate, net radiation, workload, convective heat loss, and evaporative heat loss, respectively. Further, C and E include the heat exchange caused by respiration. The unit for each term is W/m^2.

If the thermal state is not at a neutral level, a positive or negative thermal load is applied to the human body. This load amount is referred to as the human thermal load F_{load} (W/m^2) and is defined by the heat balance equation as:

$$F_{load} = M + R_{net} - W - C - E \tag{3}$$

Metabolic rate is an important determinant of the comfort or the strain resulting from the exposure to thermal environment, particularly in a hot climate [20]. The metabolic rate M refers to the heat generation by humans, and it is calculated from body surface area A_D (m^2), oxygen consumption V_{O_2} (L/min), and carbon-dioxide elimination V_{CO_2} (L/min) using the metabolic measurement system (VO2000, MGC Diagnostics, Saint Paul, MN, USA) based on Weir's formula [21].

$$M = \frac{69.735(3.9V_{O_2} + 1.1V_{CO_2})}{A_D} \tag{4}$$

The body surface area A_D can be determined using the following formula by Kurazumi et al. [22] for Japanese participants:

$$A_D = 2034.309W^{0.425}H^{0.725} \tag{5}$$

where W (kg) denotes the body weight and H (m) denotes the height.

The metabolic rate can be determined according to the type of activity and occupant. As a simplified method for outdoor experiments for practical applications, the metabolic equivalents of task (met) can be used for determining the activity level. The met is defined as the ratio of the working metabolic rate relative to the resting metabolic rate for an activity, and the list of met values for different activities is widely available (e.g., ASHRAE) [23]. To reflect individual variability, the resting metabolic rate is preliminary determined in an indoor chamber using Equation (4), and then, the metabolic rate is determined by multiplying with the met value. The met value for a person standing still is met = 1.2.

Workload W denotes the mechanical work performed by humans. Since the human participants were standing and in a resting posture in this experiment, it was assumed that $W = 0$ [23].

The net radiation R_{net} is the amount of solar and infrared radiation received by the human body and emitted from the human body, and it is calculated as:

$$R_{net} = (1 - \alpha_h)R_{sh} + \varepsilon_h\left\{R_{ln} - \sigma(T_{skin} + 273.15)^4\right\} \tag{6}$$

where α_h, ε_h, R_{sh} (W/m^2), R_{ln} (W/m^2), σ (W/(m^{2}K^4)), and T_{skin} (°C) denote the reflectance of the human body (=0.3) [24], emissivity of the human body (=0.98) [25], gain of heat from solar radiation, gain of heat from infrared radiation, the Stefan–Boltzmann constant, and the mean skin temperature, respectively. As the participants were dressed in a tight-fit garment in this experiment, α_h can be partly replaced by the reflectance of the garment α_{clo}, which was preliminarily determined using the method proposed by the authors [18]. The mean skin temperature was the area weighted and calculated as:

$$T_{skin} = \sum_i F_i T_i \tag{7}$$

where i, F_i, and T_i (°C) denote the body region, weighting factor for region i, and skin temperature for region i, respectively. The measuring sites varied from using fewer point to a large number of points, such as Ramanathan 4-points, Hardy-DuBois 7-points, and ISO9886 14-points [26]. Based on Hardy and DuBois's 7-point formula [27], the weighting factors of different body regions were determined in the present study as listed in Table 1. The skin temperature of different regions in the body was measured using thermistors (N543R, Nikkiso-Therm Co., Tokyo, Japan).

Table 1. Weighting factors for the mean skin temperature of different body regions.

Body Region i	Weighting Factor F_i
1: Forehead	0.07
2: Abdomen	0.35
3: Forearm	0.14
4: Back of hand	0.05
5: Thigh	0.19
6: Leg	0.13
7: Back of Foot	0.07

In this study, the human thermal load method was improved for assessing nonuniform environments in the up and down direction, such as on pavements. In general, global solar radiation influences the outdoor thermal environment. To evaluate the effects of climatic radiation and radiation from the ground surface separately, net radiation is divided into downward net radiation $R_{net}\downarrow$ and upward net radiation $R_{net}\uparrow$ as

$$R_{net}\downarrow = (1-\alpha_h)R_{sh}\downarrow + \varepsilon_h\left\{R_{ln}\downarrow - \sigma(T_{skin}+273.15)^4\right\} \tag{8}$$

$$R_{net}\uparrow = (1-\alpha_h)R_{sh}\uparrow + \varepsilon_h\left\{R_{ln}\uparrow - \sigma(T_{skin}+273.15)^4\right\} \tag{9}$$

where $\downarrow$ indicates the downward components and $\uparrow$ indicates the upward components. To represent the effect of the direction of irradiation and the size of the person on the net radiation, the person was simplified to a floating rectangular shape facing the sun, as shown in Figure 2. Upward and downward solar radiation and infrared radiation were then calculated as

$$R_{sh}\downarrow = \frac{A_{up}S\downarrow + A_f(S_T\cos z + \gamma S_D) + \gamma S_D(A_l + A_r + A_{bk})}{A_D} \tag{10}$$

$$R_{sh}\uparrow = \frac{A_{bt}S\uparrow + \gamma S\uparrow\left(A_f + A_l + A_r + A_{bk}\right)}{A_D} \tag{11}$$

$$R_{ln}\downarrow = \frac{A_{tp}L\downarrow + \gamma L\downarrow\left(A_f + A_l + A_r + A_{bk}\right)}{A_D} \tag{12}$$

$$R_{ln}\uparrow = \frac{A_{bt}L\uparrow + \gamma L\uparrow\left(A_f + A_l + A_r + A_{bk}\right)}{A_D} \tag{13}$$

where A_{up}, A_f, A_r, A_l, A_{bk}, and A_{bt} (m^2) denote the areas of the upper, front, right, left, back, and bottom planes, respectively. $L\downarrow$ and $L\uparrow$ (W/m^2) represent the infrared radiation from the sky and from the ground measured by the net radiometer. S_T (W/m^2) denotes the direct solar radiation and S_D (W/m^2) denotes the diffuse solar radiation; they were estimated using Udagawa's formula [28] from $S\downarrow$ measured by the net radiometer. z denotes the solar altitude angle (°), and γ represents the view factor between the human and the sky or the surface, which was assumed to be 0.5 for an open area. The area ratio for each body plane was uniformly determined for each participant: $0.3A_D$ for A_f and A_{bk}, $0.15A_D$ for A_l and A_r, $0.05A_D$ for A_{up} and A_{bt}, respectively.

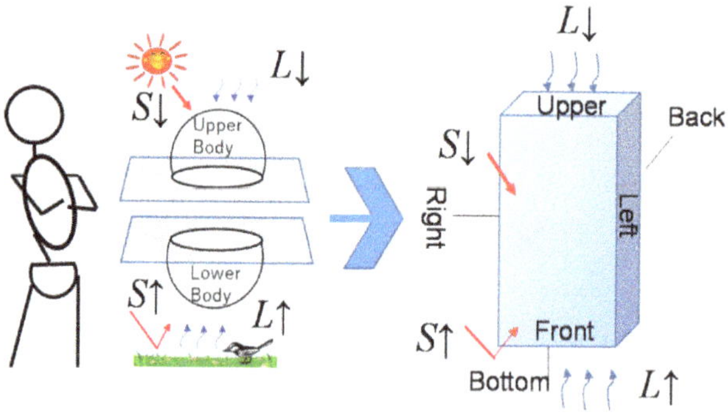

Figure 2. Abstraction of the separation of the upward and downward radiation components.

Convective heat loss C is the sum of the dry heat loss from the skin (C_{sk}), and through the respiratory system (C_{res}). C_{sk} is based on Burton and Edholm [29], and C_{res} is based on the ASHRAE model [23]; the total convective heat loss is determined as:

$$C = C_{sk} + C_{res} \tag{14}$$

$$C_{sk} = F_{cl}h_c(T_{skin} - T_{air}) \tag{15}$$

$$C_{res} = 0.0014M(34 - T_{air}), \tag{16}$$

where T_{air} (°C) denotes the temperature of the ambient air, F_{cl} represents the Burton's reduction factor, and h_c ($W/m^2/°C$) indicates the convective heat transfer coefficient [30] given as:

$$F_{cl} = \frac{1}{(1 + 0.209h_cI)} \tag{17}$$

$$h_c = 3.86 + 6.96v^{1.02} \tag{18}$$

where I (clo) denotes the clothing insulation [31] and v (m/s) denotes the wind speed.

Evaporative heat loss E is the sum of the wet heat loss from the skin (E_{sk}) and the respiratory system (E_{res}). The evaporative heat loss from the skin is a combination of thermo-regulatory sweating E_{rsw} and insensible natural diffusion E_{dif} [23].

$$E = E_{sk} + E_{res} \tag{19}$$

$$E_{sk} = E_{rsw} + E_{dif} \tag{20}$$

The amount and timing of evaporative heat loss is important; however, in practice, it is not easy to measure both in outdoor field experiments. Based on the two-node model [32,33], the evaporative heat loss is determined as

$$E_{rsw} = c_{sw}(T_b - T_{b,set})exp\left(\frac{T_{sk} - T_{sk,set}}{10.7}\right) \tag{21}$$

$$E_{dif} = 0.06\ E_{max} \tag{22}$$

$$E_{max} = LwF_{cl}h_c(P^*{}_{skin} - P_{air}) \tag{23}$$

$$E_{res} = 0.0173M(5.87 - P_{air}) \tag{24}$$

where c_{sw} ($W/(m^2K)$) is a proportional constant for sweat control (=170), T_b (°C) denotes the blood temperature, and $T_{b,set}$ and $T_{sk,}$ (°C) denote the sweating threshold for blood and skin temperature, respectively. Lw (°C/kPa) represents the Lewis ratio, $P^*{}_{skin}$ (kPa) denotes the saturated water vapor pressure at skin temperature, and P_{air} (kPa) represents the water vapor pressure in ambient air. Regardless of the situation for thermo-regulatory sweating, the ratio of the diffusion evaporative heat loss was set to 6% of E_{max}. E_{rsw} can be predicted as a function of blood temperature and mean skin temperature. The blood temperature was determined as:

$$T_b = (1 - \alpha_{sk})T_{cr} + \alpha_{sk}T_{skin} \tag{25}$$

where α_{sk} denotes the mass ratio of the skin component to the whole body. T_{cr} (°C) denotes the core temperature of the human body, and the rectal temperature was measured using a thermistor (N543R, Nikkiso-Therm Co., Tokyo, Japan).

To understand the physiological conditions of the participants, the body-weight and heartrate were measured using a precise electronic balance (GP-100K, A&D Co. Ltd., Tokyo, Japan) and a heartrate sensor (RS800CX Training Management System, Polar, Kempele, Finland). To limit the above evaporative heat loss E compared to the participant experiments, we set an upper limit E' to total evaporative heat loss based on the time-averaged sweat evaporation from the body-weight measurement as:

$$E' = E_{sk}' + E_{res}' = \frac{l}{A_D}\frac{\Delta w}{t} \tag{26}$$

where Δw (kg) denotes the body-weight change before and after the experiment, and t (s) represents the duration between the body-weight measurement before and after the experiment. When $E > E'$, E was replaced by E'.

2.2.5. Human Perceptions

As the perception of the participants' thermal comfort is a basic concept of environmental evaluation using the adaptive approach, we asked the participants to evaluate their perceptions on the thermal sensation, wettedness, and thermal comfort using the visual analog scale (VAS). Then, for quantification, each perception was scored from −3 to 3 for thermal sensation, from −2 to +2 for wettedness, and from −2 to 2 for thermal comfort, with linear interpolation based on the standards from the Architectural Institute of Japan [34]. The obtained scores correspond to the numbers listed in Table 2. Every perception was recorded on-site at 5 min intervals.

Table 2. Quantification of human thermal environment evaluated by the study participants.

Value	Thermal Sensation	Wettedness	Thermal Comfort
3	Hot	-	-
2	Warm	Wet	Comfortable
1	Slightly warm	Slightly wet	Slightly comfortable
0	Neutral	Neutral	Neutral
−1	Slightly cool	Slightly dry	Slightly uncomfortable
−2	Cool	Dry	Uncomfortable
−3	Cold	-	-

2.2.6. Data Analysis

As the experiments were conducted outdoors, the variations in the ambient atmosphere could have affected the results. The climate was relatively stable during the 30 min of the experiment, and human thermal states were almost stable. Thus, the investigation was performed using the mean value recorded during the experiments with the participants. A Student's *t*-test analysis was conducted to investigate whether the difference in pavements in the variables was significant. The effect size for a *t*-test was evaluated using Cohen's d. The items measured and accuracy of the instruments are presented in Table 3 for uncertainty evaluations. Since the uncertainty of solar and infrared radiation measurement under outdoor conditions depends on many factors, please refer to the specifications for a more complete description (https://eko-usa.com).

Table 3. Measurement items and instruments.

Parameter	Accuracy	Instrument
Air temperature	±0.3 °C (0–50 °C)	TR-73U, T&D
Relative humidity	±5% R.H. (10–95% R.H.)	TR-73U, T&D
Wind speed	±2% or 0.02 m/s	Model-6531, Kanomax
Surface temperature	±1.5 °C (−40–375 °C)	Thermocouple, J-type
Ventilatory gas	±0.1% for O_2/ ±0.2% for CO_2	VO2000, MGC Diagnostics
Body temperature	±0.2 °C (0–70 °C)	N543R, Nikkiso-Therm
Body weight	±10 g	GP-100K, A&D
Heartrate	±1% or 1 bpm	RS800CX, Polar

3. Results

3.1. Experimental Conditions

The air temperature, humidity, solar and infrared radiation, and wind speed measured at a 1.5 m height during the experiment are shown in Figure 3. The average values (mean ± s.d.) of these physical quantities measured over AS were as follows: air temperature (33.3 ± 1.4 °C), humidity (53.2 ± 7.0% R.H.), global solar radiation (583 ± 237 W/m^2), infrared radiation from the sky (520 ± 13 W/m^2), and wind speed (0.79 ± 0.18 m/s). The average maximum temperature in late August in Okayama is 32.7 °C, based on the database from the Japan Meteorological Agency. The global solar radiation showed some fluctuation because of the condition of cloud cover. However, other variables were mostly stable. Thus, the overall climatic conditions during the experiment can be considered as typical of summer weather.

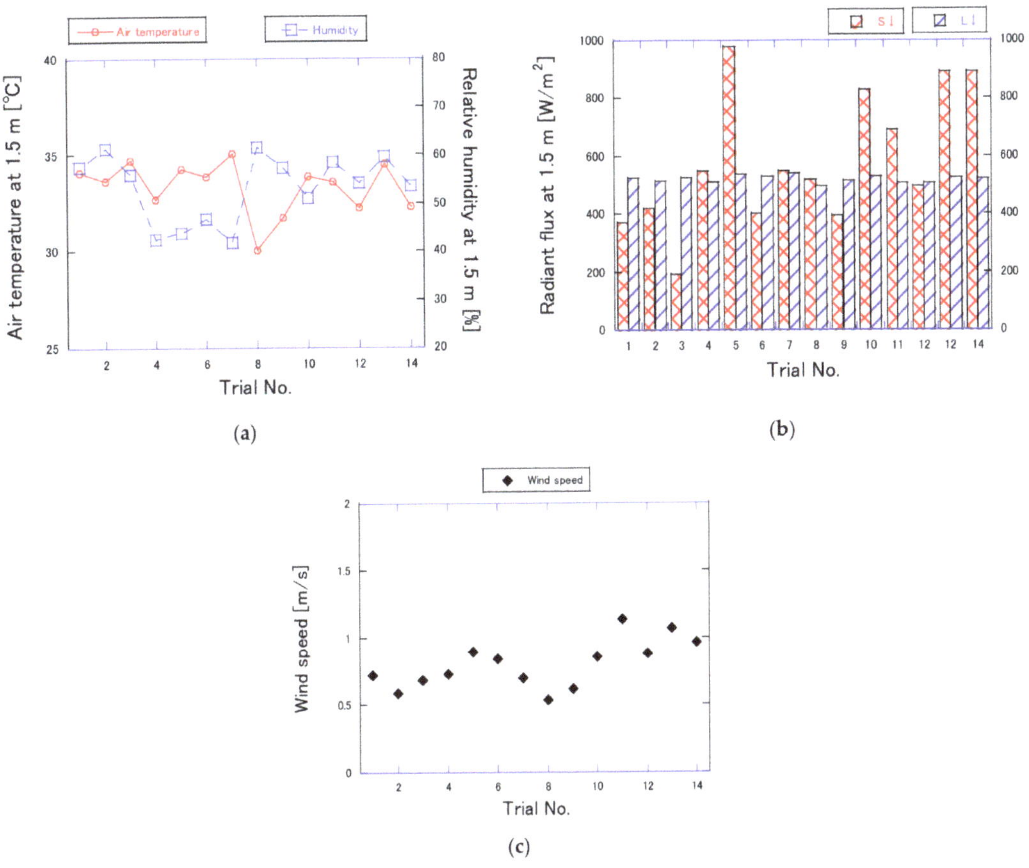

Figure 3. Climatic conditions at 1.5 m during each trial of the experiment: (**a**) air temperature and humidity, (**b**) global solar radiation and infrared radiation from the sky, and (**c**) wind speed.

3.2. Pavement Conditions

The surface temperatures of both pavements were compared using infrared images, as shown in Figure 4. The surface temperature was uniformly distributed over the surface. Using the thermocouples, the average surface temperature of AS and WR during the experiments was measured to be 50.2 ± 5.0 °C and 40.7 ± 2.9 °C, respectively, and this difference was found to be significant (p < 0.001, d = 2.38). The performance of WR is analyzed in Figure 5. The surface temperature difference varied approximately linearly with the surface temperature on AS, so the effectiveness of surface temperature reduction could be assessed completely.

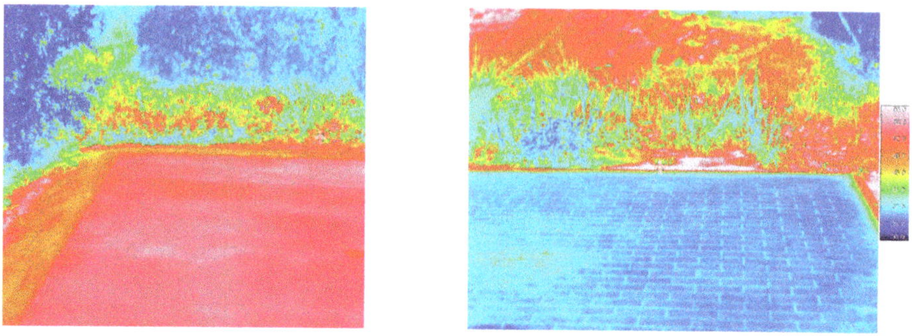

(a) (b)

Figure 4. Thermal images of both pavements (at 10:20 a.m. on 26 August 2016): (**a**) AS and (**b**) WR.

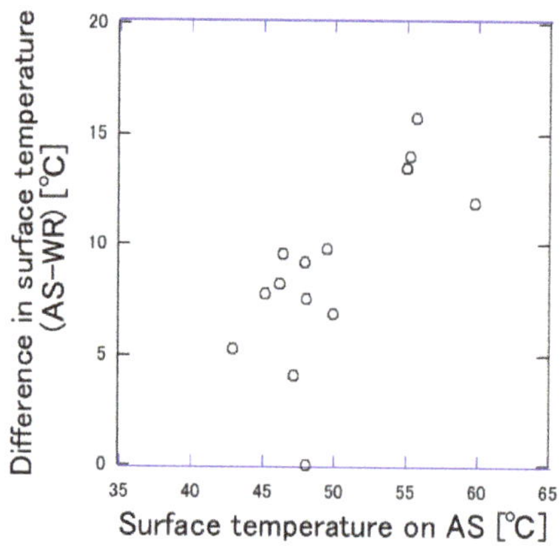

Figure 5. A comparison of surface temperatures.

The vertical temperature and humidity profiles of both pavements are shown in Figure 6. To eliminate the effect of temperature on humidity, the absolute humidity is presented in the figure. It is natural for the air temperature to be lower as the measuring point increases. The air temperature profiles showed similar values for both pavements, except on the surface. Absolute humidity fell within the range of approximately 17 g/m^3. A clear effect of water evaporation on WR and height on absolute humidity could not be observed.

The values of albedo were obtained with acceptable fluctuation, as listed in Table 4. The measurement was conducted simultaneously during participant experiments to assess the WR effect at the wet state. AS absorbs more solar energy and WR reflects more solar energy because pavements are considered as opaque and AS has a lower albedo value.

Table 4. Reflectance of pavements.

	AS	**WR**
Reflectance [1]	0.085 ± 0.010	0.253 ± 0.029

[1] Values in mean ± s.d.

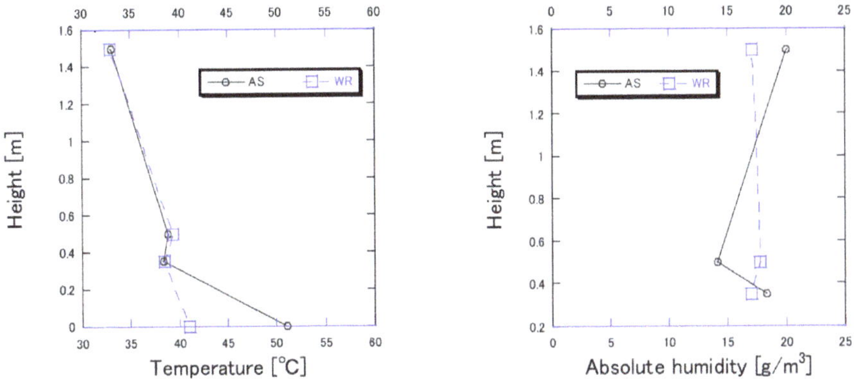

(**a**) (**b**)

Figure 6. The vertical profile over the pavements: (**a**) temperature and (**b**) absolute humidity.

3.3. *Human Thermal Condition at a Height of 1.5 m*

The human thermal loads at pedestrian height (1.5 m) for each trial, the relative contribution of $R_{net}\uparrow$ of AS and WR from $S\uparrow$ and $L\uparrow$, and the mean net radiations are compared in Figure 7, and each thermal load component is summarized in Table 5. Since $F_{load} = 0$ is thermally neutral, it is assumed to be comfort. In summer, a lower human thermal load is preferred. Since each participant experienced environments on both pavements, the metabolic rate should be similar (87 W/m^2). Air conditions such as temperature, humidity, and airflow were similar for both pavements and did not affect heat loss. Thus, other factors such as personal differences could have affected the difference in heat losses. A noticeable difference was observed in the net radiation. $R_{net}\downarrow$ is dependent on climatic radiation and was similar for both pavements, while $R_{net}\uparrow$ indicates a smaller tendency for WR ($p = 0.10$, d = 0.31). Further, the reflected solar radiation $S\uparrow$ of WR was significantly larger ($p < 0.001$, d = 2.31) and the infrared radiation $L\uparrow$ of WR was significantly smaller than that of AS ($p < 0.001$, d = 2.39). The total human thermal load was measured to be 237 ± 38 W/m^2 for AS and 215 ± 49 W/m^2 for WR. A significant difference was observed in the human thermal load between AS and WR, and it presented a small sized effect ($p = 0.01$, d = 0.47).

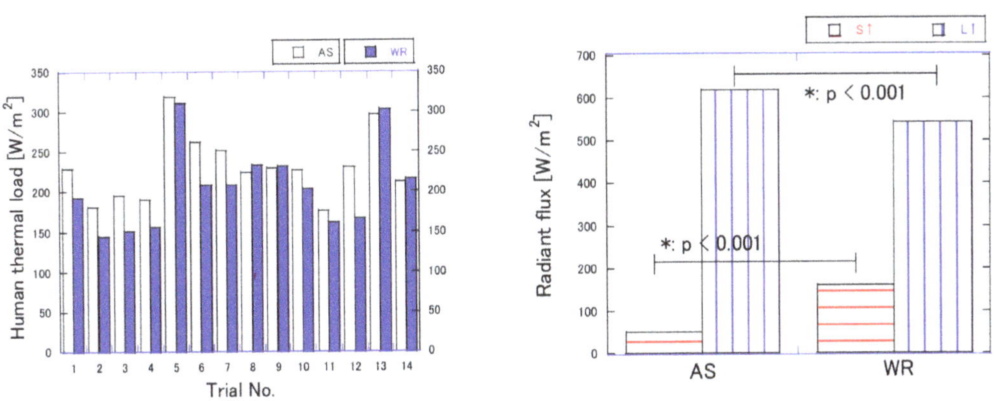

(**a**) (**b**)

Figure 7. *Cont.*

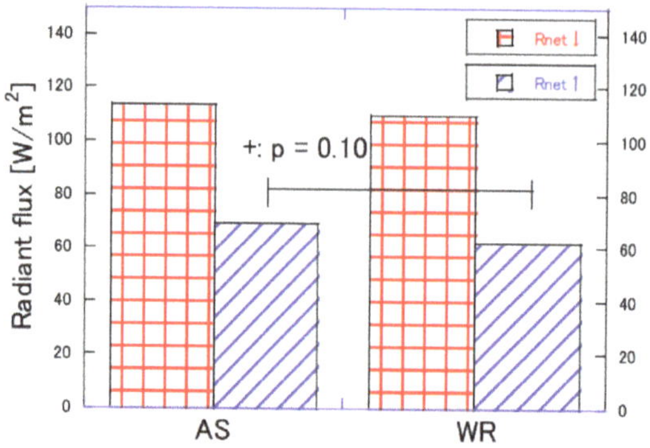

(c)

Figure 7. Comparison between AS and WR at pedestrian height (1.5 m): (**a**) human thermal load, (**b**) comparison of radiation components of AS and WR, and (**c**) net radiation.

Table 5. Components of human thermal load.

Component	AS	WR
Metabolic rate	87 ± 2	87 ± 2
Workload	0	0
Net radiation	193 ± 42	181 ± 55
Convective heal loss	15 ± 6	18 ± 4
Evaporative heat loss	29 ± 3	35 ± 5
Human thermal load *	237 ± 38	215 ± 49

* Significant difference (p = 0.01, d = 0.47).

3.4. Human Perceptions

The mean value of each perception is shown in Figure 8. In the case of AS, the thermal sensation was +0.69, wet sensation was +0.04, and comfort sensation was −0.33. In the case of WR, the thermal sensation was +0.48, wet sensation was −0.02, and comfort sensation was −0.14. The participants found that the environment was hot and uncomfortable overall. There was a trend towards reporting a cooler sensation on WR (p = 0.04, d = 0.31), and WR did not significantly affect other perceptions. Interestingly, no difference in wet sensation was induced on WR, even when the participant stood on the wetted WR surface.

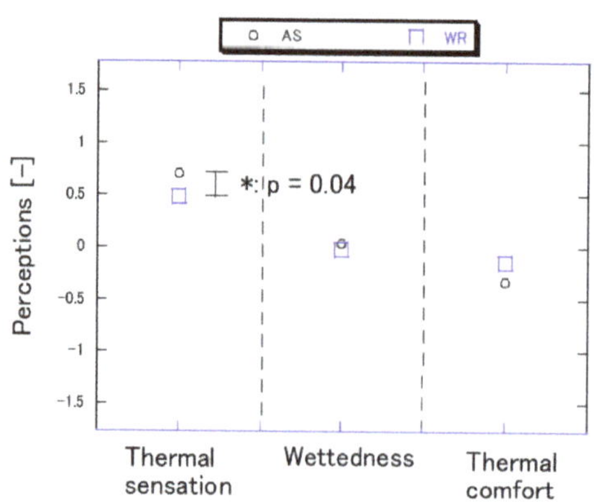

Figure 8. A comparison of the perceptions of the participants on AS and WR.

The relationships among the human thermal load, thermal sensation, and thermal comfort are plotted in Figure 9. Figure 9a shows the mean value for each variable (vertical and horizontal axis) and Figure 9b shows the line of equality ($y = x$). The human thermal load correlated with thermal sensation, and an almost linear relationship between human thermal load and thermal sensation can be observed. Overall, the plots on WR tended to be distributed in regions of smaller human thermal load and cooler perception. Further, thermal comfort had a correlation with thermal sensation (Figure 9c), as is often reported by researchers.

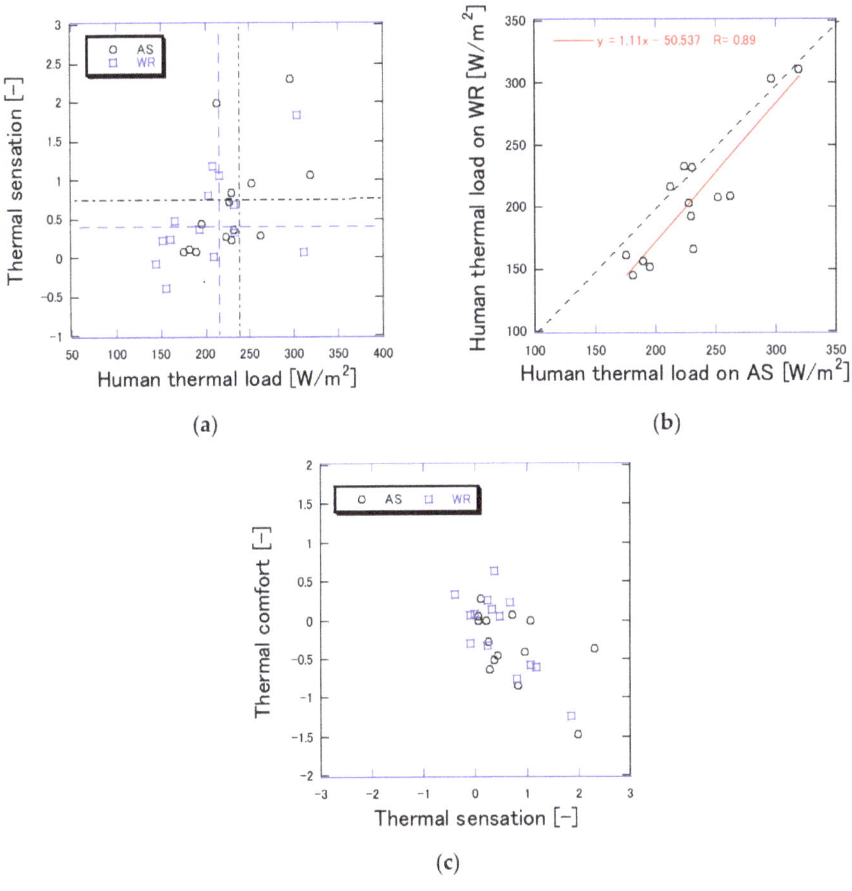

Figure 9. Relationships among the human thermal load, thermal sensation, and thermal comfort: (**a**) human thermal load and thermal sensation on AS and WR, (**b**) comparison of human thermal load of AS and WR, (**c**) thermal sensation and thermal comfort on AS and WR.

4. Discussion

4.1. Effects of Properties of WR on Human Thermal Environment

Two types of cool pavements are considered for surface improvement: reflective and evaporative. The WR in this experiment had a water-retentive function and higher reflectivity (=0.253) relative to AS (=0.085); thus, it is a hybrid cool pavement. The surface temperature on WR was significantly lower (by 9.5 °C) than conventional AS, and the potential of surface temperature reduction tended to be larger as the solar radiation became intense. However, the air temperature and humidity above both pavements were not significantly different. Numerous studies have reported on the effect of the color of the pavement materials on their surface temperature, and these effects can lead to a reduction of up to approximately 20 °C, depending on material properties, climatic conditions, and timing [35]. The air temperature was reported to be less sensitive than surface temperature [7]. There was a concern that WR would induce wet sensation; however, the evaporation above WR did not make the pedestrians on it feel wet. The areas of the test pavements used in this study may not be sufficient to influence the ambient environment, and a vast pavement may be more influential on temperature and humidity formation. If the lower surface temperature is widely spread on the pavement, it can potentially suppress the air temperature, which would be useful in mitigating the effects of UHIs. Such pavement modifications will be studied in future work.

4.2. Human Thermal States and Perceptions

The environmental measurement of pavements has been widely conducted. However, subjective evaluation of functional pavements such as WR by humans has rarely been reported previously. Using the indicator of human thermal load and its components, we compared and evaluated the thermal conditions and human effects of the WR pavement.

The sensible and evaporative heat loss did not show a significant difference because air temperature, humidity, and wind conditions experienced by the participants on both pavements were similar. The net radiation consists of the effects of solar radiation and infrared radiation. Because $R_{net}\downarrow$ is dependent on climatic radiation, $R_{net}\uparrow$ can reflect pavement effects. In Figure 7b, the reflected solar radiation $S\uparrow$ of WR was significantly larger and the infrared radiation $L\uparrow$ of WR was significantly smaller than that of AS. As the infrared radiant emission is proportional to the fourth power of the temperature, the infrared radiation from WR was lower as the surface temperature level was lower. It is likely to have had a negative impact on the participants due to the increased reflected solar radiation, especially under strong solar radiation. This suggests that an optimum reflectance value for reducing $R_{net}\uparrow$, namely human thermal load, should be examined for environmental design.

To explore the degree of adaptation to the environment based on the proposed approach, the relationships among the human thermal load, thermal sensation, and thermal comfort are examined in Figure 9. The human thermal load correlated with the thermal sensation, and the thermal sensation correlated with the thermal comfort. The plots on WR tended to be distributed in regions of smaller human thermal load and cooler perception. In addition, a linear relationship between the human thermal load on AS and WR is observed in Figure 9b. The slope of the regression line in Figure 9b is slightly larger than 1.0. This indicates the possibility that the difference in the human thermal load between AS and WR could become narrower for larger human thermal load conditions, because under these conditions, solar radiation should be intense, and thus, humans would receive more reflected solar radiation on the WR pavement. Since thermally neutral states can be considered as thermal comfort, the relationship between thermal sensation and thermal comfort shows a mound-shaped distribution. Thermal comfort has a negative correlation with thermal sensation in Figure 9c due to the warmer climate that was present during the experiment.

4.3. Limitations

This study was designed to create thermal comfort for outdoor pedestrians based on a field experiment on WR. The outdoor space varies both spatially and temporally, and thus it may be important to develop a thermal comfort evaluation method that considers the instantaneous and local effects of thermal comfort or discomfort. A line of the best regression fit for scatter plots is useful for drawing predictions. However, all experiments were conducted in a warm environment in the present study. Hence, in future work, we propose to perform additional studies in a variety of environments, such as in more neutral and cooler environments. Since the data varied because of exposure to environmental conditions caused by the outdoor participant experiments, participants potentially reacted physiologically differently, and reported a somewhat wider range of thermal perceptions. Further, it is necessary to understand the effect of human activities such as street walking on the metabolic rate and clothing functions instead of when the pedestrian is stationary, for future work. Since evaporative heat loss is an important factor considering thermal comfort under heat, it is desirable to utilize a novel on-site measuring system for sweat evaporation. Further studies are required to optimize and modify pavement performances such as reflectance by considering the relative contribution of radiation components. The properties of pavements, such as emissivity and reflection, also need to be investigated in association with conditions of pavements such as surface roughness. Although the measurement was conducted at a height of 1.5 m, the effect of the height above the pavement on the thermal environment was important to be examined. Human thermal load is a concept for quantifying the human thermal environment in the form heat flux, and the determination method for each term can be replaced, modified, or updated according to the requirement. To solidify the findings of this study, comparisons with outdoor thermal indices such as PET and UTCI can be conducted as a future work.

5. Conclusions

Currently, there are few established methods that allow easy assessment of outdoor thermal comfort specific to the evaporative-pavement-occupied pedestrian environment. In this paper, we proposed quantification techniques to study the influence of WR pavements and conventional AS pavements on the thermal environment, human body, and thermal comfort; in particular, we focused on the relative contribution of radiation components.

The key experimental observations on the thermal behavior of WR are as follows: (1) WR significantly reduced the surface temperature when compared to AS. However, the air temperature and humidity above both the pavements were not significantly different. (2) The sensible and evaporative heat loss of humans did not make a significant difference due to the similar climatic air conditions. The reflected solar radiation from WR was significantly larger and the infrared radiation of WR was significantly smaller than that of AS. Thus, WR significantly reduces human thermal load. (3) The proposed human thermal load correlates with the subjective thermal sensation. WR tends to induce a cooler sensation and does not induce an unpleasant wet sensation. As there is a strong relationship between thermal sensation and thermal comfort, we can utilize human thermal load as an environmental indicator and assessment tool for UHI adaptation. (4) It can be concluded that the thermal environment of WR is better in terms of human thermal state when compared to AS. However, there is still room for improvement of the reflectance of WR by reducing the reflective solar heat received. Experiments in more realistic scenarios, such as when walking on the street, should be performed.

Author Contributions: Conceptualization, Y.S., M.A.; methodology, validation, and investigation; Y.S., M.A., J.N.; data curation, H.O.; writing—original draft preparation, Y.S.; supervision, A.Y. All authors have read and agreed to the published version of the manuscript.

Funding: This work was partly supported by JSPS KAKENHI Grant number 26281057 (P.I. Atsumasa Yoshida).

Institutional Review Board Statement: The study was conducted according to the guidelines of the Declaration of Helsinki, and retrospective approved by the Ethics Committee of Osaka Prefecture University (Project title: The relationship between thermal environmental factors and human physiological responses (PI: Atsumasa YOSHIDA), approved date: 30 September 2016).

Informed Consent Statement: Informed consent was obtained from all subjects in-volved in the study.

Data Availability Statement: Data sharing not applicable.

Acknowledgments: The authors acknowledge Yayoi Satsumoto of Yokohama National University and Tomonori Sakoi of Shinshu University for their technical advice.

Conflicts of Interest: The authors declare no conflict of interest.

References

1. Akbari, H.; Konopacki, S.; Pomerantz, M. Cooling energy savings potential of reflective roofs for residential and commercial buildings in the United States. *Energy* **1999**, *24*, 391–407. [CrossRef]
2. Qin, Y. A review on the development of cool pavements to mitigate urban heat island effect. *Renew. Sustain. Energy Rev.* **2015**, *52*, 445–459. [CrossRef]
3. Bretz, S.; Akbari, H.; Rosenfeld, A. Practical issues for using solar-reflective materials to mitigate urban heat islands. *Atmos. Environ.* **1998**, *32*, 95–101. [CrossRef]
4. Doulos, L.; Santamouris, M.; Livada, I. Passive cooling of outdoor urban spaces. The role of materials. *Sol. Energy* **2004**, *77*, 231–249. [CrossRef]
5. Chen, J.; Wang, H.; Zhu, H. Analytical approach for evaluating temperature field of thermal modified asphalt pavement and urban heat island effect. *Appl. Therm. Eng.* **2017**, *113*, 739–748. [CrossRef]
6. Takebayashi, H.; Moriyama, M. Study on surface heat budget of various pavements for urban heat island mitigation. *Adv. Mater. Sci. Eng.* **2012**, 1–11. [CrossRef]
7. Nakayama, T.; Fujita, T. Cooling effect of water-holding pavements made of new materials on water and heat budgets in urban areas. *Landsc. Urban Plan.* **2010**, *96*, 57–67. [CrossRef]
8. Higashiyama, H.; Sano, M.; Nakanishi, F.; Takahashi, O.; Tsukuma, S. Field measurements of road surface temperature of several asphalt pavements with temperature rise reducing function. *Case Stud. Constr.* **2016**, *4*, 73–80. [CrossRef]
9. Kinoshita, S.; Yoshida, A.; Okuno, N. Evaporation performance analysis for water-retentive material-Based on outdoor heat-budget and transport properties. *J. Heat Isl. Inst. Int.* **2012**, *7*, 222–230.
10. Höppe, P. The physiological equivalent temperature—A universal index for the biometeorological assessment of the thermal environment. *Int. J. Biometeorol.* **1999**, *43*, 71–75. [CrossRef]
11. Jendritzky, G.; De Dear, R.; Havenith, G. UTCI-Why another thermal index? *Int. J. Biometeorol.* **2012**, *56*, 421–428. [CrossRef]
12. Takebayashi, H.; Okubo, M.; Danno, H. Thermal environment map in street canyon for implementing extreme high temperature measures. *Atmosphere* **2020**, *11*, 550. [CrossRef]
13. Nicol, J.F.; Humphreys, M.A. Adaptive thermal comfort and sustainable thermal standards for buildings. *Energy Build.* **2002**, *34*, 563–572. [CrossRef]
14. De Dear, R.; Xiong, J.; Kim, J.; Cao, B. A review of adaptive thermal comfort research since 1998. *Energy Build.* **2020**, *214*, 109893. [CrossRef]
15. Zhang, S.; Lin, Z. Standard effective temperature based adaptive-rational thermal comfort model. *Appl. Energy* **2020**, *246*, 114723. [CrossRef]
16. Junyan, Y.; Beixiang, S.; Geyang, X.; Qin, X.; Shi-Jie, C. Impacts of urban form on thermal environment near the surface region at pedestrian height: A case study based on high-density built-up areas of Nanjing City in China. *Sustainability* **2020**, *12*, 1737. [CrossRef]
17. Akbari, H.; Pomerantz, M.; Taha, H. Cool surfaces and shade trees to reduce energy use and improve air quality in urban areas. *Sol. Energy* **2001**, *70*, 295–310. [CrossRef]
18. Shimazaki, Y.; Yoshida, A.; Satsumoto, Y.; Taketani, S. Effect of properties of sports surface and clothing materials on human thermal load under hot environment. *Procedia Eng.* **2014**, *72*, 502–507. [CrossRef]
19. Shimazaki, Y.; Yoshida, A.; Suzuki, R.; Kawabata, T.; Imai, D.; Kinoshita, S. Application of human thermal load into unsteady condition for improvement of outdoor thermal comfort. *Build. Environ.* **2011**, *46*, 1716–1724. [CrossRef]
20. ISO. *ISO 8996:2004—Ergonomics of the Thermal Environment—Determination of Metabolic Rate*; ISO: Geneva, Switzerland, 2004.
21. Weir, J. New methods for calculating metabolic rate with special reference to protein metabolism. *J. Physiol.* **1949**, *109*, 1–9. [CrossRef]
22. Kurazumi, Y.; Horikoshi, T.; Tsuchikawa, T.; Matsubara, N. The body surface area of Japanese. *Jpn. J. Biometeorol.* **1994**, *31*, 5–29. (In Japanese)
23. American Society of Heating, Refrigerating and Air-Conditioning Engineers (ASHRAE). Chapter 9: Thermal comfort. In *2009 ASHRAE Handbook—Fundamentals*; ASHRAE: Atlanta, GA, USA, 2017.
24. Marino, C.; Nucara, A.; Pietrafesa, M.; Polimeni, E. The effect of the short-wave radiation and its reflected components on the mean radiant temperature: Modelling and preliminary experimental results. *J. Build. Eng.* **2017**, *9*, 41–51. [CrossRef]
25. Fernández-Cuevas, I.; Bouzas Marins, J.C.; Arnáiz Lastras, J.; Gómez Carmona, P.M.; Piñonosa Cano, S.; García-Concepción, M.Á.; Sillero-Quintana, M. Classification of factors influencing the use of infrared thermography in humans: A review. *Infrared Phys. Technol.* **2015**, *71*, 28–55. [CrossRef]
26. ISO. *ISO 9886:2004—Ergonomics—Evaluation of Thermal Strain by Physiological Measurements*; ISO: Geneva, Switzerland, 2004.
27. Hardy, J.D.; Du Bois, E.F.; Soderstrom, G.F. The technic of measuring radiation and convection. *J. Nutr.* **1938**, *15*, 461–475. [CrossRef]
28. Udagawa, M.; Kimura, K. The estimation of direct solar radiation from global radiation. *J. Archit. Plan.* **1978**, *267*, 83–90. (In Japanese)
29. Burton, A.C.; Edholm, O.G. Man in a cold environment—Physiological and pathological effects of exposure to low temperatures. *Q. Rev. Biol.* **1956**, *31*, 151–152.

30. Yang, J.H.; Kato, S.; Seo, J. Evaluation of the convective heat transfer coefficient of the human body using the wind tunnel and thermal manikin. *J. Asian Archit. Build. Eng.* **2009**, *8*, 563–569. [CrossRef]
31. ISO. *ISO 9920:2007 Ergonomics of the Thermal Environment—Estimation of Thermal Insulation and Water Vapor Resistance of a Clothing Ensemble*; ISO: Geneva, Switzerland, 2007.
32. Gagge, A.P.; Stolwijk, J.A.J.; Nishi, Y. An effective temperature scale based on a simple model of human physiological regulatory response. *ASHRAE Trans.* **1971**, *77*, 247–262.
33. Gagge, A.P.; Fobelets, A.P.; Berglund, L.G. A standard predictive index of human response to the thermal environment. *ASHRAE Trans.* **1986**, *92*, 709–731.
34. AIJ. *Standards for Measurement of Psychological and Physiological Responses to Thermal Environments*; Maruzen Publishing: Tokyo, Japan, 2014. (In Japanese)
35. Santamouris, M. Using cool pavements as a mitigation strategy to fight urban heat island—A review of the actual developments. *Renew. Sustain. Energy Rev.* **2013**, *26*, 224–240. [CrossRef]

atmosphere

MDPI

Article

Mobile Observation of Air Temperature and Humidity Distributions under Summer Sea Breezes in the Central Area of Osaka City

Atsumasa Yoshida, Ryusuke Yasuda * and Shinichi Kinoshita

Department of Mechanical Engineering, Osaka Prefecture University, Sakai 599-8531, Japan; ayoshida@me.osakafu-u.ac.jp (A.Y.); kinosita@me.osakafu-u.ac.jp (S.K.)

* Correspondence: yasuda@me.osakafu-u.ac.jp; Tel.: +81+72-254-9491

Received: 19 October 2020; Accepted: 11 November 2020; Published: 16 November 2020

Abstract: Thermal environment of urban areas in the summertime has become harmful to human health due to global warming and the urban heat island (UHI) effect. Mobile observations enable us to obtain the distribution of air temperature at microscale, such as urban blocks, which cannot be captured by the coarse network of meteorological sites. A series of mobile measurements was executed in the central area of Osaka City in Japan, around the Nakanoshima district which lies between two rivers, to investigate the air temperature and humidity distributions in a built-up area under sea breeze conditions. Upper wind and surface temperature of the rivers were also observed using pilot balloons and infrared thermography camera, respectively. The mean air temperature in Nakanoshima was generally lower than that of the surrounding area. Urban geometries such as building density and building height seem to affect the mean air temperature by changing the ventilation efficiency. Humidity was inversely correlated with air temperature distribution but was higher at the confluence of rivers and green parks. The depth of the sea breeze layer was found to be about 1 km. Sea breezes close to the ground surface penetrated the city along the rivers, sandwiching the Nakanoshima district. During the daytime, the surface temperatures of the rivers were lower than the air temperature observed at the nearest stationary observation point, and the difference reached approximately 2 °C.

Keywords: building density; mobile observation; sea breeze; urban heat island; urban ventilation; river island

1. Introduction

The most notable feature of the urban heat island (UHI) effect is an increase in nocturnal temperature (e.g., Oke, 1987 [1]). In Japan, the mitigation of high diurnal temperatures in midsummer has become an important issue because the extremely high temperatures not only lead to excessive power consumption [2] by air conditioning units but increase the incidence of heat stroke [3–5]. Fujibe (2013) [6] showed, from an analysis based on vital statistics data for Japan between 1909 and 2011, that there is a positive correlation between summertime temperature (averaged for July and August) and heat-related mortality.

Since the thermal environment of urban areas in the summertime is becoming harmful to human health due to the UHI effects in addition to global warming [7–9], it is becoming increasingly important to pay attention to diurnal temperatures. In 2007, the Japan Meteorological Agency (JMA) defined a new statistical index on air temperature, "extremely hot day" (EHD) (daily maximum temperature ≥35 °C), in addition to previously defined indices of "hot day" (daily maximum temperature ≥30 °C) and "hot night" (daily minimum temperature ≥25 °C). The JMA issues "extreme high-temperature forecasts" when the predicted daily maximum temperature exceeds the EHD criteria. In order to prevent heat stroke, the Ministry of the Environment publishes on its website the current and forecast values of the wet bulb globe temperature (WBGT) [10] at the meteorological observation points of Automated Meteorological Data Acquisition System (AMeDAS) which is operated by JMA.

Osaka is the most populous metropolitan area in western Japan and is well known for its severe summertime thermal environment [11]. As of July 2019, the residential populations in Osaka Prefecture are 8.82 million. In Osaka Prefecture, more than 3500 people have been transported by ambulance for heat stroke every summer (May–September) since 2015; in 2018 and 2019, the number was 7138 and 5182, respectively [12]. The local government of Osaka City has used the "ventilation path" (in Japanese, Kaze-no-Michi) concept to create a future urban plan for the area [13] that will mitigate its severe summertime thermal environment by utilizing the cooling potential of the sea breezes that come from Osaka Bay.

In the areas that sea breezes penetrate, the air temperature distribution near the ground can be used to evaluate their cooling effect. However, the thermal environment near the ground is strongly affected by surrounding features such as buildings, roads, rivers, and parks. Especially in urban areas, where buildings are numerous, the urban canopy significantly affects the heat budget at the ground surface. The ventilation efficiency of a city block will strongly depend on its geometry [14–16] because mean wind speed is weakened and/or intensified by drag forces and building-induced eddies. The cooling effect of a sea breeze may differ in magnitude even within a small area. The air temperature distribution will therefore show complex patterns. Mobile observations enable us to obtain such a microscale distribution of local air temperature, which cannot be captured by the coarse network of meteorological sites.

Mobile measurement of air temperature distribution in an urban area has been carried out by various research groups, including universities, local governments, and citizen groups. Automobiles have often been used to cover an objective area when the area was relatively large. Nabeshima et al. (2006) [17] and Mizuno et al. (2009) [18] used an automobile to make mobile observations on the Osaka plain. Sahashi (1983) [19] evaluated the various errors in air temperature measurements obtained using automobiles and made some suggestions to minimize the incidence of observational errors. Some research groups targeting relatively small areas have used bicycles for mobile observation [20–22]. Since bicycles can go through narrow paths which automobiles cannot access (e.g., back streets, narrow bridges, and pathways in green parks), they can capture a more detailed pattern of air temperature distribution. Additionally, the air temperatures measured during a mobile observation using a car can be affected by the heat radiated from the hoods, roofs, and the tailpipes of the car itself and other passing cars. The use of bicycles rather than cars circumvents this problem. Mobile observation with bicycles is therefore effective for detailed research; however, the target area of such research is often limited to a linear path or a relatively small area. There are few studies involving a large number of observers on bicycles covering different areas simultaneously because a large-scale observation campaign requires a lot of staff and equipment.

In this study, we made a detailed investigation of mean temperature and humidity distributions in urban blocks in mid-summer in Nakanoshima district and the surrounding area, a central office area in Osaka City, from dense observations obtained using bicycles. The objective area was subdivided into 13 areas (only on the first day of the observation, it was divided into eight areas) and mobile observation was executed simultaneously in every area. We also conducted an observation of the upper winds and the surface temperatures of the rivers sandwiching Nakanoshima district. The relationship between building density and the cooling effect of the sea breezes was discussed.

2. Methods

2.1. Objective Area

Figure 1a shows the location of Osaka and our study area. Since Osaka plain is bordered by mountains to the north, the east, and the south, the westerly sea breezes coming from Osaka Bay in the daytime are expected to mitigate high temperature in the plain. Nakanoshima, which means "a river island" in Japanese, is located at the center of Osaka's urban area and lies between the Dojima River and the Tosabori River to the north and the south, respectively (Figure 1b). The west side in the Nakanoshima district is occupied by large high-rise buildings, and there is a small green park on the district's eastern edge. Outer sides of the Nakanoshima district are densely built-up areas comprised of mid-rise office buildings. The study area is fairly flat and lies at an altitude between 0 and 3 m above sea level.

2.2. Mobile and Stationary Observations

Table 1 shows the outline and the instruments of the mobile observations. Figure 1b shows the observation area (enclosed in a white line), which is approximately 3.5 km E-W and 1.5 km N-S. The area was divided into 13 subareas (28 July and 31 August) (Figure 1c). However, on 1 July, due to equipment availability, the target area was divided into 8 subareas. In this case, subareas 2 and 3, 4 and 5, 10 and 11 were merged, and subareas 6 and 13 were excluded. The mobile observations were executed in all subareas simultaneously by travelling around them on bicycles. One bicycle was assigned to each area. To avoid the heat from direct solar radiation, we set a thermometer sensor (coupled with a hygrometer sensor in several cases) within a double tube made of stainless steel (Figure 2a). Steel tube has high solar reflectivity and low heat capacity, which are desirable properties to prevent the sensor from radiation heating. The double tubes were 300 mm in length. The diameter of the outer/inner tube used for the thermohygrometer coupled sensor was 50 mm/30 mm, and it was equipped with a motor fan to facilitate ventilation. The diameter of the double tube used for the thermo-only sensor was 45 mm/25 mm, and it was ventilated naturally (without a fan). The thermo-only sensors were used in subarea 1, 4, and 5 on 1 July (case of 8 subareas) and in subarea 5, 6, and 13 on 28 July and 31 August (case of 13 subareas). The thermohygrometer coupled sensors were used in the other subareas. The measurements were taken 1 m above the ground level (AGL) (Figure 2b).

Table 1. Outline and instruments of mobile observation.

Date	1 July, 28 July, 31 August 2010 (3 days)
Start Time	0900, 1100, 1400, 1600, 1800, 2000 LST (1 July and 31 August 2010) 0900, 1100. 1400, 1600, 1800 LST (28 July 2010)
Duration	30–40 min/run
Instruments	Temperature-only; RTR-52A (T&D Co.) Thermistor type, Accuracy: ±0.3 °C Temperature/Relative humidity; RTR-53A (T&D Co.) Thermistor/Polymer type, Accuracy: ±0.3 °C/±5% Latitude and Longitude; eTrex H (Garmin Co.) GPS Accuracy: <10 m r.m.s.
Record Interval	1 s

On the days of the mobile observations, the temperature, humidity, wind, and short- and longwave radiation were measured continuously between 0900 and 2100 LST (see Table 2) at point P2 shown in Figure 1b in order to understand the meteorological conditions around the objective area. Additionally, pilot balloon observations of the upper winds (~2000 m AGL) were made at P2 intermittently. These measurements can be used as validation data for future model analyses. The vertical velocity of the pilot balloon was adjusted to 150 m/min. Elevation and azimuth angles of a balloon were recorded every 20 s. Surface temperatures on the Dojima and Tosabori rivers were measured using an infrared thermography camera (TVS-200EX, Avio Co., Yokohama, Japan) from a balcony on the 19th floor of the high-rise building at P3. In addition to these data, the temperature, humidity, and wind data near ground level, observed by The General Environmental Technos Co., LTD. (Osaka, Japan) at P1 (31 August) and P4 (27 July and 31 August) and by the Osaka Meteorological Observatory at P0 (all of the observation days), were used in the following analysis.

Figure 1. Maps of the study area; (**a**) Geographical feature of Osaka area. A white line and a dark-red line are the borders of Osaka Prefecture and Osaka City, respectively. (**b**) Study area (enclosed by the white line) and stationary observation points. Stationary observation of air temperature, relative humidity, and wind at ground level were executed at P2. The same elements were observed at P0 by the Osaka Meteorological Observatory and at P1 and P4 by The General Environmental Technos Co., LTD. Upper winds were observed at P2 using pilot balloons. Surface temperature on the rivers were observed from a building at P3. Yellow arrows indicate mean wind directions in the daytime at P1 and P4, respectively. (**c**) Subareas for mobile observations (13 divisions. 28 July. and 31 August., 2010). Serial number in the figure indicates the ID of each subarea. Area of the Nakanoshima district corresponds to subarea 7 and 8.

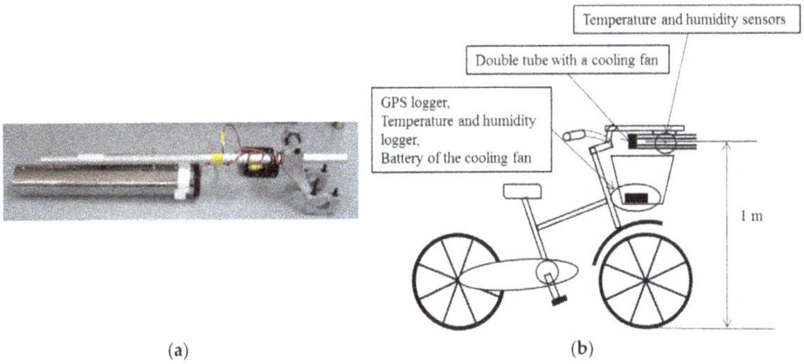

Figure 2. Equipment for the mobile observation. (**a**) A double tube with a cooling fan; (**b**) Arrangement of the instruments on a bicycle.

Table 2. Outline and instruments of stationary observation at P2.

Date	1 July, 28 July, 31 August 2010 (3 days)
Time	0900-2100 LST
Height	1.5 m AGL
Instruments	Temperature/Relative humidity; HMT100 (Visala Co.) Pt/Polymer type in a double tube with a ventilation fan Accuracy: ±0.3 °C at 30 °C/±1.7% RH Wind; USA-1 (EKO Co.) Ultrasonic type, Accuracy: ±0.01 m/s Short- and Long-wave radiations; CNR1-10 (Field pro Co.) Accuracy: daily total ±10%
Record Interval	60 s

2.3. Data Corrections and Mapping

During a mobile observation, the latitudes and longitudes of the sampling course in each area were recorded by a GPS logger. All trajectories of the moving observations were plotted on a digital map (Numerical Map 2500—Spatial Data Base by GSI) using ArcGIS (ESRI Inc., Redlands, CA, USA) and their behavior was examined. Some courses in the areas in which there were many high-rise buildings or elevated roads showed pathological routes (in a river, inside a building, extraordinarily long jump, etc.). These may have been caused by disturbances to GPS signals. Such unrealistic routes were corrected manually by referencing a geographical map. Data that jumped to remote areas were deleted. Data passing through buildings and crossing rivers other than bridges were assigned to the nearby roads and bridges, respectively. Data whose behavior could not be understood were deleted. Figure 3 shows a sample of observation paths from the run started at 1400 on 31 August.

Instrumental errors in the thermometer sensors and thermometer/hygrometer coupled sensors used for the mobile observation were calibrated by adjusting them to the reading of the instrument used at the stationary observation point (P2). The thermometer/hygrometer coupled sensors and the thermo-only sensors used in our mobile observations are relatively slow to respond, which means that the observed value recorded at each 1 s interval does not reflect the instantaneous environment at the point indicated by the GPS logger and is influenced by the local environment along the path already passed through. To retrieve instantaneous fluctuation data from the temperature and humidity records, we assumed that the response of the sensors could be approximated by a first-order delay system [23] and estimated the instantaneous values using the following equation:

$$T_e^n = \frac{T_s^{n+1} - rT_s^n}{1-r},\ r = \exp\left(-\frac{\Delta t}{\tau}\right) \quad (1)$$

where, T is air temperature or relative humidity, τ is a time constant of a thermometer or a thermometer/hygrometer sensor, Δt is the time interval of the data sampling (= 1 s), and n is the time index of the raw data. The subscript "s" indicates the value recorded by a data logger and "e" indicates the instantaneous value at the sampling point. From the results of our laboratory measurement made in 3 m/s wind conditions, we used a time constant of 15 s for the thermometer sensor (52A, 53A) and 120 s for the hygrometer sensor (53A). Since the magnitude of the instantaneous fluctuation was overestimated due to the relatively large time constants, the retrieval fluctuations were smoothed with a moving average of 9 s for temperature and 75 s for humidity. After this, data points obtained at a traveling speed (5 s average) less than 5 m/s were excluded from subsequent analyses to prevent disturbance to the data arising from radiation heating caused by insufficient ventilation.

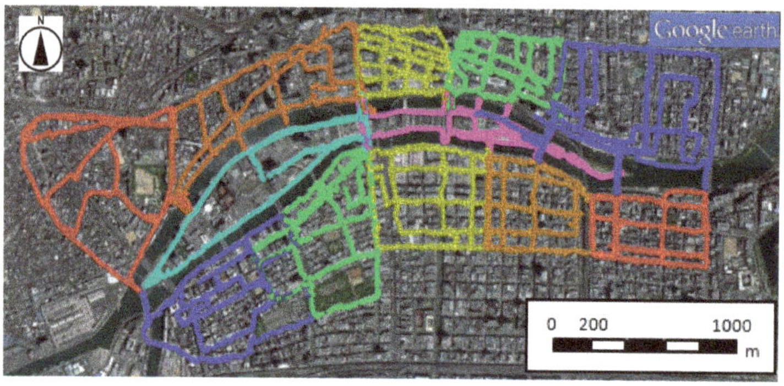

Figure 3. Travelling routes of the mobile observation started at 1400 LST, 31 August 2010. Different colors were used to distinguish adjacent subareas. The same color was used for different subareas if it was not confusing (see Figure 1c).

A time correction was also required for the temperature and humidity data obtained by the mobile observations in order to eliminate the diurnal change during travelling [17]. For the correction, we used the observation data obtained at P2 and P0. In addition to these data, the observation data at P1 and P4 were used when they were available. In each run (0900/1100/1400/1800/2000 LST) the correction amount of the diurnal change in temperature measurements (same for relative humidity) at every mobile point was estimated as follows.

$$\Delta T_p(t) = \hat{T}_p(t) - \hat{T}_p(t_0) \tag{2}$$

$$\Delta T_m(t) = \sum_p w_p(t)\Delta T_p(t) / \sum_p w_p(t) \tag{3}$$

$$w_p(t) = 1/\left(\left|x_m(t) - x_p\right|^2 + \left|y_m(t) - y_p\right|^2\right) \tag{4}$$

$$\widetilde{T}_m(t_0) = T_m(t) - \Delta T_m(t) \tag{5}$$

where T is a temperature, (x,y) is the position of a measurement point at an observation time, t, t_0 is the starting time of the run. Subscript m and p indicate values at a mobile and stationary points, respectively. (x_m, y_m) is time dependent and (x_p, y_p) is constant. $\hat{T}_p(t)$ is the air temperature at each stationary point (p). $\hat{T}_p(t)$ was given as a 3rd order polynomial function using 10 min-averaged temperatures observed during the run. The time-correction amount of temperature at a mobile point, $\Delta T_m(t)$, was estimated by averaging the temperature differences at each stationary points, $\Delta T_p(t)$, with a weight function $w_p(t)$. A guess value of temperature at the start time, $\widetilde{T}_m(t_0)$, was obtained by subtracting $\Delta T_m(t)$ from the instantaneous value, $T_m(t)$.

3. Results and Discussion

3.1. Overview of Weather on the Observation Dates

Except for intermittent light rain (0.0 mm/h at Osaka Meteorological Observatory, P0 in Figure 1b) after 1600 LST on 28 July, the weather on each observation day was fine. The data for 2000 LST on 28 July is therefore missing. The daily maximum temperature at the Osaka Meteorological Observatory on each day were 32.2 °C (1510 LST, 1 July), 33.3 °C (1401 LST, 28 July) and 35.5 °C (1547 LST, 31 August). As an example, meteorological data obtained at the Osaka Meteorological Observatory (P0) and the stationary points (P1, P2, and P4) on 31 August are shown in Figure 4. Figure 4a shows the global solar radiations at P2. P2 is suitable for monitoring overall changes of weather in the objective area because there are few obstacles (such as buildings and trees) around it. The air temperature at P4 was lower than that at P1 and P2 during the daytime (0900–1800 LST) on 31 August. In these hours, relative humidity observed at P4 was higher than that at P1 and P2 (Figure 4b), which was reasonable if water vapor pressures at these points were nearly constant.

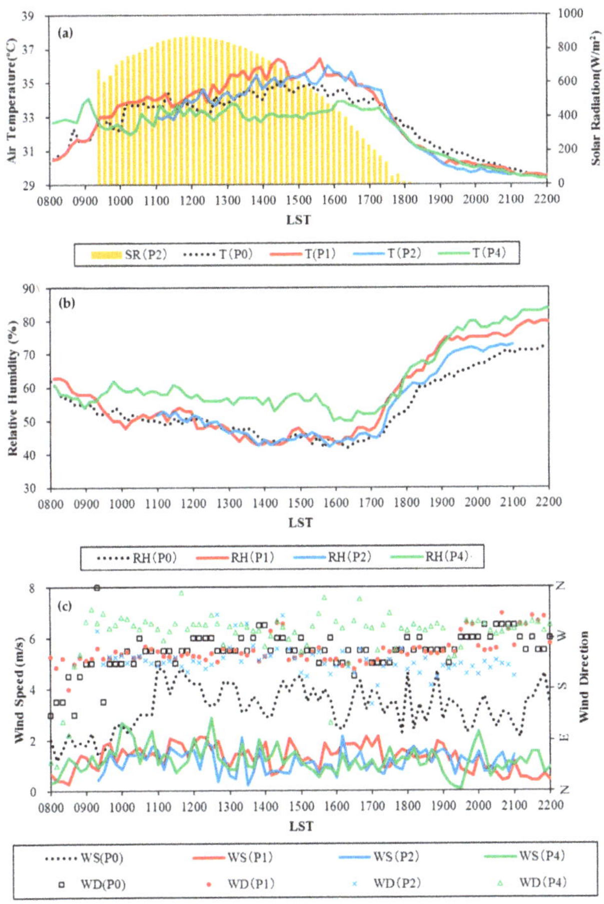

Figure 4. Time variations of meteorological elements on 31 August 2010. (**a**) SR: global solar radiation, T: Air temperature, (**b**) RH: Relative humidity. (**c**) WS: Wind speed, WD: Wind direction. P1, P2, and P4 are the stationary points. P0 is the Osaka Meteorological Observatory nearby the study area (see Figure 1b).

The wind direction during each observation period was stable and ranged between SW and W. Sea breezes from Osaka Bay penetrated into the study area on every observation day. Land breezes were not observed during the period of our mobile observation. As shown in Figure 4c, wind direction during the daytime was SW at P1 on the western edge of Nakanoshima district and at P2 in the central area; however, wind direction at P4 on the eastern edge was around WNW (shown by arrows in Figure 1b). Since these directions almost coincide with the alignment (but in the opposite direction) of the two rivers sandwiching Nakanoshima, we conclude that the sea breezes near ground level flow along the rivers in this area.

3.2. Temperature and Humidity Distributions in the Objective Area

As an example of mobile measurements, time series of the air temperature and the relative humidity variations on 31 August in the three subareas (1, 7, and 8) are shown in Figure 5, with the measurements obtained at the stationary point included in the subarea. Note that the record interval is 10 min at the stationary points and 1 s for the mobile observations. These data are raw measurements. No corrections have been applied. The time variations of the mobile observation data almost correspond to those obtained at stationary points. The temperature in subarea 1 is lower during the daytime (1100–1600 LST) and higher in the evening than the P1 data. This relationship also applies between subarea 7 and P2. The results obtained by traveling in the street canyons seem to exhibit urban canopy effect more intensely than the data of the stationary points located in relatively open places. On the other hand, in subarea 8, the mobile measurements of air temperature in the daytime are about 2 °C higher and the relative humidity is about 10% lower consistently than those observed at P4. The reason for this difference may be that P4 is in a humid place, in a green park near the river junction, while mobile measurements are obtained on paved streets.

After the data corrections for GPS route, instrumental error, time difference, and time responses, we made distribution maps of air temperature and relative humidity from the mobile measurements by interpolating them using the inverse distance weighted (IDW) method in ArcGIS.

In general, the air temperature variation in the study area is large in the daytime and small in the nighttime. However, the temperature distribution patterns are similar across days and hours. Here we show the characteristics of the distribution based on the results on 31 August as the examples. Figure 6 shows the air temperature distribution data obtained on 31 August. Air temperatures were generally lower within the Nakanoshima district (subareas 7 and 8) than in the outer areas. Outside the district, the temperatures in riverside areas along the Dojima and Tosabori Rivers were also relatively low. The mean temperature roughly increased from the west to the east, except in the most northwestern area (subarea 1). Relatively high temperatures were observed in the northern part of subarea 5 and 6 and in the southern part of subareas 11–13. These areas are crowded with office buildings and are distant from the rivers.

Figure 7 shows the statistical distribution of the air temperature at 1400 and 1800 LST on 31 August by box plots. The whiskers indicate the minimum and the maximum values excluding outliers, respectively. The lines of the box show lower quarter(Q1), median (Q2), upper quarter(Q3).

Q3–Q1 is called the IQR (interquartile range) which is 1.35 times the standard deviation (σ) in the case of the normal distribution. Data smaller than Q1-1.5 × IQR or larger than Q3 + 1.5 × IQR are regarded as outliers. In Figure 7, outlier plots are omitted to avoid visual complications. Mean values are shown in asterisks. The differences between Q2 and the mean value are small, less than 0.1 °C, for all runs of the observations. IQR/σ ranges 0.91–1.49 (1 July), 0.98–1.65 (29 July), and 1.16–1.50 (31 August). Clear relationship between IQR/σ and subarea or run is not found. At all times, Q2 in each subarea is near the midpoint of Q1 and Q3, which means that the skewness of the statistical distribution is not prominent.

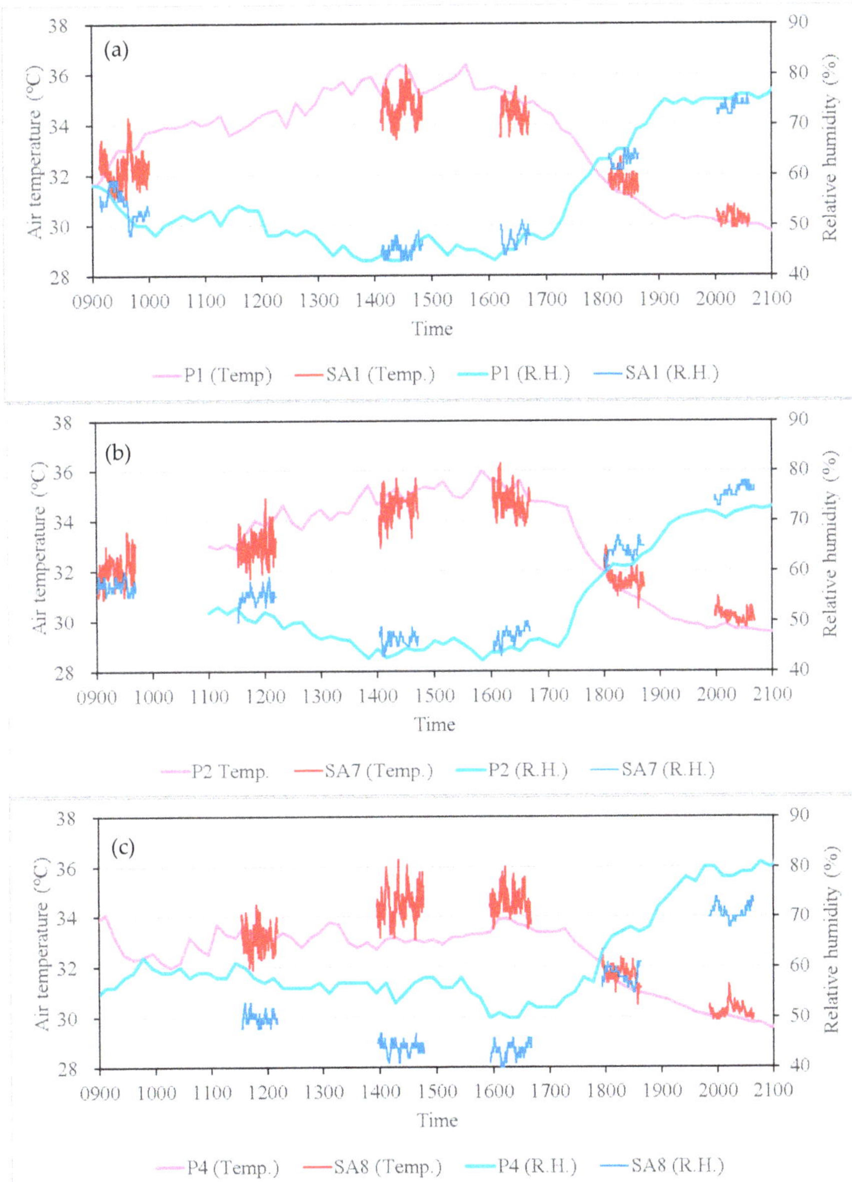

Figure 5. Time series variations of the air temperature and the relative humidity obtained at a stationary point and mobile observations in the subarea on 31 August. (**a**) P1 and Subarea-1, (**b**) P2 and Subarea-7, (**c**) P4 and Subarea-8.

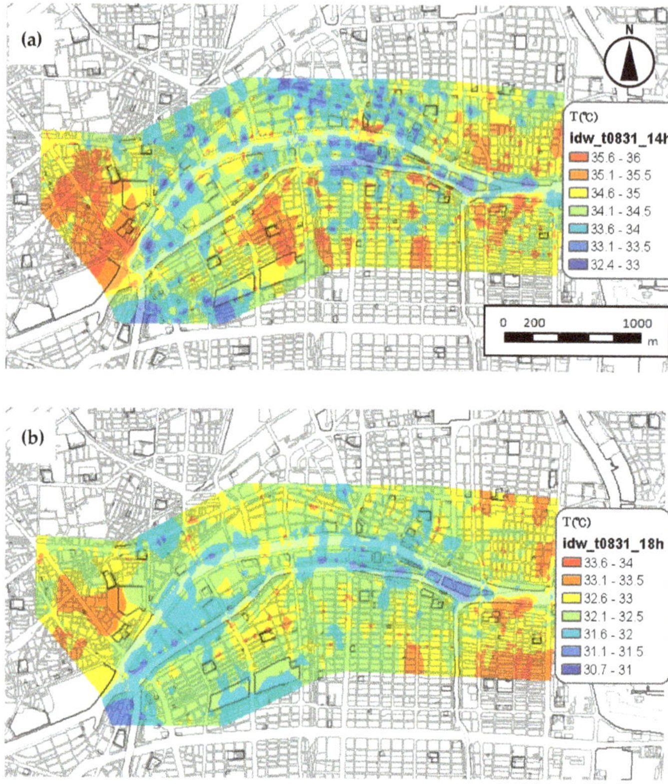

Figure 6. Horizontal distribution of air temperature on 31 Aug. 2010. (**a**) 1400 LST, (**b**) 1800 LST.

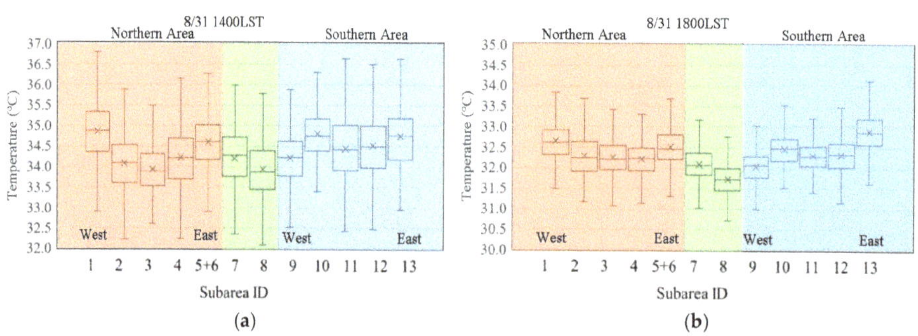

Figure 7. Box plots of the air temperatures in each subarea on 31 August 2010. (**a**) 1400 LST, (**b**) 1800 LST. Subarea IDs are shown in Figure 1c. Green background shows subareas in the Nakanoshima district. Subarea 5 and 6 were merged on this day.

The IQR in each subarea was relatively large (0.8–1.1 °C at 1400 and 0.5–0.7 °C at 1800), considering that the range of difference in median value of the temperature between the subareas was around 1 °C (1.0 °C at 1400 and 1.2 °C at 1800). These results suggest that the geometric characteristics of the urban block in a subarea (such as building density, building height, and road width) affect the mean temperature to some extent. We therefore investigated the relationship between the mean temperatures and geometric characteristics of the blocks. As a temperature index for an area, the deviation of the mean temperature of that area from the average value of all subareas was used. As geometrical indices of the urban blocks, the gross building coverage ratio and the mean number of stories of the buildings were used. Gross building coverage ratio (GBC), which represents building density, is defined as the ratio of the total area covered by buildings to the total ground area, and we estimated it from "Osaka City Mesh Data (2005)" provided by Planning and Coordination Bureau of Osaka City.

As shown in Figure 8a, mean temperatures tend to increase with GBC in both daytime and nighttime. Therefore, it seems that the low temperatures in the Nakanoshima district are due to the area's low building density; the ventilation efficiency of an area promotes sea breeze cooling in the daytime, which reduces heat storage in the ground and the bodies of buildings and leads to low temperatures in the nighttime. Additionally, as shown in Figure 8b, the mean temperature in a subarea tends to decrease as the mean number of stories of buildings (S) increases. One of the reasons for this correlation is the effect of shadows during the daytime. Another possible reason is the increase in wind speed at ground level around tall buildings, which is induced by turbulent eddies. Both of these effects will reduce heat storage in the daytime. Figure 8c shows the correlations between the s.t.d. of air temperature (σ_T) and GBC. From the daytime to the evening, σ_T are reduced in all subareas. The decrease in σ_T in the outer subarea of the Nakanoshima district is smaller than in the inner subarea. In the outer subareas, effect of diurnal variation in radiation on air temperature may be relatively small because urban canopy effect is prominent by their high building density. In a low GBC area, such as the inner subareas of Nakanoshima district, open spaces occupy a large portion of it. In there, radiative cooling develops in the evening, and it uniformly lowers the air temperature on the open surface. Therefore, the σ_T decreases significantly from daytime to the nighttime. As a result, the correlation between GBCs and σ_T is inverted between the daytime and the nighttime. The relationship between the σ_T and S were shown in Figure 8d. As in the case of the mean temperature difference, the correlation with S shows the opposite tendency to the case in GBC. Note that the correlations between the geometrical indices and the deviation of mean temperature are significant. On the other hand, the relationships between the geometrical indices and σ_T are not significant.

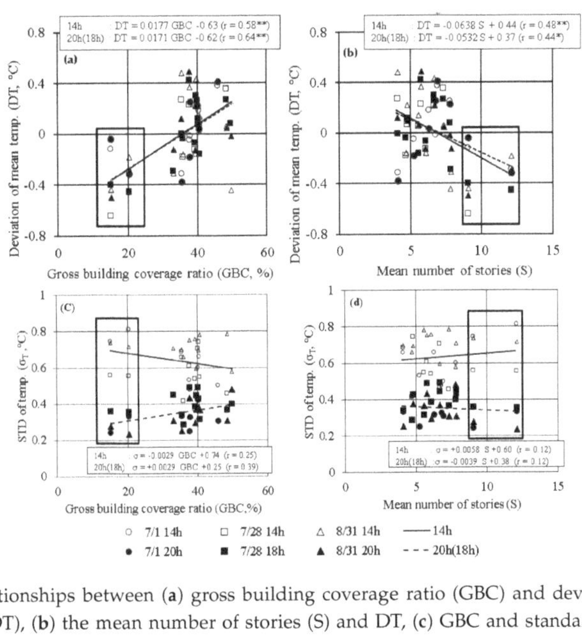

Figure 8. Relationships between (**a**) gross building coverage ratio (GBC) and deviation of mean temperature (DT), (**b**) the mean number of stories (S) and DT, (**c**) GBC and standard deviation of temperature in a subarea (σ_T), and (**d**) S and σ_T. The symbols in a box of thick line are the data within the Nakanoshima district (subarea 7 and 8). **: $p < 0.01$, *: $p < 0.05$.

Figure 9 shows the distribution of relative humidity on 31 August. Because thermo-only sensors were used in subarea 6 and 13, relative humidity in the most eastern part of the study area was not obtained. The distribution pattern of specific humidity is similar to that of relative humidity, possibly due to the temperature difference in the objective area being relatively small. In general, the pattern of humidity distribution correlated inversely with that of temperature; high humidity was found in the riverside zones of the Dojima and the Tosabori Rivers. Figure 10 shows the mean relative humidity in each subarea. The main features of the relative humidity distribution, as opposed to the temperature distribution, can be summarized as follows: both temperature and humidity were higher on the western side of the Nakanoshima district (subarea 7), which is near a confluence of the rivers, than on the eastern side of the district (subarea 8); a high humidity zone can be seen at the southern edge of the observation area, where there is a large green park.

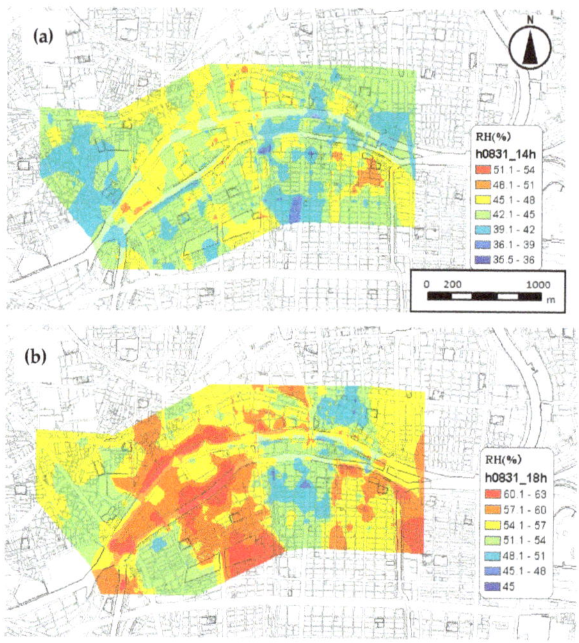

Figure 9. Horizontal distribution of relative humidity on 31 August 2010. (**a**) 1400 LST, (**b**) 1800 LST.

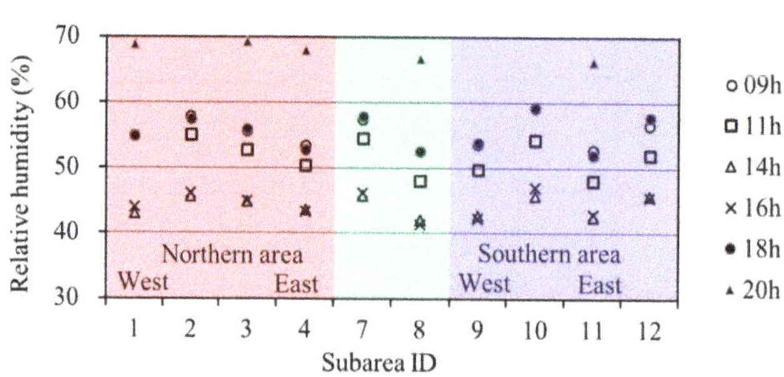

Figure 10. Mean relative humidity in each subarea on 31 Aug. 2010. Subarea IDs are shown in Figure 1c. Data of subarea 5, 6, and 13 does not exist because observations in these subareas were made with thermo-only sensors. Green background shows subareas in the Nakanoshima district.

3.3. Characteristic of Upper Wind and River Surface Temperature

Figure 11a shows an example of the upper wind profiles obtained on 31 August. Some discontinuous changes in wind direction and wind speed (e.g., wind speed at 1112 LST, both wind direction and wind speed at 1820 LST) were observed at around 1 km AGL. In the lower layer, the wind direction was between S and W, and the wind speed was light to moderate (3–6 m/s), which suggests that the depth of the sea breeze layer during this observation period was approximately 1 km. As shown by the trajectories of pilot balloons (Figure 11b), the direction of the sea breeze, unlike the surface wind direction, does not necessarily coincide with the paths of the rivers sandwiching the Nakanoshima district (see Section 3.1). On the other hand, wind direction above the sea breeze layer was variable; it is likely to depend on the synoptic conditions. Wind speed above the sea breeze layer was often lower than that in the layer because our observation was executed under typical anticyclone conditions. These features of the wind profiles were often observed on other days and hours during the observation period.

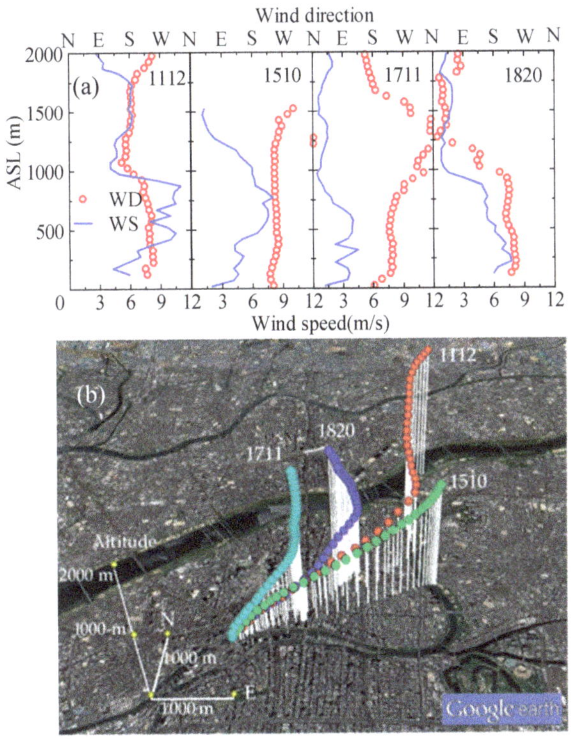

Figure 11. Upper wind around the study area on 31 August 2010. (**a**) Vertical profiles of the horizontal winds. The numbers in the figures are starting time (hhmm LST) of the observations. (**b**) Three-dimensional trajectories of the pilot balloons. The circular mark indicates balloon position of every 20 s (vertical interval is approx. 50 m). The starting time of each observation is shown near the end point of the trajectory.

Figure 12 shows the diurnal variations in the surface temperatures of the rivers sandwiching Nakanoshima, as obtained using thermography on 31 August. The surface temperatures of the rivers increase with air temperature, though the increment of the surface temperature is smaller than that in air temperature, and the rivers therefore act as heat sinks in the daytime. The surface temperatures of the rivers were generally lower than the air temperature at P2 from 0900 to 1700 LST, and the maximum difference is about 2 °C.

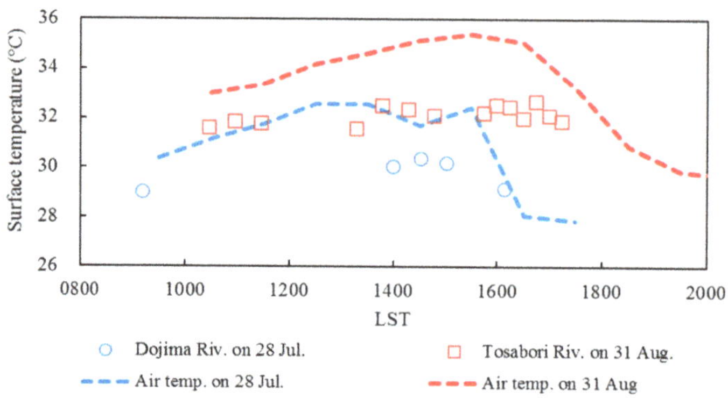

Figure 12. Surface temperature of the rivers and air temperature at ground level. The surface tempera-tures of the rivers were observed from a balcony of a hi-rise building at P3, using an IR thermography camera (see Figure 1b). The air temperature was observed at P2.

4. Conclusions

In the observation area, large temperature differences were found even within a relatively small area. The building density seems to affect the mean air temperature because it influences the ventilation efficiency of the area. The temperature of the Nakanoshima district is generally lower than that of its surrounding area because the building density is relatively low. The relative humidity distribution was inversely correlated with the air temperature distribution in general; however, both temperature and relative humidity were high at the confluence of the rivers and green parks. The depth of the sea breeze layer penetrating the Nakanoshima district was found to be about 1 km. The sea breeze direction near the ground is aligned with but opposite to the direction of the rivers surrounding the Nakanoshima district. During the daytime, 0900 to 1700 LST, the surface temperature of the rivers sandwiching the district was lower than the air temperature observed at the nearest stationary observation point, and the difference reached approximately 2 °C.

Author Contributions: Conceptualization, A.Y.; Methodology, A.Y.; Software, R.Y.; Validation, R.Y. and S.K.; Formal analysis, R.Y.; Investigation, S.K. and R.Y.; Resources, R.Y. and A.Y.; Data curation, R.Y.; Writing—original draft preparation, R.Y.; Writing—review and editing, A.Y. and S.K.; Visualization, R.Y.; Supervision, A.Y.; Project administration, S.K.; Funding acquisition, A.Y. All authors have read and agreed to the published version of the manuscript.

Funding: This study was partially funded by a Grant-in-Aid for Scientific Research (17H00802, Principal Investigator, Atsumasa Yoshida) granted by the Japanese Society for Promotion of Science.

Acknowledgments: The authors acknowledge the provision of stationary observation data by The General Environmental Technos Co., LTD., the provision of the instruments for the pilot balloon observations by the Japan Weather Association, and the cooperation on the mobile observations of H. Miyazaki and the members of "Netsu-zemi" in Kansai University. The authors wish to give special thanks to K. Takenaka and the other students of our research group who took part in the mobile observations and in the data preprocessing.

Conflicts of Interest: The authors declare no conflict of interest.

References

1. Oke, T.R. *Boundary Layer Climates*, 2nd ed.; Routledge: London, UK, 1987.
2. Komatsu, H. Evaluation of recent hot summer in Kansai: Basic information for supply and demand of electricity. *Jpn. J. Biometeorol.* **2014**, *51*, 37–47.
3. Ono, M. Chapter 3-Heat stroke in the summer of 2010. In *Meteorological Research Note*; The Meteorological Society of Japan: Tokyo, Japan, 2012; Volume 225, pp. 29–35.
4. Yokoyama, T.; Fukuoka, T. Regional tendency of heat disorders in Japan. *Jpn. J. Biometeorol.* **2006**, *43*, 145–151.

5. Hoshi, A.; Nakai, S.; Kaneda, E.; Yamamoto, T.; Inaba, Y. Regional characteristics in the death due to heat disorders in Japan. *Jpn. J. Biometeorol.* **2010**, *47*, 175–184.
6. Fujibe, F. Long-term variations in heat mortality and summer temperature in Japan. *Tenki* **2013**, *60*, 371–381.
7. Fujibe, F. Evaluation of background and urban warming trends based on centennial temperature data in Japan. *Pap. Meteorol. Geophys.* **2012**, *63*, 43–56. [CrossRef]
8. Ono, M.; Ueda, K. Chapter 16—Prediction of heat stroke damage by global warming. In *Meteorological Research Note*; The Meteorological Society of Japan: Tokyo, Japan, 2012; Volume 225, pp. 177–182.
9. Adachi, A.S.; Kimura, F.; Kusaka, H.; Inoue, T.; Ueda, H. Comparison of the impact of global climate changes and urbanization on summertime future climate in the Tokyo metropolitan area. *J. Appl. Meteorol.* **2012**, *51*, 1441–1454. [CrossRef]
10. Ministry of the Environment, Government of Japan. Heat Stress Index: WBGT, Heat Illness Prevention Information (Online). Available online: http://www.wbgt.env.go.jp/en/ (accessed on 1 July 2020).
11. Kono, H.; Narumi, D.; Shimoda, Y. *Osaka, Klimaatlas for Urban Environment*; Gyosei Co.: Tokyo, Japan, 2000; pp. 53–65.
12. Fire and Disaster Management Agency. Heatstroke Information. Available online: https://www.fdma.go.jp/disaster/heatstroke/post1.html (accessed on 1 October 2020).
13. Osaka City, "Kaze-no-Michi" Vision [Basic Policy]. 2011. Available online: http://www.city.osaka.lg.jp/kankyo/page/0000123906.html (accessed on 1 July 2020).
14. Kubota, T.; Miura, M.; Tominaga, Y.; Mochida, A. Wind tunnel tests on the nature of regional wind flow in the 270 m square residential area, using the real model, Effects of arrangement and structural patterns of buildings on the nature of regional wind flow Part 1. *J. Archit. Plan. Environ. Eng.* **2000**, *529*, 109–116.
15. Takebayashi, H.; Moriyama, M.; Miyake, K. Analysis on the relationship between properties of urban block and wind path in the street canyon for the use of the wind as the climate resources. *J. Environ. Eng.* **2009**, *74*, 77–82. [CrossRef]
16. Wong, M.S.; Nichol, J.E.; To, P.H.; Wang, J. A simple method for designation of urban ventilation corridors and its application to urban heat island analysis. *Build. Environ.* **2010**, *45*, 1880–1889. [CrossRef]
17. Nabeshima, M.; Kozaki, Y.; Nakao, M.; Nishioka, M. Experimental study of heat island phenomena by mobile measurement system, Horizontal air temperature distribution at night in the Osaka plain. *J. Heat Island Inst. Int.* **2006**, *1*, 23–29.
18. Mizuno, M.; Nabeshima, M.; Nakao, M.; Nishioka, M. Study on temperature distribution of an urbanized area using a vehicle with mobile measurement system: Analysis on characteristics of the 330 horizontal distribution using semivariogram and kriging. *J. Environ. Eng.* **2009**, *74*, 1179–1185. [CrossRef]
19. Sahashi, K. Errors in the air temperature observation by traveling method with the automobiles: Effects of the automobiles. *Tenki* **1983**, *30*, 509–514.
20. Melhuish, E.; Pedder, M. Observing an urban heat island by bicycle. *Weather* **1998**, *53*, 121–128. [CrossRef]
21. Takayama, N.; Yamamoto, H.; Zhang, J.; Iwaya, K.; Wang, F.; Harada, Y.; Kaneishi, A.; Tsuchiya, Y.; Mori, H. Development and application of mobile observation method of urban climate to analyze occurrence of heat-island in Changchung city of China. *Archit. Inst. Jpn. Technol. Des.* **2009**, *15*, 827–832. [CrossRef]
22. Brandsma, T.; Wolters, D. Measurement and statistical modeling of the urban heat island of the city of Utrecht (the Netherlands). *J. Appl. Meteorol. Climatol.* **2012**, *51*, 1046–1060. [CrossRef]
23. Nakao, M.; Nabeshima, M.; Hasegawa, M.; Nishioka, M. Continuous mobile measurement method to estimate thermal environmental distribution in street. In *Architectural Institute of Japan, Summaries of Technical Papers of Annual Meeting (Chugoku)*; Architectural Institute of Japan: Tokyo, Japan, 2008; pp. 981–982.

Publisher's Note: MDPI stays neutral with regard to jurisdictional claims in published maps and institutional affiliations.

atmosphere

MDPI

Article

Indicators Framework for Sustainable Urban Design

Andrea López Chao [1,*], Amparo Casares Gallego [1], Vicente Lopez-Chao [1] and Alberto Alvarellos [2]

1 School of Architecture, University of A Coruña, 15008 A Coruña, Spain; amparo.casares@udc.es (A.C.G.); v.lchao@udc.es (V.L.-C.)

2 RNASA-IMEDIR, Computer Science Faculty, University of A Coruna, 15071 A Coruña, Spain; alberto.alvarellos@udc.es

* Correspondence: andrea.lopez.chao@udc.es

Received: 25 September 2020; Accepted: 20 October 2020; Published: 22 October 2020

Abstract: Climate change and sustainability have recently been object of study due to the impact on the planet and on human activity of the first and the benefits that could derive from the efficiency of the second. Particularly, urban environments are locations that represent a high percentage of emissions of gases, waste, resources use and so forth. However, they are places where great changes can be made, in an attempt to accomplish the urgent challenge to adapt to current and projected rates of climate change. Research has shown that a fruitful approach to urban sustainability is to describe indicators that measure the effectiveness of current processes of urban infrastructures, analyze areas in need of improvement and measure the effect of any actions taken. The significant feature of this research relies on its global approach, considering both major worldwide used and less widely-spread frameworks and the analysis of the 32 selected tools and guidelines, including over 2000 indicators. The result is a proposed structure of 14 categories and 48 indicators, easily applicable in urban areas, that tries to fulfill basic aspects to obtain a general diagnosis of the sustainable nature of the urban environment, which can serve as support to detect the strongest and weakest areas in terms of their sustainability.

Keywords: architecture; assessment; tool; indicators; sustainability; cities

1. Introduction

Climate change is one of the most important environmental issues of our time. This concern reflects the reality that not only climate change is sensitive to human activity but also that much of human activity is sensitive to climate change and that adapting to current and projected rates of climate change could be very challenging. Although both natural processes and human activities can cause climate change, recent global warming has been largely attributed to human activity. However, there are proven energy and infrastructure solutions that can be put into use as of today to provide a more sustainable development.

The conversion of Earth's land surface to urban uses is one of the most irreversible human impacts on the global biosphere and the environmental impacts of urban expansion are undeniable [1]. But cities are also places of great efficiency, innovation and can be, especially in dense cities, great places of energy savings, too, although such savings do not appear automatically—an effort to accomplish improvements in these matters is mandatory. As the United Nations stated, managing urban areas has become one of the most important development challenges of the 21st century.

Cities are responsible for a great part of our energy expenses and harmful greenhouse gases. According to UN-Habitat, cities consume 78% of the world's energy and produce more than 60% of greenhouse gas emissions and 68% of the world population is projected to live in urban areas by 2050 [2]. But they are also the places where great efficiencies can potentially be made in terms of sustainability, through careful management of urban development and optimized resource use efficiency [3].

This makes it necessary to understand the form and content of urbanization to implement a reduction of our footprint. Understanding the contribution of cities to climate change will help intervene at a local level. The right approach to delivering and maintaining transport, housing, energy, water and communication infrastructure is essential to create a strong and competitive economy and provide social services.

Sustainable urban infrastructure is an infrastructure that facilitates a place's progress towards the goal of sustainable living. Attention also needs to be paid to technological and government policy, which enables urban planning for sustainable initiatives. Challenges resulting from increasing population growth generate a need for sustainable infrastructure that is high performing, cost-effective, resource-efficient and environmentally friendly.

Research has shown that an accurate way to take a fruitful approach to urban sustainability is to describe the most appropriate indicators to measure the effectiveness of current processes of urban infrastructures, as well as to be able to measure the effect of taken actions and therefore develop qualitative and quantitative descriptors of urban environments. Urban sustainability indicators are greatly important instruments for assessing the performance of cities and they can be of major help to improve it.

Different public actions have been used during the past few decades to manage city infrastructures considering several aspects considered within urban sustainability, both at a micro and macro level. Governments at all levels are currently taking into account the concepts of sustainable and smart and developing programs that involve both notions towards a better quality of life for citizens while maintaining economic growth.

Since 2014, 34 Organization for Economic Cooperation and Development (OECD) countries have been trying to collect data about people's well-being several times a year. Comparisons have been made using different criteria, such as access to services, civic engagement, the environment, individual incomes, employment and education—with open data being made available to researchers and citizens.

In the same line of action, the UN Sustainable Development Goals (SDGs) were unanimously adopted by the 193 UN Member States in 2015. They are comprised of 17 Global Goals and 169 targets intended to balance the economic, social and environmental dimensions of sustainable development. The SDGs and associated targets, indicators and evaluation metrics represent an internationally accepted framework for the evaluation of sustainability at a global level.

Also, in Europe, certain organizations are trying to identify the best frameworks to evaluate urban sustainability, which we will go into detail later on.

It is of big importance to dispose of a framework that includes indicators which are not only useful per se because of the information they provide but also as easy to obtain and measure as possible and preferably publicly available.

During the past few years there has been a certain level of research on said indicators, mostly by world-wide organisms in charge of actions against climate change (UN, World Bank, etc.) but oftentimes centered in a certain aspect of urban planning (water, energy, green spaces, roads, ...) or including a really wide, general range of topics. Different guides have been developed to support the achievement of the SDGs, yet the alignment between the SDG targets and indicators and the great number of individually focused frameworks developed to evaluate specific actions at small scale remains unclear [3].

It would be highly useful to dispose of a framework that summaries a series of indicators that provide information about different aspects of the city which can result in a proper global image of its sustainability state, with those indicators being easy to measure (they should be formulated so they can be easily incorporated into an on-going program of gathering statistics) and easy to made widely available for online public access.

Our objective is to review the more relevant existent frameworks, models, assessments and so forth and if appropriate, define a new holistic assessment framework that can be applied both to new and existing settlements, to create more efficient urban settlements and improve existing ones with certain measures depending on the results obtained with the framework, as regional and city planners do not count yet on a valid, widely accepted tool that enables their better assessment concerning sustainable urban planning and cities/towns/regions/countries do not count on a tool that helps them measure and improve their urban aspects enabling them an easier reach of UN goals.

The goal of this research is to support current cities' transformation into sustainable smart cities of the future, through the measurement of certain aspects, the implementation of actions to improve them and the monitoring of the results over time. Ideally, with a model that could even be used to compare cities, carefully applied.

While this research could happen at different levels (continents, countries, regions, cities, small settlements ...), we will be focusing cities, as they are considered by literature the most suitable as a self-sufficient spatial unit for showing redevelopment results [4] although the framework could easily be adapted to smaller settlements, bigger regions or even countries.

The proposed framework (whether it is an existing one or a new one) would fundamentally enable smart cities to strengthen their strategic planning efforts and measure their progress towards a more sustainable progress and reaching the UN SDGs, among others.

2. Urban Sustainability and Sustainability Indicators

Since the emergence of the term Sustainable Development [5], the appearance of tools and assessments that attempt to measure the level of sustainability of an urban system through indicators has been quite fruitful (OECD, 2014) and to date, there is a heterogeneous amount of options to assess sustainability in an urban context, taking into account different aspects: the geographical scope of the selected area of study, the main interest of study and so forth. However, not all the sustainability criteria are always contemplated by each assessment or even those included in each one have not been thoroughly analyzed. There are also tools that include an overwhelming number of indicators, which are not easy to measure or obtain and are not necessarily relevant in every case.

According to Kennedy et al. [6], a sustainable city can only be one for which the inflow of material and energy resources and the disposal of wastes, do not exceed the capacity of the city's surrounding environment. As stated before, research shows that sustainability depends on social, economic, environmental and governance factors, which can only be achieved through great effort from a holistic point of view.

With respect to cities, indicators are understood to provide a barometer of how various aspects and parts of a city are unfolding and performing. These trends are most often revealed by charting them as graphs or maps or inserted into models which enable a general view of the current situation or a future possibility [7].

To create a sustainable urban environment, it is necessary to measure, assess policies and measure again in order to check the effectiveness of the applied policies. Indicators are an effective tool to pursue this objective, as properly selected, they allow the measurement and comparison of any aspect.

The selection of appropriate indicators is the base of any relevant framework/tool and should also be the first step of databases. Every indicator should be selected to detect both the current state and measure the progress in meeting the objectives of sustainability understood in its broadest sense.

An indicator (understood as a measurement or a value that provides relevant information) should be, according to existing literature (p. 31) [8]:

- Specific (must clearly relate to outcomes)
- Measurable (implies that it must be a quantitative indicator).
- Usable (practical).
- Sensitive (must readily change as circumstances change).

- Available (it must be relatively straightforward to collect the necessary data for the indicator).
- Cost-effective (it should not be a very expensive task to access the necessary data).

Concerning the aspects of sustainability, the three-pillar conception of sustainability (social, economic and environmental) has become widely represented [9]. The inclusion of the three initial pillars did not occur at the same time but as a result of needs that appeared as the notion of sustainability matured. However, we find it necessary to include a fourth aspect, governance, as it is key in the implementation, management and reporting processes of overall sustainability.

Therefore, the range of indicators when it comes to our field of work should cover environmental, economic, social and governance areas, as sustainability should be obtained in every aspect without neglecting any of them. The challenge appears when defining a compound of indicators that conform a tool which is easy to implement and worth the effort.

BREEAM (UK) is one of the first tools to be have been introduced in sustainability assessments focused on buildings. Ever since its appearance, various tools have been developed worldwide, as the case of LEED in the US, CASBEE in Japan, ECOCITY in Europe or HQE in France.

Different governmental and non-governmental scientific organizations, such as SDEWES center, financial institutions like the World Bank, resulting in frameworks, toolkits or guides such as Urban Sustainability Framework (The World Bank) or Reference Framework for European sustainable cities.

Some projects concerning sustainability and indicators have also been funded by the different EU programs or funds and developed obtaining key performance indicators and data collection procedures, such as CITYkeys, ECO-city and KITCASP. There are also initiatives of the European Union to implement sustainable, clean and efficient urban transport measures, as CIVITAS.

Certain studies that have compared some of those tools [3], while others have applied some of them to compare their results in a practical case and some studies have even compared the tools, set a new tool based on their criteria and applied it [4].

The significant feature of this paper relies on its global approach, considering both major world-wide used and less widely-spread frameworks. In this study, indicators were selected in first place on the basis of their frequency, based on the analyzed data, consisting of 32 indicator sets, 224 categories and 2.060 indicators and in second place taking into account the following criteria: objectivity (clear, easy to understand, precise and unambiguous); relevance, measurability and reproductivity (quantitative, systematic observable); validity; comparability; and accessibility (available databases, use of existing data), as we will go into detail throughout the article.

3. Materials and Methods

The criteria for inclusion in this systematic review were studies, frameworks, assessments, tools, that evaluated or considered sustainability indicators in environmental, social, economic, and/or governance aspects. Those included papers, books, reviews, theoretical comparisons and tools with certain level of relevance. Case studies or studies that implemented frameworks were not included in the final selection, even though they were taken into consideration in the initial selection, among proceedings and gray literature.

The development of different search strategies for the each used database (Scopus, Web of Science, ...), gray literature in Google Scholar and research by relevant Organizations, allowed us to examine an extensive number of references and choose a significant amount that met our criteria for consequent analysis. We managed all references using Zotero desktop software.

Study selection consisted of two phases. The first phase included a brief review of all the titles and abstracts of all identified references in the consulted databases and the discard of all articles, tools, assessments and so forth, that did not meet the eligibility criteria. This phase also included the selection of several frameworks and most highly used worldwide tools to assess urban or buildings sustainability, as well as the review of gray literature which was not to be included as reference but could provide valuable information.

The second phase consisted of an extensive analysis of every selected papers and tools/assessments and the extraction of the necessary data concerning indicators that would then be classified and analyzed.

The characteristics collected from the selected studies and assessment tools included the following: authors and year of publication in the case of papers, geographical scope, the objective of the study or assessment, indicator(s) evaluated, classification of the indicator(s) if existed (category, subcategory and others).

Considering the literature review done, the final selection consisted of 32 indicators sets, which we proceeded to analyze and classify.

Table 1 shows the list of the included sets and how they are structured into categories. This in-depth analysis of each tool allowed the identification of the structure of indicators contained in each set to evaluate the sustainability of an urban area. In our analysis, we took into account aspects such as the geographical scope of each set, its original classification into categories and subcategories and the description of each indicator.

Table 1. Frameworks, Assessment Tools and Guidelines object of study.

N	Framework/Assessment Tool/Guideline	N. Categories	N. Indicators
1	SDEWES	7	35
2	LEED v4 for Neighborhood Development	5	59
3	BREEAM Communities	9	62
4	CITYKeys	5	99
5	CIVITAS	9	28
6	Ecocity	5	186
7	Le Modele Indi-Ru 2005	5	73
8	KITCASP	4	20
9	LB	7	93
10	SMIS	6	39
11	CGYM	8	52
12	SEV	7	44
13	BCN	4	13
14	BIL	12	34
15	VERDE	6	42
16	DGNB	6	63
17	HQE	14	158
18	White Paper on Sustainability of Spanish Urban Planning	7	154
19	European Smart Cities Ranking	6	64
20	REPLICATE PROJECT	7	56
21	Smart City PROFILES	5	21
22	City Protocol	9	190
23	ISO 37120 Sustainable development of communities	17	100
24	Reference framework for European sustainable cities (RFSC)	5	30
25	Sustainability Tools And Targets for the Urban Thematic Strategy project	8	46
26	UN Habitat indicators	5	42
27	CASBEE for Urban Development (CASBEE-UD)	7	82
28	Urban Sustainability Framework (The World Bank)	9	43
29	The bridge project	3	28
30	CESBA tool: new buildings	5	22
31	European green city index	8	30
32	System of indicators and conditions for large and medium-sized cities	4	52
Total		224	2060

Before comparing the indicators used in each tool, it was necessary to define a common structure as each tool uses a different classification system and a distinct nomenclature. To solve this aspect, the first step was to create a two-level structure of categories and subcategories with their corresponding objective.

The method used to obtain this was a thorough review of the indicator's structures proposed in the existent literature and a frequency analysis for categories, subcategories and indicators of each tool selected for this study. It then appeared that generally, categories are defined in such a way that each can include a wide range of indicators with diverse objectives.

An important observation related to this is that many indicators and categories are closely related, which would make it possible for an indicator or subcategory to be considered under different categories or even under different aspects (environmental, social, economic or governance).

4. Results

The proposed structure of categories, subcategories and indicators is based on their frequency among the 32 indicator sets object of this study. Figure 1 shows the word frequency of the 2060 indicators contained in the 32 selected indicator sets.

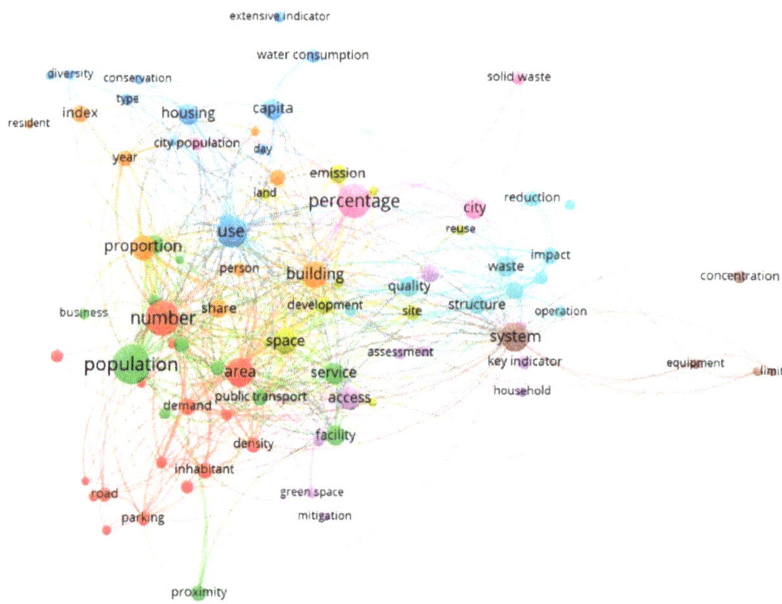

Figure 1. Visualization of the cluster network from the terms of sustainable urban indicators shown in Table 1 with a lower limit of co-occurrence of 7.

Indicators were analyzed through a content analysis, which implies the codification of the meaning that allows its categorization, for which the analysis codes were established; the codification of the meaning, where a keyword is given to a part of the text and facilitate its identification; condensation of meaning, in which the meanings expressed in short formulations are summarized; and interpretation of meaning, in which the text is interpreted. As the meaning and the language are intermingled, this type of analysis allows attention to the linguistic features of an interview, being able to contribute to generate and verify the meaning of the statements, as well as to improve the precision of the questions in the interview [10].

Additionally, cluster analysis was used as support and verification of the correlations made in the previous analysis. This method, highly used in bibliometric environments, allows assigning relevance values to the terms based on the frequency of appearance in this case in the indicators, to generate categories. The method allows to set a minimum repetition value of the term to be entered in order to test different grouping results based on the relevance of the terms, so that, with a high number of frequencies, it would only include highly repeated terms and with a lower value of frequencies, low strength relationships are established.

Figure 1 is the result of concurrency 7, which included 90 terms, repeated at least 7 times. It resulted in 10 groups, which provided support for the analysis of categorization of meaning carried out in the traditional way.

Color identifies different categories and the importance of the term is given by the size of the circle. The closeness between terms indicates a greater relationship between them. In this sense, the result confirms the need for an interdisciplinary vision of sustainability in urban environments and the difficulty of generating a measurement tool that encompasses issues that cover such diverse and specific topics.

Table 2 shows the results of the proposed category structure based on the analysis of the meaning of the indicators supported by the cluster analysis: environmental, social, governance, economy aspects, resulting in 14 categories:

Table 2. Proposed category structure.

ESGE	Area	Category
Environmental	Resources	Energy Waste Water Pollution
	Territory management	Mobility & transport Plan & design Site and land
Social	Social	Social aspects
Governance	Governance	Management City Planning & Innovation Transparency
Economy	Economy	Local aspect Labor Finances

From the analysis carried out regarding aspects considered by each of the tools, we can conclude that all the frameworks include indicators related to the environment, around 80% consider the economic aspect, although in a generally low proportion, 90% consider the social aspect (this being the second aspect with the highest total number of indicators, after the environmental aspect) and only around 70% consider some aspect related to the government in some way. Figure 2 shows the number of indicators related to each aspect for each framework (the numbering of the frameworks corresponds to that shown in Table 1):

Figure 2. Distribution of considered aspects by framework.

4.1. Environmental Aspects

Environmental sustainability ensures fundamental needs such as the preservation and improvement of natural resources and the protection and restoration of environment and habitat. A sustainable environment implies that an ecosystem is able to maintain its functionality over time. The environmental sustainability proposed structure consists of seven categories, widely classified in two big groups, considering the area of action they fall into: Resources and Territory Management.

Figure 3 shows the distribution of environmental aspects considered by each analyzed framework, according to the proposed categories within this section: water, waste, site and land, pollution, plan and design, mobility and transport and energy.

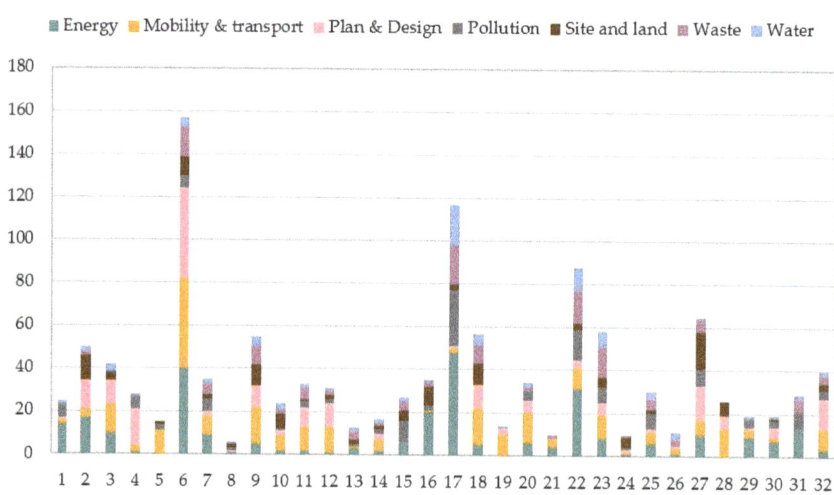

Figure 3. Distribution of environmental aspects considered by framework.

4.1.1. Resources

Resources include 4 categories—Energy, Waste, Water and Pollution (see Table 3). The four of them were present in more than half of the analyzed assessment tools (17/32) and at least 2 of these categories were present on every framework, which came as no surprise, as natural resources and pollution are some of the most concerning aspects of environment conservation.

Table 3. Proposed indicators for environmental aspects (resources area).

Category	Subcategory	N	Indicator
Energy	Indoor environment quality	1	Proportion of buildings certified by an environmental quality sign
	Consumption	2	Total energy consumption
	Efficiency	3	Proportion of energy self-sufficiency based on renewable energies use
		4	Infrastructure and Buildings Energy Efficiency
	Buildings	5	Housing density
	Conservation	6	Building age index
		7	Investment in restoration and conservation projects
	Emissions	8	Greenhouse gases emissions
Waste	Waste water treatment	9	Use of systems to reuse/treat wastewater
	Recycled/treated solid waste	10	Percentage of municipal solid waste recycled.
	Waste management	11	Generation and waste management
Water	Access	12	Percentage of population with sustainable access to an improved water source
	Efficiency	13	Efficiency of water usage
	Consumption	14	Water consumption
Pollution	Light pollution	15	Reduce light pollution
	Noise pollution	16	Reduce noise pollution
	Gas pollution	17	Reduce gas pollution
	Air quality	18	Improve air quality

Energy category represented almost a 25% of the environmental indicators object of study, which gives a good view of its importance when it comes to sustainability and the need for its proper representation in any assessment model [11].

Energy consumption and efficiency are two closely related concepts. Vast quantities of energy are consumed as economy [12] and people's expectations of comfort grow [13], which makes it vital to move towards more sufficient, renewable-based, pollutant-mitigating consumption policies [14].

Buildings constitute one of the main consumers of energy in cities, besides industry and transportation systems. It is important that buildings are constructed to be efficient and self-sufficient, according to their geographical situation [15] and use. As a result of this, a considerable number of certifications is available nowadays specifically for measuring sustainability of buildings based on those factors (use, location, etc.) [16].

Indoor environment quality (IEQ) involves many factors, as it refers to the quality of the interior spaces of a building in relation to the wellbeing of those that make use out of it. It includes aspects such as air quality, comfortable temperature, air quality and adequate humidity levels. Research has shown that IEQ has a relevant impact in the performance of those who occupy the space [17–19]. IEQ is a value that indicates to what extent the energy used for the construction and use of a building is useful or efficient.

Conservation is another key in urban sustainable development, as they are concepts that share common ground [20]. Successful architectural conservation and the adaptative reuse of heritage buildings to meet efficient resources management criteria could be beneficial [21], however it is a tricky subject, as it has been shown that despite its possible positive impact when it comes to sustainability, it could create certain issues [22].

All energy sources have certain impact on the environment; fossil fuels do substantially more harm than renewable energy sources by most measures, which makes it indispensable to promote aspects such as efficiency, self-sufficiency, percentage of renewable energies use and to reduce emissions [23].

Waste: the environmental effects of waste can be the cause of serious issues, both contributing to the greenhouse effect and affecting health. Gases coming from the incineration of waste may produce air pollution and contribute to acid rain if not treated properly [24]. Waste water is by definition the process that removes contaminants from waste water and sewage water so it can be returned to the water cycle or reused for certain purposes without a relevant impact on the environment [25,26]. Treatments may consist of chemical, biological or physical processes or even a combination of them and water can be treated to obtain any desired level of quality. However, the level of quality comes at a cost, hence the need to select appropriate outcome levels according to the future use of the treated water or its return back to the water cycle [27].

Water: population growth, the movement of large numbers of people to certain areas and the consequent high density of population in said areas, the demands for higher living standards, are some aspects that have affected the use of water resources. Once the basic need for water access [28] and water quality is accomplished, it is necessary to look into an efficient use of water resources [29,30]. Water consumption in urban areas is essential, hence being able to measure it becomes crucial. Different models for predicting urban water consumption have been developed during the past years, which can enable an improvement in the performance of water distribution systems [31]. This leads us to water efficiency, which should be approached both in households [32,33] and considering the whole urban water infrastructure [34].

Pollution: it is defined as "the contamination of the physical and biological components of the earth/atmosphere system to such an extent that normal environmental processes are adversely affected" and it may poison soils and waterways, lead to chronic diseases [35]. A holistic approach to resource management could be applied to start the way towards a solution of increasing worldwide pollution issues [36]. Even though gas pollution and air quality might be some of the most relevant and more studied factors within pollution, noise and light [37–39] pollution have been shown to be related to population wellbeing.

4.1.2. Territory Management

This area includes 3 categories—Mobility and Transport, Plan and Design and Site and Land (see Table 4).

Table 4. Proposed indicators for environmental aspects (territory management area).

Category	Subcategory	N	Indicator
Mobility & transport	Proximity to services	19	Distance to basic services
	Traffic	20	Surface of public road for automobile traffic and public transport.
	Accessibility	21	Accessibility to key services
	Public transport	22 23 24	Access to public transport network Quality of public transport network Smart public transport network
	Pedestrian-oriented urban structures	25 26	Pedestrianized streets Enlarged pedestrian and pedestrian-priority areas
	Cycling network availability	27	Quality cycling network
Plan & design	Open/green areas	28 29	Percentage of open/green public areas Number of trees / Kilometer urban road
	Roads	30	Length and surface or urban roads
	Urban structure	31	Pedestrian-oriented urban structures with short distances
Site and land	Land use	32	Up-to-date sustainable land use
	Conservation	33	Maintain or re-cultivate/restore green and water elements
	Reuse	34	Prioritize the use of existing sites

Mobility and Transport: transport is an important factor in the context of sustainability, as generally it contributes highly to contamination and the greenhouse effect. An efficient transport system that provides smart, sustainable mobility structures is key for the health of both the environment and economy [40].

Accessibility and proximity to services have lately been considered as concepts of "walkable cities" or "15 minutes' city" [41]. Proximity to basic services has a wide range of positive impacts on the environment, from the reduction of vehicle emissions to the wellbeing of citizens [42]. Traffic management is another approach that can be taken to move towards sustainability in cities, which research that proposes different ways to achieve it [43].

ICT and sustainability connect to every service provided to communities in smart cities and play an important role in smart city planning, including in public transportation [44]. An efficient public network, with adequately positioned stops, can benefit largely from real-time information. An integrated cycling network availability that develops the concept of smart velomobility allows citizens to use an active, sustainable and intelligent way of transportation [45].

Pedestrian-oriented urban structures imply pedestrianized streets and enlarge pedestrian and pedestrian-priority areas, to improve the comfort citizens experiment in their walking environment, always considering aspects such as accessibility and safety, among others [46,47].

Plan and Design: environment can be integrated into urban planning, contemplating environmental priorities and strategies into development plans, promoting a smart mobility structure, a pedestrian-oriented urban structure and other aspects like the conservation and creation of public green areas [48,49].

Sustainable road network development should meet the usual sustainability criteria and reduces its impact on climate change, human health and biological diversity [50]. Concepts like "green roads" have been studied as an alternative for more sustainable civil engineering constructing practices [51].

Urban structure represents another factor that has been proven to affect city sustainability since it determines the location of pollution emission sources and traffic patterns [52], as well as it is related to mobility possibilities [53].

Site and Land: land use is closely related to Mobility and Transport and Plan and Design aspects. Changes in land use and the prioritization of the use of existing sites may reduce the need for transportation and hence reduce emissions [54]. Conservation of site implies the maintenance or recultivation/restoration of green and water elements of urban areas, which play an important role in citizen's sense of place with their environment [55,56].

The reuse land has become a strong policy aim in certain governments in their attempt for sustainable planning strategies [57,58]. The reuse of existing sites is also considered in the concepts of green urbanism [59].

4.2. Social Aspects

The third pillar of sustainability, its social aspect represents the contribution to a more equitable, diverse future. The social part of the overall sustainability index has received little attention compared to the previous aspects until recently [60], even though it provides the level of social sustainability performance and its areas of improvement, which would enhance a holistic better development and quality of life. Better social futures require anticipation, which could be obtained via relationally real structures [61,62].

The proposed social sustainability index consists of three sub-categories, which are housing, public spaces quality and health and wellbeing (see Table 5).

Housing and public quality: these two first sub-categories have been a concern historically and taken into account in City Planning since decades ago. Initially, the sole purpose of social housing was to overcome the need for shelter for socially vulnerable citizens. However, studies have shown that the low level of quality, which used to be common in this type of housing, had a negative impact both socially and environmentally which makes it important to consider these aspects in the planning approach to social housing [63,64].

Table 5. Proposed indicators for social aspects.

Category	Subcategory	N	Indicator
Social aspects	Housing	35	Proportion of social housing in the neighborhood
	Public spaces quality	36	Integrate public spaces and streetscapes of high spatial quality
	Health and wellbeing	37	Access to basic services
		38	Encouraging a healthy lifestyle
		39	Ensuring healthy outdoor spaces and healthy atmosphere in indoor spaces

Accessibility to public greenspaces and open spaces of quality has been recommended as an indicator for public health [65]. The availability of these kind of spaces has proven to be beneficial not only for the environment but for citizen's wellness [66].

However, the concept of wellness and wellbeing is a growing concern that has been intensified in the past years, provoking a growing interest in processes that create places that promote wellbeing, resulting in the appearance of new assessment tools and certifications that focus on this sustainability aspect, like WELL certification. City livability is also closely related to the concept of urban wellbeing and to other environmental aspects mentioned before like urban pollution [67].

4.3. Governance Aspects

Governance helps implement sustainability strategies, manage goal-setting and reporting processes and ensure overall accountability. However, despite the fact that the relationship between sustainability and governance is fundamental in the development of the former, the work including governance in sustainability frameworks is still limited [68].

Governance indicators are key to ensure the functioning of the whole society and when used efficiently, informed by science, they can be a critical component of sustainable development. They are usually overseen and poorly included in indicator frameworks and assessments, which is incoherent, as public institutions should be leading the change towards a more sustainable territory, both through legislation and through actions. From Management to City Planning and even Innovation, every aspect matters and should be taken into account. Ultimately, indicators must reflect political reality, information availability and a relevant scale of analysis [69].

United Nations Development Program (UNDP) published "A Users' Guide to Measuring Local Governance" where they specified a total of 22 governance frameworks [70] which are adopted by various countries across the world in order to evaluate governance [71]. Since nowadays transparency should be part of every public organism, the contemplated data should be easily obtained. Governance indicators have been categorized in four categories based on their representation of a common goal, as follows (see Table 6).

Management: it is necessary to ensure objectivity during the process of diagnosing, decision making, drafting and approving urban plans and also while integrating Agenda 21 [72].

City Planning and innovation: the effects of urban planning indicators on the urban regime can provide useful tools to urban planners and policymakers for the evaluation of the impact of an urban transformation of the city at different scales: green areas and how they affect the intra-urban thermal regime [73] the availability and accessibility of public transport and so forth. Innovation positively evaluates the implementation of innovative solutions in different aspects of urban sustainability. Innovation is the key to progress, particularly in times of crisis [74].

Table 6. Proposed indicators for governance aspects.

Category	Subcategory	N	Indicator
Management	Monitoring and reporting	40	Implementation of monitoring and reporting systems
	Integrating agenda 21	41	Integration of Agenda 21 into urban planning
City planning and innovation	Smart city policy development	42	Transition towards smart city (implementation of sensors, open data, ...)
		43	Citizens access to ICT information
Transparency	Transparency	44	Transparency and open Government

Transparency: this aspect is often credited with generating government accountability, which would lead to reduction in government corruption, bribery and other malfeasance. However, even in democratic societies, public officials have incentives to pursue secrecy, as it provides fertile ground for special interests and undermines the ability to provide an effective check against the actions of government. It is fundamental for government to provide transparency in sustainability as in all matters, acting in the interest of citizens and their territory [68].

4.4. Economic Aspects

The economic pillar of sustainability implies understanding that the measures of unsustainability arising from a consumer led culture that treats finite resources as an income leads to an inevitable. This economic aspect is used to define strategies that promote de use of resources optimally, considering an equitable, efficient distribution of resources.

The number of proposed economic indicators in literature is almost endless, has economy has been historically studied to very large extents. However, if we take into account the 32 indicators sets considered in this article, economic aspects do not represent a high percentage (nearly 5%) of the total of unified categories. The fact that the sets are mostly focused on environmental factors could explain this, as they contain principally basic economy indicators, while only a few of them go more in depth into economy issues. The economic sustainability index consists of three sub-categories, which are the local aspect of economy, labor and finances (see Table 7).

Table 7. Proposed indicators for economic aspects.

Category	Subcategory	N	Indicator
Local aspect	Local resources	45	Encourage use of local resources
	Labor and skills	46	Strengths and local specifics of the labor force including the availability of workers with different qualifications
Labor	Employment	47	Employment rate
Finances	Household income	48	Average household income

Local resources: one of the objectives of sustainability is to establish local economies that are environmentally friendly, socially responsible and of course economically viable. The use of local resources would be in favor for those three aspects, considering it could enhance the creation of jobs locally, reduce emissions due to transportation and contribute to the area's economy [75].

Labor: as previously stated, climate change is sensitive to human activity but also that much of human activity is sensitive to climate change, hence the emergence of concepts such as green economy and economic sustainability [76].

Finances: average house income is present in most of economic indicator-based assessments, as it can be used as an indicator for the monetary well-being of a country's citizens. Low-income households are a matter of debate due to its possible consequences for sustainability [77].

Finally, in Figure 4 the developed structure has been graphically represented, which makes it easier to check visually the characteristics of a sustainable urban design at a glance. The graphic represents the proposed areas, categories and subcategories.

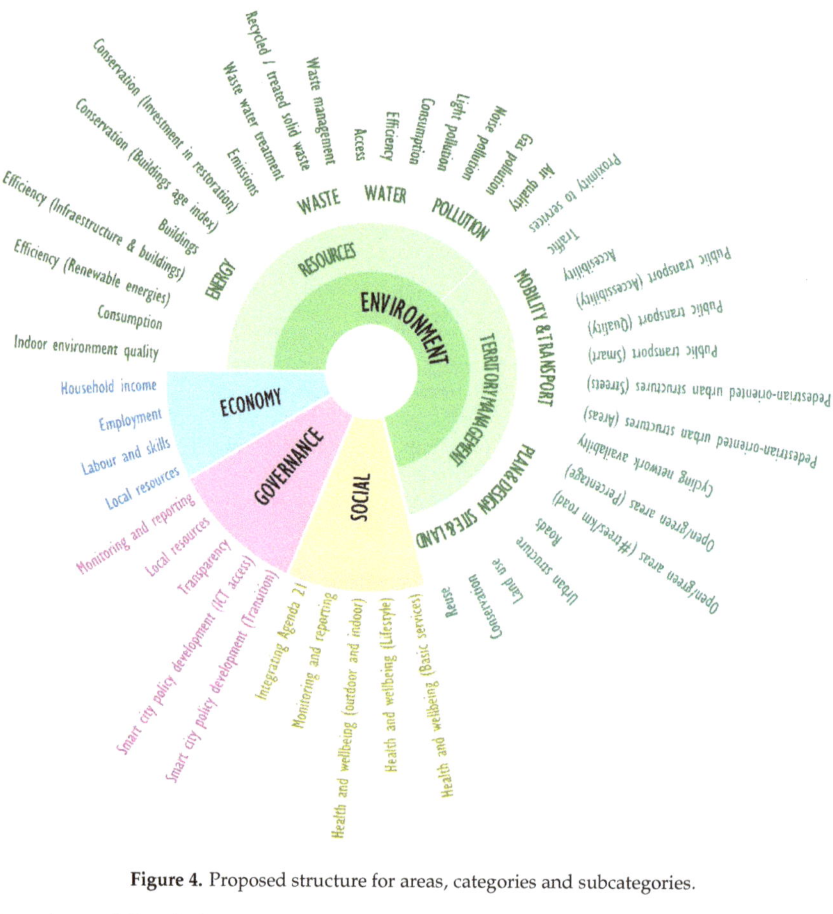

Figure 4. Proposed structure for areas, categories and subcategories.

5. Discussion and Conclusions

Sustainability has recently been object of study due to its impact on the planet and on human activity. Particularly urban environments are locations that represent a high percentage of emissions of gases, waste, energy and so forth. Numerous authors have developed tools for measuring the sustainability of buildings and urban environments to address this global challenge issue. The importance of assessment tools for the measure of sustainability in urban areas is clear, as they are a proven method for driving sustainable urban developed, which shows considering the increasing number of frameworks being developed in a short amount of time. However, while the pursued aim is often the same, the way to reach it varies from one to another and a common criterion has not yet been set in that sense, resulting in diverse, heterogeneous frameworks.

While some frameworks or guidelines offer a very wide range of almost 200 hundred indicators, it has not been proven that such a high number of information compensates the effort that it takes to obtain it. On the other hand, assessment tools with a very small number of indicators may result in poor overall information. It is then recommended to use enough indicators to get an overall view of the sustainability aspects of an urban area, without compromising its total cost.

The approaches may lie in certain aspects of the cities due to cultural and social characteristics or the great differences in terms of the level of development and industry, among others. Therefore, there are very specific tools for certain problems (such as City Blueprint), without a totally holistic solution.

This study proposes a response to this situation, considering 32 international frameworks and tools of diverse natures, resulting in a structure consisting of 14 categories (Energy, Waste, Water, Pollution; Mobility & Transport, Plan & Design, Site & Land; Social aspects, Governance Management, City Planning & Innovation, Transparency; Economy's Local Aspect, Labor and Finances) and 48 indicators.

This has been possible through a thorough review of the indicator's structures proposed in the existent literature and a frequency analysis for categories, subcategories and indicators of each tool selected for this study, supported by a cluster analysis used for the verification of the preciously made correlations.

The result provides a list of basic indicators that tries to fulfill all due aspects to obtain a general diagnosis of the sustainable nature of the urban environment, which can serve as support to detect the strongest and weakest areas in terms of their sustainability of the urban environment. Subsequently, other more specific existing tools such as those mentioned above could be applied to measure and improve said specific aspect.

The list of proposed indicators is not final, as it should evolve as knowledge and data availability improve and it is definitely arguable, as any attempt of proposal of a set of indicators would be, however similar the objectives could be. Ultimately, the list is expected to include key indicators for highly different yet interconnected issues such as resources and territory management and social, governance and economy factors.

Author Contributions: Conceptualization, A.L.C., A.C.G. and V.L.-C.; Data curation, A.L.C.; Formal analysis, A.L.C.; Methodology, A.L.C. and V.L.-C.; Software, A.L.C., V.L.-C. and A.A.; Supervision, A.C.G., V.L.-C. and A.A.; Visualization, V.L.-C.; Writing—original draft, A.L.C.; Writing—review & editing, V.L.-C. and A.A. All authors have read and agreed to the published version of the manuscript.

Funding: This research received no external funding.

Conflicts of Interest: The authors declare no conflict of interest.

References

1. Seto, K.C.; Fragkias, M.; Güneralp, B.; Reilly, M.K. A Meta-Analysis of Global Urban Land Expansion. *PLoS ONE* **2011**, *6*, e23777. [CrossRef] [PubMed]
2. United Nations, Department of Economic and Social Affairs, Population Division. *World Urbanization Prospects: The 2018 Revision*; United Nations, Department of Economic and Social Affairs, Population Division: New York, NY, USA, 2016.
3. Wendling, L.A.; Huovila, A.; zu Castell-Rüdenhausen, M.; Hukkalainen, M.; Airaksinen, M. Benchmarking Nature-Based Solution and Smart City Assessment Schemes Against the Sustainable Development Goal Indicator Framework. *Front. Environ. Sci.* **2018**, *6*, 69. [CrossRef]
4. Garau, C.; Pavan, V. Evaluating Urban Quality: Indicators and Assessment Tools for Smart Sustainable Cities. *Sustainability* **2018**, *10*, 575. [CrossRef]
5. Brundtland, G.H. Our Common Future—Call for Action. *Environ. Conserv.* **1987**, *14*, 291–294. [CrossRef]
6. Kennedy, C.; Cuddihy, J.; Engel-Yan, J. The Changing Metabolism of Cities. *J. Ind. Ecol.* **2007**, *11*, 43–59. [CrossRef]
7. Kitchin, R.; Lauriault, T.P.; McArdle, G. Knowing and governing cities through urban indicators, city benchmarking and real-time dashboards. *Reg. Stud. Reg. Sci.* **2015**, *2*, 6–28. [CrossRef]
8. Bell, S.; Morse, S. *Measuring Sustainability: Learning by Doing*; Earthscan Publications Ltd.: London, UK, 2003; ISBN 978-1-85383-839-2.

9. Purvis, B.; Mao, Y.; Robinson, D. Three pillars of sustainability: In search of conceptual origins. *Sustain. Sci.* **2019**, *14*, 681–695. [CrossRef]
10. López-Chao, V. El impacto del Diseño del Espacio y Otras Variables Socio-Físicas en el Proceso de Enseñanza-Aprendizaje. Ph.D. Thesis, Universidade da Coruña, A Coruña, Spain, 2017.
11. Grigoroudis, E.; Kouikoglou, V.S.; Phillis, Y.A.; Kanellos, F.D. Energy sustainability: A definition and assessment model. *Oper. Res. Int. J.* **2019**. [CrossRef]
12. Apergis, N.; Payne, J.E. Energy consumption and economic growth: Evidence from the Commonwealth of Independent States. *Energy Econ.* **2009**, *31*, 641–647. [CrossRef]
13. Saint Akadiri, S.; Alola, A.A.; Akadiri, A.C.; Alola, U.V. Renewable energy consumption in EU-28 countries: Policy toward pollution mitigation and economic sustainability. *Energy Policy* **2019**, *132*, 803–810. [CrossRef]
14. Chappells, H.; Shove, E. Debating the future of comfort: Environmental sustainability, energy consumption and the indoor environment. *Build. Res. Inf.* **2005**, *33*, 32–40. [CrossRef]
15. Yılmaz, Z. Evaluation of energy efficient design strategies for different climatic zones: Comparison of thermal performance of buildings in temperate-humid and hot-dry climate. *Energy Build.* **2007**, *39*, 306–316. [CrossRef]
16. Berardi, U. Sustainability Assessment in the Construction Sector: Rating Systems and Rated Buildings: Sustainability Assessment in the Construction Sector. *Sust. Dev.* **2012**, *20*, 411–424. [CrossRef]
17. López-Chao, V.; Amado Lorenzo, A.; Martín-Gutiérrez, J. Architectural Indoor Analysis: A Holistic Approach to Understand the Relation of Higher Education Classrooms and Academic Performance. *Sustainability* **2019**, *11*, 6558. [CrossRef]
18. Baeza Moyano, D.; Fernández, M.S.J.; González Lezcano, R.A. Towards a Sustainable Indoor Lighting Design: Effects of Artificial Light on the Emotional State of Adolescents in the Classroom. *Sustainability* **2020**, *12*, 4263. [CrossRef]
19. López-Chao, V.; Amado Lorenzo, A.; Saorín, J.L.; de la Torre-Cantero, J.; Melián-Díaz, D. Classroom Indoor Environment Assessment through Architectural Analysis for the Design of Efficient Schools. *Sustainability* **2020**, *12*, 2020. [CrossRef]
20. Rodwell, D. *Conservation and Sustainability in Historic Cities*; Publishing House of Electronics Industry: Beijing, China, 2015; ISBN 978 7 121 27256 1.
21. Ma, Z.; Cooper, P.; Daly, D.; Ledo, L. Existing building retrofits: Methodology and state-of-the-art. *Energy Build.* **2012**, *55*, 889–902. [CrossRef]
22. Bullen, P.A.; Love, P.E.D. Adaptive reuse of heritage buildings. *Struct. Surv.* **2011**, *29*, 411–421. [CrossRef]
23. Song, J.; Feng, Q.; Wang, X.; Fu, H.; Jiang, W.; Chen, B. Spatial Association and Effect Evaluation of CO_2 Emission in the Chengdu-Chongqing Urban Agglomeration: Quantitative Evidence from Social Network. *Anal. Sustain.* **2018**, *11*, 1. [CrossRef]
24. Liu, M.; Woudstra, T.; Promes, E.J.O.; Restrepo, S.Y.G.; Aravind, P.V. System development and self-sustainability analysis for upgrading human waste to power. *Energy* **2014**, *68*, 377–384. [CrossRef]
25. Tchobanoglous, G.; Burton, F.L.; Stensel, H.D. *Wastewater Engineering: Treatment and Reuse*; McGraw-Hill series in civil and environmental engineering; McGraw-Hill: Boston, MA, USA, 2002; ISBN 978-0-07-124140-3.
26. Leslie, G.P.G., Jr.; Glen, T.D.; Nancy, G.L.; Carlos, D.M.F. *Biological Wastewater Treatment*, 3rd ed.; IWA Publishing: London, UK, 2011; ISBN 978-1-4200-0963-7.
27. Englande, A.J.; Krenkel, P.; Shamas, J. Wastewater Treatment &Water Reclamation. In *Reference Module in Earth Systems and Environmental Sciences*; Elsevier: Amsterdam, The Netherlands, 2015; ISBN 978-0-12-409548-9.
28. Aiga, H.; Umenai, T. Standardisation of the definition of access to safe water. *Lancet* **2003**, *361*, 2156. [CrossRef]
29. Yang, W.; Hyndman, D.W.; Winkler, J.A.; Viña, A.; Deines, J.M.; Lupi, F.; Luo, L.; Li, Y.; Basso, B.; Zheng, C.; et al. Urban water sustainability: Framework and application. *Ecol. Soc.* **2016**, *21*, art4. [CrossRef]
30. Milman, A.; Short, A. Incorporating resilience into sustainability indicators: An example for the urban water sector. *Glob. Environ. Chang.* **2008**, *18*, 758–767. [CrossRef]
31. Cutore, P.; Campisano, A.; Kapelan, Z.; Modica, C.; Savic, D. Probabilistic prediction of urban water consumption using the SCEM-UA algorithm. *Urban Water J.* **2008**, *5*, 125–132. [CrossRef]
32. Lee, M.; Tansel, B.; Balbin, M. Urban Sustainability Incentives for Residential Water Conservation: Adoption of Multiple High Efficiency Appliances. *Water Resour Manag.* **2013**, *27*, 2531–2540. [CrossRef]
33. Lee, M.; Tansel, B.; Balbin, M. Influence of residential water use efficiency measures on household water demand: A four year longitudinal study. *Resour. Conserv. Recycl.* **2011**, *56*, 1–6. [CrossRef]

34. Bell, S. Renegotiating urban water. *Prog. Plan.* **2015**, *96*, 1–28. [CrossRef]
35. Finco, A.; Nijkamp, P. Pathways to urban sustainability. *J. Environ. Policy Plan.* **2001**, *3*, 289–302. [CrossRef]
36. Niemczynowicz, J. New aspects of urban drainage and pollution reduction towards sustainability. *Water Sci. Technol.* **1994**, *30*, 269–277. [CrossRef]
37. Brunekreef, B.; Holgate, S.T. Air pollution and health. *Lancet* **2002**, *360*, 1233–1242. [CrossRef]
38. Goines, L.; Hagler, L. Noise Pollution: A Modem Plague. *South Med. J.* **2007**, *100*, 287–294. [CrossRef]
39. Chepesiuk, R. Missing the Dark: Health Effects of Light Pollution. *Environ. Health Perspect.* **2009**, *117*. [CrossRef] [PubMed]
40. Lopez-Carreiro, I.; Monzon, A. Evaluating sustainability and innovation of mobility patterns in Spanish cities. Analysis by size and urban typology. Sustain. *Cities Soc.* **2018**, *38*, 684–696. [CrossRef]
41. Marquet, O.; Miralles-Guasch, C. The Walkable city and the importance of the proximity environments for Barcelona's everyday mobility. *Cities* **2015**, *42*, 258–266. [CrossRef]
42. de Nazelle, A.; Morton, B.J.; Jerrett, M.; Crawford-Brown, D. Short trips: An opportunity for reducing mobile-source emissions? *Transp. Res. Part D Transp. Environ.* **2010**, *15*, 451–457. [CrossRef]
43. Boltze, M.; Tuan, V.A. Approaches to Achieve Sustainability in Traffic Management. *Procedia Eng.* **2016**, *142*, 205–212. [CrossRef]
44. Bifulco, F.; Tregua, M.; Amitrano, C.C.; D'Auria, A. ICT and sustainability in smart cities management. *Int. J. Public Sect. Manag.* **2016**, *29*, 132–147. [CrossRef]
45. Behrendt, F. Why cycling matters for Smart Cities. Internet of Bicycles for Intelligent Transport. *J. Transp. Geograph.* **2016**, *56*, 157–164. [CrossRef]
46. Forsyth, A.; Southworth, M. Cities Afoot—Pedestrians, Walkability and Urban Design. *J. Urban Des.* **2008**, *13*, 1–3. [CrossRef]
47. Ovstedal, L.; Ryeng, E.O. Who Is The Most Pleased Pedestrian? In Proceedings of the 3rd International Conference, Steps towards Liveable Cities, Donastia-San Sebastián, Spain, 8–11 May 2002.
48. Newman, P.W.G. Sustainability and cities: Extending the metabolism model. *Landsc. Urban Plan.* **1999**, *44*, 219–226. [CrossRef]
49. Yiftachel, O.; Hedgcock, D. Urban social sustainability. *Cities* **1993**, *10*, 139–157. [CrossRef]
50. Puodziukas, V.; Svarpliene, A.; Braga, A. Measures for Sustainable Development of Road Network. *Transp. Res. Procedia* **2016**, *14*, 965–972. [CrossRef]
51. Muench, S.T.; Anderson, J.; Bevan, T. Green Roads: A Sustainability Rating System for Roadways. *Int. J. Pavement Res. Technol.* **2010**, *3*, 270–279.
52. Borrego, C.; Martins, H.; Tchepel, O.; Salmim, L.; Monteiro, A.; Miranda, A.I. How urban structure can affect city sustainability from an air quality perspective. *Environ. Model. Softw.* **2006**, *21*, 461–467. [CrossRef]
53. Madlener, R.; Sunak, Y. Impacts of urbanization on urban structures and energy demand: What can we learn for urban energy planning and urbanization management? *Sustain. Cities Soc.* **2011**, *1*, 45–53. [CrossRef]
54. Pauleit, S.; Duhme, F. Assessing the environmental performance of land cover types for urban planning. *Landsc. Urban Plan.* **2000**, *52*, 1–20. [CrossRef]
55. Colley, K.; Craig, T. Natural places: Perceptions of wildness and attachment to local greenspace. *J. Environ. Psychol.* **2019**, *61*, 71–78. [CrossRef]
56. Carrus, G.; Lafortezza, R.; Colangelo, G.; Dentamaro, I.; Scopelliti, M.; Sanesi, G. Relations between naturalness and perceived restorativeness of different urban green spaces. *Psyecology* **2013**, *4*, 227–244. [CrossRef]
57. Williams, K.; Dair, C. A framework for assessing the sustainability of brownfield developments. *J. Environ. Plan. Manag.* **2007**, *50*, 23–40. [CrossRef]
58. Dair, C.M.; Williams, K. Sustainable Land Reuse: The Influence of Different Stakeholders in Achieving Sustainable Brownfield Developments in England. *Environ. Plan. A* **2006**, *38*, 1345–1366. [CrossRef]
59. Lehmann, S. Transforming the City for Sustainability: The Principles of Green Urbanism. *J. Green Build.* **2011**, *6*, 104–113. [CrossRef]
60. Baffoe, G.; Mutisya, E. Social sustainability: A review of indicators and empirical application. *Environ. Manag. Sustain. Dev.* **2015**, *4*. [CrossRef]
61. Hale, J.; Legun, K.; Campbell, H.; Carolan, M. Social sustainability indicators as performance. *Geoforum* **2019**, *103*, 47–55. [CrossRef]

62. Popovic, T.; Barbosa-Póvoa, A.; Kraslawski, A.; Carvalho, A. Quantitative indicators for social sustainability assessment of supply chains. *J. Clean. Prod.* **2018**, *180*, 748–768. [CrossRef]
63. Petković-Grozdanović, N.; Stoiljković, B.; Shubenkov, M. Location Criteria Relevant for Sustainability of Social Housing Model. *MATEC Web Conf.* **2016**, *73*, 06001. [CrossRef]
64. Hall, M.; Purchase, D. Building or bodging? Attitudes to sustainability in UK public sector housing construction development. *Sust. Dev.* **2006**, *14*, 205–218. [CrossRef]
65. Annerstedt van den Bosch, M.; Mudu, P.; Uscila, V.; Barrdahl, M.; Kulinkina, A.; Staatsen, B.; Swart, W.; Kruize, H.; Zurlyte, I.; Egorov, A.I. Development of an urban green space indicator and the public health rationale. *Scand. J. Public Health* **2016**, *44*, 159–167. [CrossRef]
66. Lee, A.C.K.; Maheswaran, R. The health benefits of urban green spaces: A review of the evidence. *J. Public Health* **2011**, *33*, 212–222. [CrossRef]
67. del Mar Martínez-Bravo, M.; Martínez-del-Río, J.; Antolín-López, R. Trade-offs among urban sustainability, pollution and livability in European cities. *J. Clean. Prod.* **2019**, *224*, 651–660. [CrossRef]
68. Aras, G.; Crowther, D. Governance and sustainability: An investigation into the relationship between corporate governance and corporate sustainability. *Manag. Decis.* **2008**, *46*, 433–448. [CrossRef]
69. Kaufmann, D.; Kraay, A.; Zoido-Lobatón, P. Aggregating Governance Indicators. Policy Research Working Paper. Available online: https://info.worldbank.org/governance/wgi/pdf/govind.pdf (accessed on 2 May 2020).
70. Wilde, A.; Narang, S.; Laberge, M.; Moretto, L.; UNDP Oslo Governance Centre. A Users' Guide to Measuring Local Governance. Available online: https://www.undp.org/content/dam/aplaws/publication/en/publications/democratic-governance/dg-publications-for-website/a-users-guide-to-measuring-local-governance-/LG%20Guide.pdf (accessed on 24 March 2020).
71. Biswas, R.; Jana, A.; Arya, K.; Ramamritham, K. A good-governance framework for urban management. *J. Urban Manag.* **2019**, *8*, 225–236. [CrossRef]
72. Braulio-Gonzalo, M.; Bovea, M.D.; Ruá, M.J. Sustainability on the urban scale: Proposal of a structure of indicators for the Spanish context. *Environ. Impact Assess. Rev.* **2015**, *53*, 16–30. [CrossRef]
73. Petralli, M.; Massetti, L.; Brandani, G.; Orlandini, S. Urban planning indicators: Useful tools to measure the effect of urbanization and vegetation on summer air temperatures: Urban indicators and summer intra-urban air temperatures. *Int. J. Climatol.* **2014**, *34*, 1236–1244. [CrossRef]
74. Crielaard, J.P.; Omta, S. Innovation and sustainability: Create the innovative organisation. *Eng. Bus.* **2008**, *1*, 1–35.
75. Curtis, F. Eco-localism and sustainability. *Ecol. Econ.* **2003**, *46*, 83–102. [CrossRef]
76. Prinz, L.; Pegels, A. The role of labour power in sustainability transitions: Insights from comparative political economy on Germany's electricity transition. *Energy Res. Soc. Sci.* **2018**, *41*, 210–219. [CrossRef]
77. Lettenmeier, M.; Hirvilammi, T.; Laakso, S.; Lähteenoja, S.; Aalto, K. Material Footprint of Low-Income Households in Finland—Consequences for the Sustainability Debate. *Sustainability* **2012**, *4*, 1426–1447. [CrossRef]

Publisher's Note: MDPI stays neutral with regard to jurisdictional claims in published maps and institutional affiliations.

atmosphere

MDPI

Article

Concentrations of Four Major Air Pollutants among Ecological Functional Zones in Shenyang, Northeast China

Liguang Li *, Ziqi Zhao, Hongbo Wang, Yangfeng Wang, Ningwei Liu, Xiaolan Li and Yanjun Ma *

Institute of Atmospheric Environment, China Meteorological Administration, Shenyang 110166, China; ziqizhao@yahoo.com (Z.Z.); hb_wang@foxmail.com (H.W.); wyf-7818@163.com (Y.W.); ningweiliu@163.com (N.L.); leexl.ouc@163.com (X.L.)

* Correspondence: liliguangyjs@foxmail.com (L.L.); myj@sina.com (Y.M.); Tel.: +86-24-83893253 (L.L.); +86-24-83893249 (Y.M.)

Received: 14 September 2020; Accepted: 4 October 2020; Published: 8 October 2020

Abstract: Air pollution is a critical urban environmental issue in China; however, the relationships between air pollutants and ecological functional zones in urban areas are poorly understood. Therefore, we analyzed the spatiotemporal characteristics of four major air pollutants (particulate matter less than or equal to 2.5 μm ($PM_{2.5}$) and 10 μm (PM_{10}) in diameter, SO_2, and NO_2) concentrations over five ecological functional zones in Shenyang, Liaoning Province, at hourly, seasonal, and annual scales using data collected from 11 monitoring stations over 2 years. We further assessed the relationships between these pollutants and meteorological conditions and land-use types at the local scale. Peaks in PM, SO_2, and NO_2 concentrations occurred at 08:00–09:00 and 23:00 in all five zones. Daytime PM concentrations were highest in the industrial zone, and those of SO_2 and NO_2 were highest in residential areas. All four air pollutants reached their highest concentrations in winter and lowest in summer. The highest mean seasonal PM concentrations were found in the industrial zone, and the highest SO_2 and NO_2 concentrations were found in residential areas. The mean annual PM and SO_2 concentrations decreased in 2017 in all zones, while that of NO_2 increased in all zones excluding the cultural zone. The natural reserve zone had the lowest concentrations of all pollutants at all temporal scales. Pollutant concentrations of $PM_{2.5}$, PM_{10}, SO_2, and NO_2 were correlated with visibility, and their correlation coefficients are 0.675, 0.579, 0.475, and 0.477. Land coverage with buildings and natural vegetation negatively and positively influence air pollutant concentrations, respectively.

Keywords: ecological functional zone; hourly concentration; land-use type; pollutant

1. Introduction

Air pollution has detrimental effects on human health [1–3], atmospheric visibility [4,5], air quality [6,7], and regional and global climate change [8,9]. Seven of the 10 most-polluted cities in the world are in China [10], and air quality in China has been extremely poor in recent decades. To combat this problem, environmental monitoring stations have been established in most Chinese cities to monitor pollution and inform control measures. These data have been used extensively to describe the spatiotemporal characteristics of air pollutants [11–15]. Data collected from 2006 to 2014 at 24 stations indicated that particulate matter (PM) concentrations were highest in Xi'an, Zhengzhou, and Gucheng, followed by cities in Northeast and South China [16]. Zhao et al. [14] found that, within China, cities located in the North China Plain were the most severely polluted, and SO_2 concentrations were highest in the northern part of the country. As for NO_2, however, there was no significant difference between northern and southern Chinese cities [12]. Generally, air pollution is higher in winter and lower in summer [17–20]. However, pollution characteristics vary among cities. For instance, particulate matter with a diameter < 2.5 μm ($PM_{2.5}$) concentrations in northern Chinese cities exhibited bimodal annual trends [14], whereas no pattern was observed in southern Chinese cities [14].

Although the emissions are the main cause of high concentrations, the influence of meteorological conditions and human activities cannot be ignored because both are important for the dispersion and transformation of air pollutants. Generally, strong wind speed favors the dispersion of air pollutants [21]. Additionally, temperature [22], relative humidity [23], wind direction [24], temperature inversion [25], and so on are related to dispersion and transformation of air pollutants. On the other hand, the intricate feedback loop between urban heat island, temperature-driven chemistry and urban pollution, urbanization also plays an important role for the air quality [26]. Land cover [27], high population density [28], and heavy traffic [29] all can affect pollutants' concentrations. Rendón et al. [30] simulated the impact of increasing urban land cover from 0% (rural) to 100% (urbanscape) on urban pollutants, and it indicated that the 40% urban cover was the most penalized in terms of air quality due to the combined effect of high urban emissions and temperature inversion persistence [31].

Shenyang, the capital of Liaoning Province, Northeast China, is undergoing rapid economic expansion and urban population growth. Air pollution is a major concern, with frequent air pollution events that can be severe. Famously, in November 2015, the maximum hourly $PM_{2.5}$ concentration exceeded 1000 $\mu g \cdot m^{-3}$ (http://news.sina.com.cn/c/nd/2015-11-08/doc-ifxknutf1607479) in Shenyang. Studies designed to inform the development of control methods, including pollution monitoring [32,33], transport paths [34], and compounding factors [29], have been carried out in Shenyang. Although numerous studies have investigated air pollution in the city, there are few analyses of pollutant characteristics across ecological functional zones [35,36].

To address this limitation, we classified the ecological functional zones at 11 established monitoring stations in Shenyang and compared the variations in four major air pollutants ($PM_{2.5}$, PM_{10}, SO_2, and NO_2) among these zones over three time scales (hourly, seasonal, and annual). The aims of this research were to (1) analyze temporal variations in PM, SO_2, and NO_2 concentrations in five ecological functional zones in Shenyang; (2) understand differences in PM, SO_2, and NO_2 concentrations among these zones; (3) determine local relationships of air pollutants with meteorological elements and land-use types, with the goal of providing the local government with information and suggestions for future pollution control measures.

2. Methods

2.1. Study Area

Shenyang, the capital city of Liaoning Province in the northeastern China, is an industrial city with a population of more than 8 million [37,38]. The topography is generally flat, and the city sits at < 50 m above sea level. The Hunhe River flows across the southern urbanized area. The climate is subhumid continental with a hot, rainy season from June to August [39].

2.2. Air Pollution, Meteorological, Urbanization, and Imaging Data

Hourly measurements of $PM_{2.5}$, PM_{10}, SO_2, and NO_2 concentrations at all stations collected over a 2-year period (1 January 2016 to 31 December 2017) were obtained from Shenyang Environment Monitoring Center for use in this study. Corresponding hourly meteorological data from the Shenyang weather station (Figure 1) were obtained from the Liaoning Provincial Meteorological Information Center, including relative humidity, wind speed, and visibility. Urbanization data, such as population and number of motor vehicles, were obtained from various internet sources (https://tieba.baidu.com/p/4936711038?red_tag=2949498218).

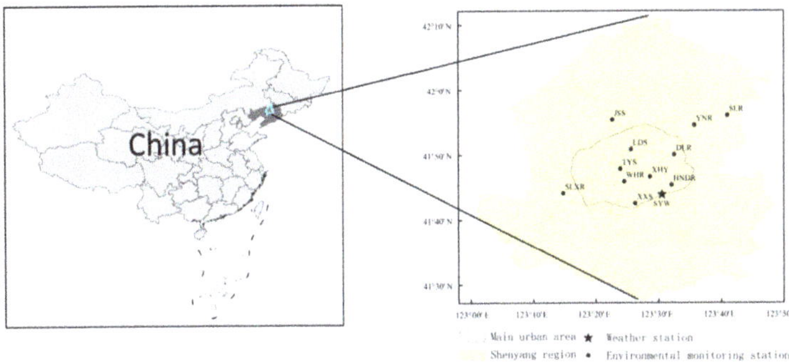

Figure 1. Locations of the 11 environmental monitoring stations and one weather station in the study area.

We classified all land-use types surrounding the monitoring stations based on Landsat Thematic Mapper images from 2015 with a resolution of 30 m. Images were obtained from the United States Geological Survey (http://glovis.usgs.gov). We used a preprocessing method for the Landsat images as described in Li et al. [37]. The ERDAS Imagine software and the ArcGIS software were also used in this study. Data used for classification included remotely sensed image, road maps, and field surveys. At last, land use was categorized into six types: buildings, roads, grassland, cropland, water, and bare land, with a combination of visual interpretation and supervised classification methods.

2.3. Ecological Function Zones

There were 11 pre-existing air quality monitoring sites in Shenyang (Figure 1, Table 1). An ecological function zone was assigned to each station based on characteristics of the surrounding environment and geographical position when it was first established. The ecological function zone in this paper refers to the ecological function of the area where the monitoring site is located. The specific process how to determine ecological functional zone is as follows: (1) we made a 3 km circular buffer around each site, and then we counted the areas of main land use types and typical ecological function within each buffer (Table 1). (2) The 11 monitoring sites were divided into five ecological function zones. Firstly, the natural reserve and non-natural reserve were determined based the percentages of main land use areas within 3 km buffer of each station. The Senlinlu (SLR) station is located in the natural reserve and its green land area accounted for 99% within the buffer, so we determined it as the natural reserve zone (Z5). Secondly, the other stations were divided into residential–commercial zone (Z1), residential zone (Z2), cultural zone (Z3), and industrial zone (Z4) according to whether there was a flourishing business street, a concentrated residential block, a local famous university, industrial park in the corresponding zone, respectively. The classification results are shown in Table 1.

Table 1. Eleven air quality monitoring sites in Shenyang, Liaoning Province. Height refers to the height of the sampling port from the ground. The number in the brackets is the percentage of land use area within 3 km buffer.

Number	Name (abb.)	Height (m)	Main Land use Types and Typical Feature Sites within 3 km Buffer						Ecological Functional Zone (abb.)
			Main Land Use Types	Commercial Block	Industrial Park	Local Universities	Concentrated Residential Blocks	Natural Reserve	
1	Taiyuanjie (TYS)	10	Buildings (69%)	√					Residential–commercial zone (Z1)
2	Xiaoheyan (XHY)	10	Buildings (76%)				√		Residential zone (Z2)
3	Wenhualu (WHR)	20	Buildings (54%)				√		Residential zone (Z2)
4	Lingdongjie (LDS)	10	Buildings (66%), green land (19%)			√			Cultural zone (Z3)
5	Senlinlu (SLR)	5	Green land (99%)					√	Natural reserve zone (Z5)
6	Donglinglu (DLR)	10	Buildings (26%), green land (35%)			√			Cultural zone (Z3)
7	Xinxiujie (XXS)	20	Buildings (53%)		√				Industrial zone (Z4)
8	Jingshenjie (JSS)	20	Buildings (49%), green land (15%)	√					Residential–commercial zone (Z1)
9	Hunnandonglu (HNDR)	20	Buildings (45%), green land (12%)	√					Residential–commercial zone (Z1)
10	Shenliaoxilu (SLXR)	20	Buildings (66%)		√				Industrial zone (Z4)
11	Yunonglu (YNR)	6	Green land (52%)			√			Cultural zone (Z3)

3. Results and Discussion

3.1. Diurnal Variations in Four Major Air Pollutants among Ecological Functional Zones

The hourly air pollutant concentrations in all five ecological functional zones are shown in Figure 2. Hourly $PM_{2.5}$ concentrations were lowest in the natural reserve zone (Z5), and were 4.96–19.47 $\mu g{\cdot}m^{-3}$ lower than the concentrations found in other zones (Figure 2a,b). $PM_{2.5}$ concentrations were consistently higher in the industrial zone (Z4) relative to the other zones. Feizizadeh et al. [28] also indicated that the highest PM concentration occurred in the petrochemical industrial site of Tabriz of Iranian. Among the four anthropogenic zones, the cultural zone (Z3) had the lowest levels of all pollutants. The variations in pollutant concentrations among the zones indicated that pollution was more problematic and variable in zones that contained a source of anthropogenic emissions (i.e., Zones 1–4) than in the zone without an anthropogenic emissions source (Z5).

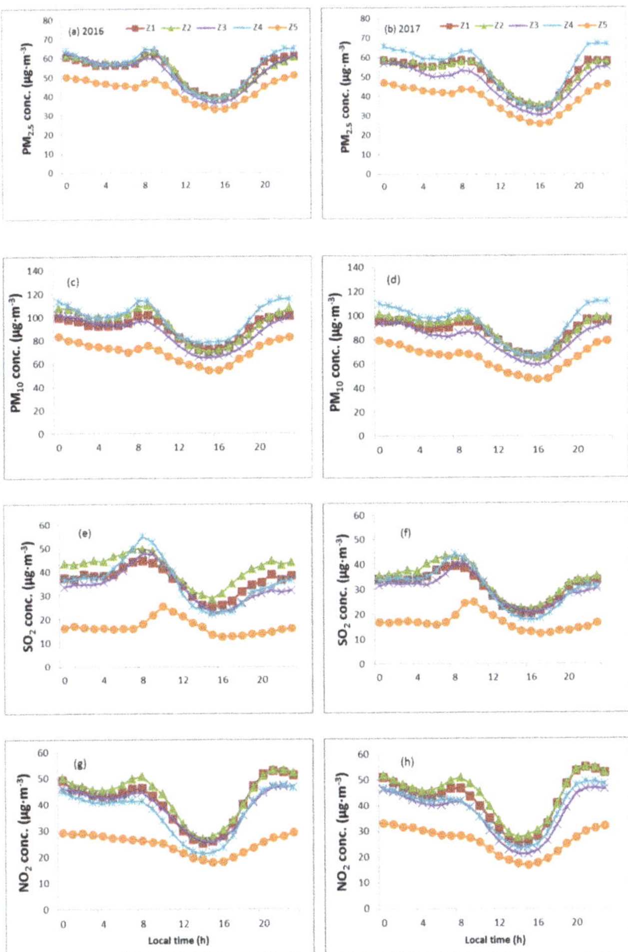

Figure 2. Average hourly concentrations of (**a**) $PM_{2.5}$ in 2016, (**b**) $PM_{2.5}$ in 2017, (**c**) PM_{10} in 2016, (**d**) PM_{10} in 2017, (**e**) SO_2 in 2016, (**f**) SO_2 in 2017, (**g**) NO_2 in 2016, and (**h**) NO_2 in 2017 in five ecological functional zones in Shenyang.

Average hourly $PM_{2.5}$ concentration exhibited a bimodal distribution in all five zones in both 2016 and 2017 (Figure 2a,b). This is consistent with observations in other cities in Northeast China such as Siping and Anshan, but not with cities in the south such as Changde and Panyu [16]. Peaks in $PM_{2.5}$ concentrations were less obvious in Z5 relative to the other zones, but peak values in all zones occurred between 08:00 and 09:00. Lows (valleys) in concentrations were generally observed at 15:00 in 2016 and at 16:00 in 2017 in all zones. These hours correspond to the variation in the diurnal boundary layer and to morning and evening rush hours [29]. The trends in hourly PM_{10} concentration across all zones in both years were consistent with those observed for $PM_{2.5}$, excluding some small local fluctuations (Figure 2c,d).

Among all zones, average diurnal SO_2 concentrations were highest in the residential zone (Z2), excluding a few hours in the morning when the average concentration was highest in Z4 (Figure 2e,f), which suggested that residential heating and industrial emission were the SO_2 main pollution source. Again, Z5 had the lowest concentrations among all zones. The greatest fluctuations in SO_2 concentrations were observed in Z4, where values ranged from 12.81 to 55.05 $\mu g{\cdot}m^{-3}$ in 2016 and 12.30 to 44.38 $\mu g{\cdot}m^{-3}$ in 2017. It means that the industrial emission affects significantly SO_2 concentrations.

Pronounced diurnal variations in SO_2 concentration were observed in all five zones. Zones with sources of anthropogenic emissions had similar trends, with two peaks and one valley. This is different from most cities and regions of China such as in Lanzhou [40], Guangzhou [41], and Sichuan Basin [42] where there is only one peak for variation of diurnal SO_2 concentrations. However, in Z5, the only zone without a source of anthropogenic emissions, only one obvious peak was observed (Figure 2e,f). The above variation characteristics mean that SO_2 pollution situation is complex in Shenyang that it is an old and famous base for heavy industry. In Zones 1–4, the first peak in SO_2 concentrations occurred at 08:00, whereas concentrations peaked at 10:00 in Z5. This suggests that SO_2 emitted in urban areas (Zones 1–4) takes up to 2 h to spread to suburban areas (Z5).

The average hourly variations in NO_2 showed a similar pattern to PM concentrations (Figure 2g,h), including the timing of peaks and valleys. However, the greatest NO_2 concentrations were observed in Z2 due to high vehicle traffic with increasing private cars, as opposed to Z4 for PM concentrations. There was no observed spike in NO_2 concentrations during the day in Z5, likely because Z5 has very low vehicle traffic.

3.2. Seasonal Variations in Four Major Air Pollutants among Ecological Functional Zones

Average seasonal $PM_{2.5}$ concentrations were highest in Z4 (Figure 3a), suggesting that industrial emissions were the main source of $PM_{2.5}$ in Shenyang. Z3 had the lowest concentrations among zones with sources of anthropogenic emissions. In 2016, concentrations were highest in Z4 in summer and autumn, and in the residential–commercial zone (Z1) in spring and winter. Concentrations were substantially lower in Z5 than in all other zones over both years. We observed differences between zones with sources of anthropogenic emissions; the within-season variations were relatively small in 2016 (differences < 5.7 $\mu g{\cdot}m^{-3}$) and greater in 2017 (differences < 10.3 $\mu g{\cdot}m^{-3}$).

Mean seasonal $PM_{2.5}$ concentrations were highest in winter (mean ± SE, 71.86 ± 12.47 $\mu g{\cdot}m^{-3}$ in 2016 and 70.61 ± 11.58 $\mu g{\cdot}m^{-3}$ in 2017) and lowest in summer (33.27 ± 5.68 $\mu g{\cdot}m^{-3}$ in 2016 and 28.28 ± 4.24 $\mu g{\cdot}m^{-3}$ in 2017), which is related to the boundary layer height. The higher boundary layer height in summer is favorable to pollutants dispersion, while the opposite is true in winter [43]. The difference in mean concentration between winter and summer was 23.39 $\mu g{\cdot}m^{-3}$ in 2016 and 16.48 $\mu g{\cdot}m^{-3}$ in 2017, indicating a large range in $PM_{2.5}$ concentrations among seasons. $PM_{2.5}$ concentrations decreased between years for all seasons excluding spring, where it increased by an average of 5.65 $\mu g{\cdot}m^{-3}$ in 2017. The seasonal variations and patterns in PM_{10} concentrations across all seasons, years, and zones were similar to those observed for $PM_{2.5}$ (Figure 3b).

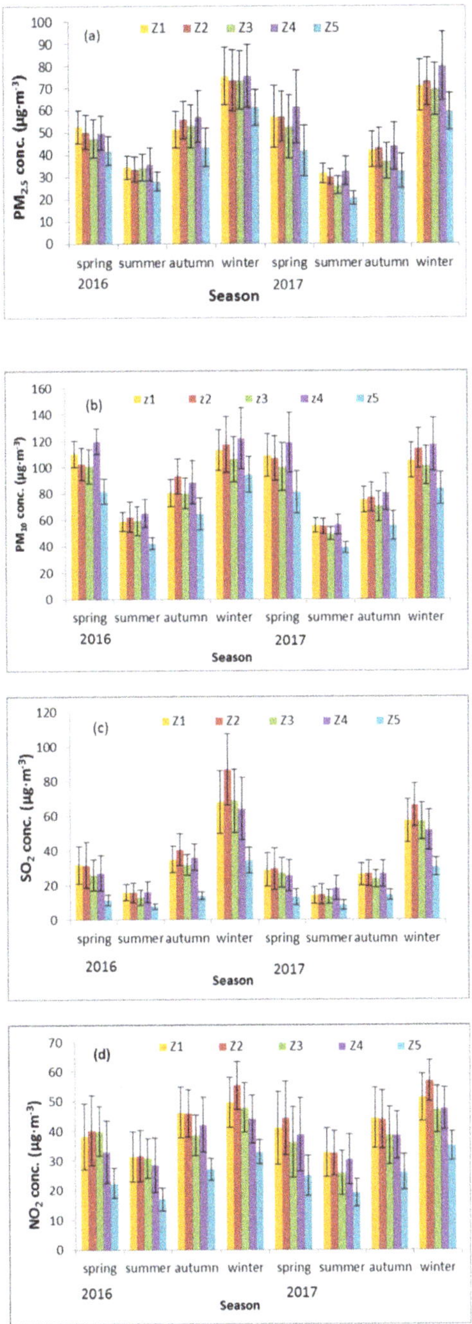

Figure 3. Mean (± SE) seasonal concentrations of (**a**) $PM_{2.5}$, (**b**) PM_{10}, (**c**) SO_2, and (**d**) NO_2 in five ecological functional zones in Shenyang from January 2016 to December 2017.

Average seasonal SO_2 concentrations were highest in Z2, excluding the spring and summer of 2016, and consistently lowest in Z5 across both years (Figure 3c). Concentrations were highest in winter (2016: 73.70 ± 15.32 $\mu g \cdot m^{-3}$; 2017: 58.48 ± 9.70 $\mu g \cdot m^{-3}$) and more than double the average concentrations observed in spring and autumn (Figure 3c). SO_2 concentrations were lowest in the summer of both years (2016: 14.08 ± 4.22 $\mu g \cdot m^{-3}$; 2017: 14.13 ± 4.63 $\mu g \cdot m^{-3}$). Higher SO_2 concentrations in winter are a result of coal combustion during the heating season, which is consistent with most cities of north China [14]. The local government stipulates that each household must have an indoor temperature of at least 18 °C from November 1 to March 31 in Shenyang.

The variations in mean seasonal NO_2 concentrations were similar to those observed for PM and SO_2, i.e., highest in winter and lowest in summer (Figure 3d). In both years, the highest NO_2 concentrations were found in Z2 during winter. Concentrations were often high in Z2; this is potentially related to an increase in the number of family cars owned by residents of Shenyang in recent years (https://tieba.baidu.com/p/4936711038?red_tag=2949498218).

Figure 4 presents the hourly variation coupled with seasonal variation for all five zones and four air pollution components over both study years. Z5 had the lowest concentrations of all air pollutants. Z1 and Z2 showed similar concentrations across all time scales. Trends in the diurnal variation of $PM_{2.5}$ were similar among zones (Figure 4a,b). Peaks in $PM_{2.5}$ mostly occurred between 07:00 and 09:00 and between 23:00 and 00:00, excluding the winter and spring of 2017. Valleys occurred late at night and in the afternoon at approximately 14:00. Overall, hourly concentrations were smoother in the summer relative to the other seasons, which had sharp increases and decreases. Excluding spring, variations in PM_{10} (Figure 4c,d) concentrations were consistent with those of $PM_{2.5}$.

Z2 had the highest hourly SO_2 concentrations, especially during winter, when the maximum concentration reached 137.78 $\mu g \cdot m^{-3}$ (Figure 4e,f). Mean hourly SO_2 concentrations followed a bimodal pattern in Zones 1–4 during both spring and winter, but no discernable trend was observed in autumn. The first peak in the bimodal trend occurred at 08:00 and the second at 00:00. During summer, SO_2 concentrations differed most among the four zones with a source of anthropogenic emissions between 09:00 and 20:00.

Mean hourly NO_2 concentrations were consistently highest in Z2 in winter (Figure 4g,h). Concentrations showed a bimodal distribution in all seasons (excluding Z5 in winter), with the highest values occurring in the first peak in spring, autumn, and winter, and relatively similar peak concentrations in summer.

Broadly, most hourly air pollutant concentrations showed a bimodal distribution, with peaks during morning rush hour (09:00) and between 23:00 and 00:00. The lowest values were observed in the afternoon around 16:00. Peaks and valleys in concentrations are related to human activities and the height of the atmospheric boundary layer [16]. The time elapsed between the morning peak to the afternoon valley was 7 h, as was the gap between the afternoon valley and the midnight peak. However, only 5 h elapsed between the midnight peak and the morning rush hour peak the following day. This suggests that the diffusion lengths of pollutants were approximately 7 and 5 h under increased and decreased anthropogenic activity, respectively. These diffusion lengths were likely related to meteorological conditions. Wang et al. [16] suggested that meteorological factors play a major role in the short-term variation of air pollutant concentrations. In this study, most hourly mean concentration curves followed a unimodal pattern in summer. These findings are consistent with those of Li et al. [29], which suggested that unimodal distributions are linked to high relative humidity and low wind speeds at night in the summer months.

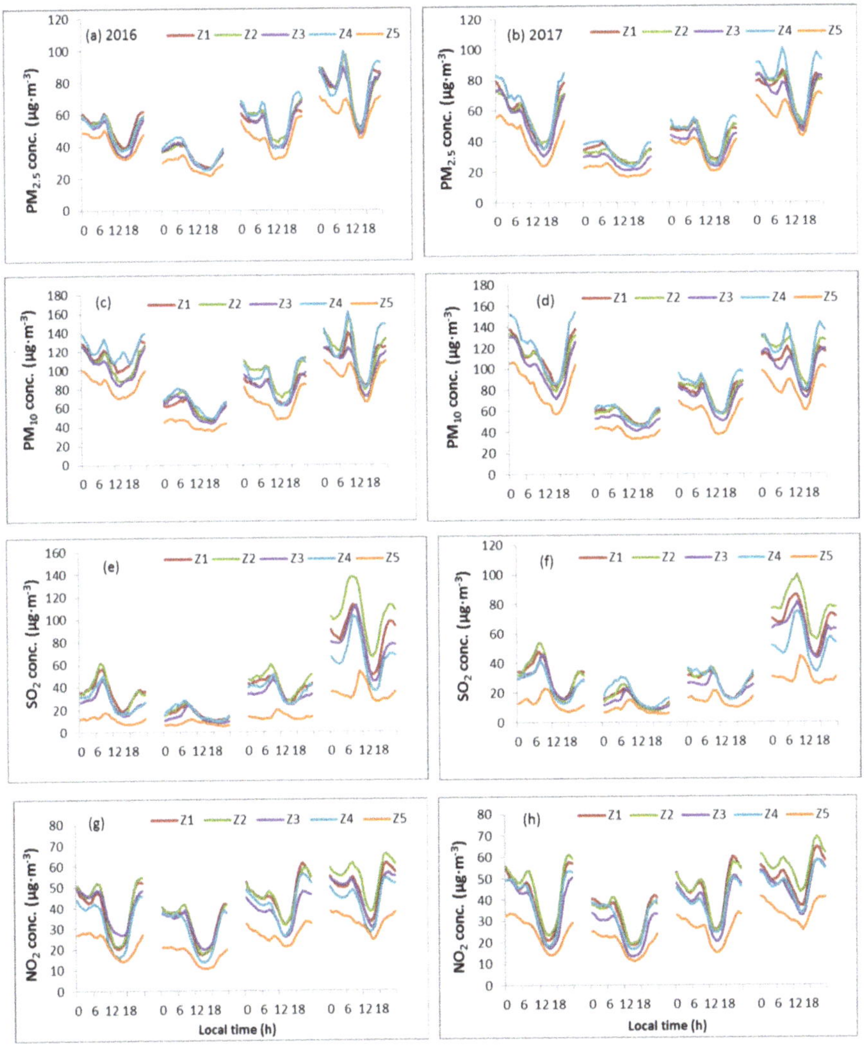

Figure 4. Hourly and season variation in concentrations of (**a**) $PM_{2.5}$ in 2016, (**b**) $PM_{2.5}$ in 2017, (**c**) PM_{10} in 2016, (**d**) PM_{10} in 2017, (**e**) SO_2 in 2016, (**f**) SO_2 in 2017, (**g**) NO_2 in 2016, and (**h**) NO_2 in 2017 across four seasons and five ecological functional zones in Shenyang from January 2016 to December 2017.

3.3. Annual Variations in Four Major Air Pollutants among Ecological Functional Zones

The mean annual $PM_{2.5}$ concentrations in Shenyang were 53.18 ± 7.76 $\mu g \cdot m^{-3}$ in 2016 and 48.18 ± 8.83 $\mu g \cdot m^{-3}$ in 2017. In 2016, among all five ecological functional zones, the highest $PM_{2.5}$ concentration (54.55 ± 9.08 $\mu g \cdot m^{-3}$) was observed in Z4, followed closely by Z1 (53.37 ± 7.71 $\mu g \cdot m^{-3}$) and Z2 (53.31 ± 7.90 $\mu g \cdot m^{-3}$) (Table 2). This pattern was consistent in 2017. The range in $PM_{2.5}$ concentrations was greater in 2017 than 2016 (2016: 33.24–64.93 $\mu g \cdot m^{-3}$; 2017: 25.71–66.87 $\mu g \cdot m^{-3}$).

Overall, $PM_{2.5}$ concentrations decreased significantly from 2016 to 2017 in all five functional zones (Table 2), particularly in Z3 (a decrease of 5.54 $\mu g \cdot m^{-3}$) and Z5 (5.14 $\mu g \cdot m^{-3}$). Only a slight decrease was observed in Z4, the industrial zone (0.15 $\mu g \cdot m^{-3}$). This suggests that the local emission reduction policies such as Implementation plan of Shenyang Blue Sky Defense War work well, including controlling motor vehicle exhaust emissions and coal-fired boilers, prohibiting crop straw burning and so on.

Table 2. Mean (±SE) annual concentrations of four major air pollutants among five ecological functional zones in Shenyang from January 2016 to December 2017. Concentrations are shown in μg·m^{-3}.

Pollutant	Year	Z1	Z2	Z3	Z4	Z5
$PM_{2.5}$	2016	53.37 ± 7.71	53.31 ± 7.90	51.97 ± 8.77	54.55 ± 9.08	43.71 ± 5.72
	2017	50.64 ± 8.76	50.86 ± 8.28	46.43 ± 8.96	54.40 ± 11.47	38.57 ± 6.99
PM_{10}	2016	90.51 ± 9.87	94.07 ± 13.34	86.65 ± 12.51	98.84 ± 13.06	70.59 ± 8.83
	2017	86.31 ± 10.51	88.35 ± 12.15	80.27 ± 11.53	93.34 ± 15.18	64.85 ± 10.23
SO_2	2016	36.11 ± 5.62	41.35 ± 6.05	34.00 ± 7.00	35.90 ± 9.20	16.83 ± 3.23
	2017	30.98 ± 6.03	33.53 ± 6.88	29.85 ± 5.61	30.57 ± 7.83	16.64 ± 3.32
NO_2	2016	41.27 ± 8.67	43.26 ± 8.49	39.26 ± 7.25	36.85 ± 8.71	24.92 ± 3.75
	2017	42.24 ± 9.23	44.33 ± 8.74	36.88 ± 8.82	38.77 ± 8.59	26.23 ± 5.29

The mean hourly SO_2 concentrations in Shenyang were 32.18 ± 5.83 μg·m^{-3} in 2016 and 28.31 ± 5.67 μg·m^{-3} in 2017. Among the five zones, SO_2 concentrations were highest in Z2 (2016: 41.35 ± 6.05 μg·m^{-3}; 2017: 33.53 ± 6.88 μg·m^{-3}), and were very similar among Z1, Z3, and Z4 (Table 2). SO_2 concentrations decreased markedly from 2016 to 2017, particularly in Z2 (a decrease of 7.81 μg·m^{-3}), although concentrations were relatively stable between years in Z5.

The mean annual NO_2 concentrations were 37.11 ± 7.28 μg·m^{-3} in 2016 and 37.69 ± 8.04 μg·m^{-3} in 2017. Z2 had the highest observed concentrations in both years. Generally, NO_2 pollution was severe in Z2 and only slight in Z5. The mean concentrations decreased in Z3 between 2016 and 2017, but increased in the remaining four zones (Table 2).

The observed decreases in major air pollutants in Shenyang, a typical industrial city, are a result of recent emission reduction policies such as the Plan of Resistance to Haze and the Blue Sky Protection Campaign. Wang et al. [44] suggested that PM concentrations are decreasing due to the progress in emissions control made in China. Although our results indicate that industrial emissions can be controlled in Shenyang, NO_2 concentrations increased from 2016 to 2017, a product of the increased use of motor vehicles, which are a source of NO_2. In the past 10 years, government policies have been developed to encourage the purchase of family cars. The number of motor vehicles in Shenyang doubled from 2010 to 2017, and exceeded 2 million by 2017 (https://tieba.baidu.com/p/4936711038?red_tag=2949498218).

3.4. Meteorological Factors and Major Air Pollutants

Meteorological factors play an important role in the diffusion of air pollutants [45–48]. The 11 monitoring stations used in this study were all in close proximity (< 20 km apart), excluding the Senlin Road station (SLR). At this scale, the local microclimate affects pollution concentrations recorded at different stations. Given that no corresponding microclimate data were available for each station, we selected the closest environmental monitoring station (Hunnandong road station), located in Z1, to examine the relationships among meteorological factors and air pollutant concentrations.

The relationships between visibility and mean daily PM_{10}, $PM_{2.5}$, and NO_2 concentrations were best represented by negative exponential curves (Figure 5). However, the relationship between visibility and the mean daily SO_2 concentration was linear, with a correlation coefficient of 0.512 (Figure 5c). Li et al. [29] indicated that fine particles can strongly constrain visibility via the atmospheric extinction effect, which is supported by these results (Figure 5d).

The relationships between pollutant concentrations and wind speed were similar to those with visibility, but the correlation coefficients (a measure of fit) of these relationships were weaker (Table 3). The relationships between pollutant concentrations and relative humidity were best represented by binomial curves, suggesting complex relationships. Wang et al. [49] reported that the contribution of relative humidity to visibility becomes increasingly important with increasing $PM_{2.5}$ concentrations, suggestive of interactive effects.

Table 3. Correlation coefficients (*R*) of the relationships between four major air pollutants and three meteorological elements ($n = 731$).

Correlation Coefficient	$PM_{2.5}$	PM_{10}	SO_2	NO_2
Wind speed	0.234	0.061	0.166	0.512
Visibility	0.675	0.579	0.475	0.477
Relative humidity	0.202	0.189	0.307	0.236

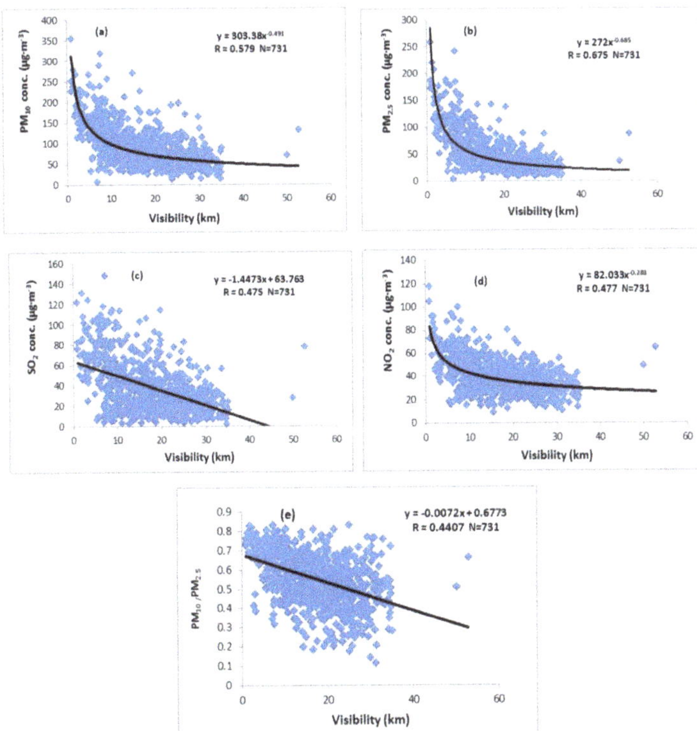

Figure 5. Scatter plots of mean daily (**a**) PM_{10}, (**b**) $PM_{2.5}$, (**c**) SO_2, (**d**) NO_2, and (**e**) PM (PM_{10} and $PM_{2.5}$) concentrations and visibility recorded at the Hunnandonglu meteorological station in Shenyang from 1 January 2016 to 31 December 2017.

3.5. Land-Use Types and Major Air Pollutants

The land-use types of the area around each station were determined by creating circular buffers with a 3-km radius around each of the 11 monitoring stations. We classified the land-use types and calculated the area covered by each type within the buffer area (Figure 6). The station SLR was located in a natural reserve, and the only land-use type around this station was green land (i.e., grassland or cropland). Therefore, we excluded this station from analyses of the relationships between land-use type and air pollutant concentration.

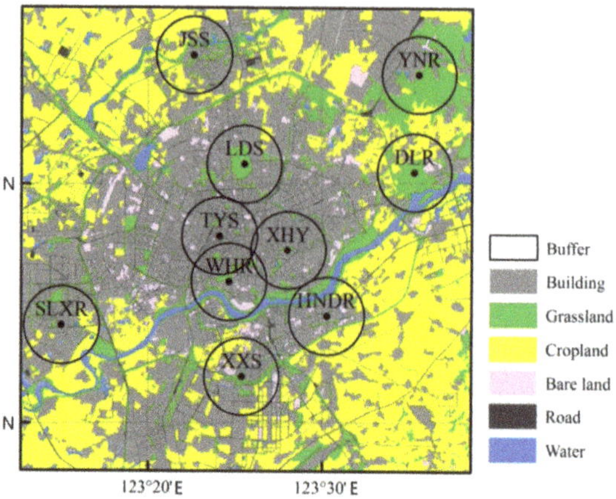

Figure 6. Land-use types surrounding 10 environmental monitoring stations in Shenyang. All buffer areas had a radius of 3 km.

At local scale, anthropogenic activities and their urban distribution play an important role for urban pollution [26]. Junk et al. [50] indicated that land use and its interaction with other factors can influence air quality. Thus, we analyzed the relationship between land-use types and air pollutants. For all stations used in this analysis, an increase in the area of buildings within the buffer area was correlated with higher air pollutant concentrations, regardless of the ecological functional zone. This is a result of a blocking effect, where higher buildings or a greater amount of buildings reduce air pollutant diffusion. There were significant positive correlations ($p < 0.01$) between built-up area and the mean annual concentrations of PM_{10}, $PM_{2.5}$, SO_2, and NO_2 in 2016, with correlation coefficients of 0.546, 0.758, 0.621, and 0.650, respectively (Figure 7).

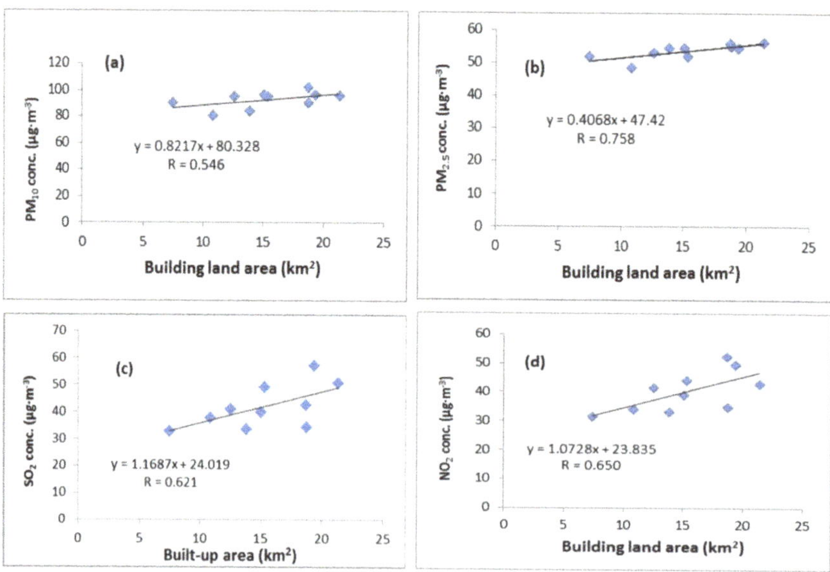

Figure 7. *Cont.*

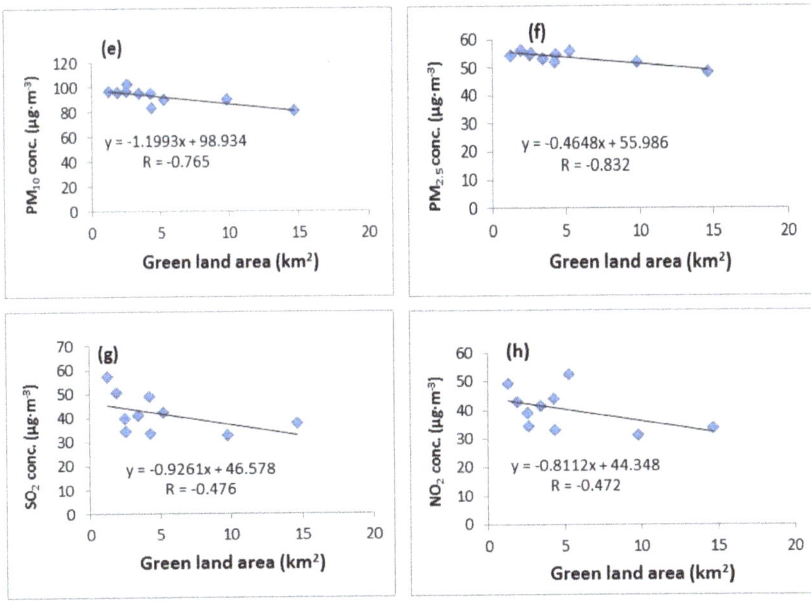

Figure 7. Scatter pots of mean annual concentrations of (**a**) PM_{10}, (**b**) $PM_{2.5}$, (**c**) SO_2, and (**d**) NO_2, and the amount of area covered by buildings in a 3-km-radius buffer, and (**e**) PM_{10}, (**f**) $PM_{2.5}$, (**g**) SO_2, and (**h**) NO_2, and the amount of area covered by green land (i.e., grassland or cropland), for 10 environmental monitoring stations in Shenyang.

In contrast, green land area was significantly negatively correlated with air pollutant concentrations ($p < 0.01$ for all four pollutants), meaning that green land can reduce air pollution to some extent. This is a product of the capacity of plants to absorb air pollutants. The concentrations of air pollutants measured at SLR, located in Z5, were the lowest among all 11 environmental monitoring stations, providing a clear example of this relationship.

4. Conclusions

We categorized 11 environmental monitoring stations into five ecological functional types (zones): mixed commercial–residential (Z1), residential (Z2), cultural (Z3), industrial (Z4), and natural reserve (Z5). Then, we compared hourly, seasonal, and annual concentrations of $PM_{2.5}$, PM_{10}, SO_2, and NO_2 from 1 January 2016 to 31 November 2017, for all five zones. We further determined the relationships between these four air pollutants with meteorological elements and land-use types.

Four of the five zones contained sources of anthropogenic emissions. Among these, the temporal variations in the four air pollutants were similar; however, Z5, which did not contain an emission source, displayed some contrary trends. PM_{10} and $PM_{2.5}$ concentrations were highest in Z4, and SO_2 and NO_2 concentrations were highest in Z2. The lowest observed values for all pollutants were found in Z5. Morning rush hour was associated with peaks in all four pollutants, and lows were observed around 16:00. Pollutant concentrations were typically highest in winter and lowest in summer. Mean annual concentrations of $PM_{2.5}$, PM_{10}, and SO_2 decreased from 2016 to 2017, while NO_2 increased. We suggest that the atmospheric conditions in Shenyang are improving annually in both developed areas and within reserve zones.

Visibility was exponentially and negatively related with PM_{10}, $PM_{2.5}$, and NO_2, as was wind speed, albeit to a weaker degree. Visibility and SO_2 had a linear relationship. Binomial relationships were observed between relative humidity and air pollutants. Government policy has had a significant influence on preventing and controlling air pollution. On a small scale, land-use type also has an important role in the diffusion and transportation of pollutants.

Based on the daily fluctuations in air pollutant concentrations, we suggest that morning and evening exercise may lead to greater pollutant inhalation, which may be unfavorable to human health. Therefore, we propose that 16:00 is the best time to exercise in Shenyang. Future city planning may impact air quality at local scales by increasing the amount of green space within urban areas, and by reducing the height and density of large buildings in urban areas.

Author Contributions: Conceptualization, L.L. and Y.M.; methodology, L.L. and Z.Z.; software, L.L. and H.W.; validation, L.L. and X.L.; formal analysis, L.L., X.L. and Y.M.; investigation, L.L., X.L. and Y.W.; resources, L.L., N.L. and Y.W.; data curation, L.L. and H.W.; writing—original draft preparation, L.L.; writing—review and editing, L.L., Z.Z., H.W., X.L., Y.W., N.L. and Y.M.; project administration, L.L. and Y.M. All authors have read and agreed to the published version of the manuscript.

Funding: This research was funded by the Key Project of the National Natural Science Foundation of China (grant number: 41730647).

Acknowledgments: We express our gratitude to the Shenyang Environmental Monitoring Center for providing us with air quality monitoring data. The English in this document has been checked by two professional editors, both native speakers of English.

Conflicts of Interest: The authors declare no conflict of interest.

References

1. Adachi, K.; Tainosho, Y. Characterization of heavy metal particles embedded in tire dust. *Environ. Int.* **2004**, *30*, 1009–1017. [CrossRef] [PubMed]
2. Kampa, M.; Castanas, E. Human health effects of air pollution. *Environ. Pollut.* **2008**, *151*, 362–367. [CrossRef] [PubMed]
3. Richardson, E.A.; Pearce, J.; Tunstall, H.; Mitchell, R.; Shortt, N.K. Particulate air pollution and health inequalities: A Europe-wide ecological analysis. *Int. J. Health Geogr.* **2013**, *12*, 1–10. [CrossRef] [PubMed]
4. Chen, J.; Zhao, C.S.; Ma, N.; Liu, P.F.; Göbel, T.; Hallbauer, E.; Deng, Z.Z.; Ran, L.; Xu, W.Y.; Liu, H.J.; et al. A parameterization of low visibilities for hazy days in the North China Plain. *Atmos. Chem. Phys.* **2012**, *12*, 4935–4950. [CrossRef]
5. Fan, G.F.; Ma, H.; Zhang, X.W.; Liu, Y. Impacts of relative humidity and $PM_{2.5}$ concentration on atmospheric visibility: A comparative study of hourly observations of multiple stations. *Acta Meteor. Sinica.* **2016**, *74*, 959–973.
6. Liu, Z.; Hu, B.; Wang, L.; Wu, F.; Gao, W.; Wang, Y. Seasonal and diurnal variation in particulate matter (PM_{10} and $PM_{2.5}$) at an urban site of Beijing: Analyses from a 9-year study. *Environ. Sci. Pollut. R.* **2015**, *22*, 627–642. [CrossRef]
7. Yu, M.F.; Zhu, Y.; Lin, C.J.; Wang, S.X.; Xing, J.; Jang, C.; Huang, J.Z.; Huang, J.Y.; Jin, J.B.; Yu, L. Effects of pollution control measures on air quality improvement in Guangzhou, China. *J. Environ. Man.* **2019**, *244*, 127–137. [CrossRef] [PubMed]
8. He, B.-J. Potentials of meteorological characteristics and synoptic conditions to mitigate urban heat island effects. *Urban Clim.* **2018**, *24*, 26–33. [CrossRef]
9. Huang, J.; Wang, T.; Wang, W.; Li, Z.Q.; Yan, H.R. Climate effects of dust aerosols over East Asian arid and semiarid regions. *J. Geophys. Res.* **2014**, *119*, 11398–11416. [CrossRef]
10. Bapna, M. Toward an environmentally sustainable future: Country environmental analysis of the People's Republic of China. The Asian Development Bank: Manila, Philippines, 2012.
11. Zheng, S.; Wang, M.; Wang, S.; Tao, Y.; Shang, K. Short-term effects of gaseous pollutants and particulate matter on daily hospital admissions for cardio-cerebrovascular disease in Lanzhou: Evidence from a heavily polluted city in China. *Inter. J. Environ. Res. Public Health* **2013**, *10*, 462–477. [CrossRef]
12. Chai, F.H.; Gao, J.; Chen, Z.X.; Wang, S.L. Spatial and temporal variation of particulate matter and gaseous pollutants in 26 cities in China. *J. Environ. Sci.* **2014**, *26*, 75–82. [CrossRef]
13. Chen, Y.; Xie, S. Characteristics and formation mechanism of a heavy air pollution episode caused by biomass burning in Chengdu, southwest China. *Sci. Total Environ.* **2014**, *473–474*, 507–517. [CrossRef] [PubMed]

14. Zhao, S.P.; Yu, Y.; Yin, D.Y.; He, J.J.; Qu, J.J.; Xiao, J.H. Annual and diurnal variations of gaseous and particulate pollutants in 31 provincial capital cities based on in situ air quality monitoring data from China National Environmental Monitoring Center. *Environ. Int.* **2016**, *86*, 92–106. [CrossRef] [PubMed]
15. Lv, B.L.; Cai, J.; Xu, B.; Bai, Y. Understanding the rising phase of the $PM_{2.5}$ concentration evolution in large China cities. *Sci. Rep.* **2017**, *7*, 46456. [CrossRef]
16. Wang, Y.Q.; Zhang, X.Y.; Sun, J.Y.; Zhang, X.C.; Che, H.Z.; Li, Y. Spatial and temporal variations of the concentrations of PM_{10}, $PM_{2.5}$ and $PM_{1.0}$ in China. *Atmos. Chem. Phys.* **2015**, *15*, 13585–13598. [CrossRef]
17. Zhang, Y.L.; Cao, F. Fine particulate matter ($PM_{2.5}$) in China at a city level. *Sci. Rep.* **2015**, *5*, 14884. [CrossRef]
18. Zhang, H.; Wang, Y.; Hu, J.; Ying, Q.; Hu, X.M. Relationships between meteorological parameters and criteria air pollutants in three megacities in China. *Environ. Res.* **2015**, *140*, 242–254. [CrossRef]
19. Kota, H.S.; Guo, H.; Myllyvirta, L.; Hu, J.; Sahu, S.K.; Garage, R.; Ying, Q.; Gao, A.; Dahiya, S.; Wang, Y.; et al. Year-long simulation of gaseous and air particulate in India. *Atmos. Environ.* **2018**, *180*, 244–255. [CrossRef]
20. Meng, X.Y.; Wu, Y.P.; Pan, Z.H.; Wang, H.; Yin, G.; Zhao, H.G. Seasonal characteristics and particle-size distributions of particulate air pollutants in Urumqi. *Int. J. Environ. Res. Pubic Health* **2019**, *16*, 396. [CrossRef]
21. Zhao, H.; Che, H.; Zhang, X.; Ma, Y.; Wang, Y.; Wang, H.; Wang, Y. Characteristics of visibility and particulate matter (PM) in an urban area of Northeast China. *Atmos. Pollut. Res.* **2013**, *4*, 427–434. [CrossRef]
22. Bhaskar, B.V.; Mehta, V.M. Atmospheric particulate pollutants and their relationship with meteorology in Ahmedabad. *Aerosol Air Qual. Res.* **2010**, *10*, 301–315. [CrossRef]
23. Cheng, Y.; He, K.B.; Du, Z.Y.; Zheng, M.; Duan, F.K.; Ma, Y.L. Humidity plays an important role in the $PM_{2.5}$ pollution in Beijing. *Environ. Pollut.* **2015**, *197*, 68–75. [CrossRef] [PubMed]
24. Fallmann, J.; Forkel, R.; Emeis, S. Secondary effects of urban heat island mitigation measures on air quality. *Atmos. Environ.* **2016**, *125*, 199–211. [CrossRef]
25. Malek, E.; Davis, T.; Martin, R.S.; Silva, P.J. Meteorological and environmental aspects of one of the worst national air pollution episodes (January, 2004) in Logan, Cache Valley, Utah, USA. *Atmos. Res.* **2006**, *79*, 108–122. [CrossRef]
26. Ulpiani, G. On the linkage between urban heat island and urban pollution island: Three-decade literature review towards a conceptual framework. *Sci. Total Environ.* **2020**. [CrossRef]
27. Romero, H.; Ihl, M.; Rivera, A.; Zalazar, P.; Azocar, P. Rapid urban growth, land-use changes and air pollution in Santiago, Chile. *Atmos. Environ.* **1999**, *33*, 4039–4047. [CrossRef]
28. Feizizadeh, B.; Blaschke, T. Examining urban heat island relations to land use and air pollution: Multiple endmemeber spectral mixture analysis or thermal remote sensing. *IEEE J. Sel. Top. Appl. Earth Obs. Remote Sens.* **2013**, *6*, 1749–1756. [CrossRef]
29. Li, X.L.; Ma, Y.J.; Wang, Y.F.; Liu, N.W.; Hong, Y. Temporal and spatial analyses of particulate matter (PM_{10} and $PM_{2.5}$) and its relationship with meteorological parameters over an urban city in northeast China. *Atmos. Res.* **2017**, *198*, 185–193. [CrossRef]
30. Rendón, A.M.; Salazar, J.F.; Palacio, C.A.; Wirth, V.; Brötz, B. Effects of urbanization on the temperature inversion breakup in a mountain valley with implications for air quality. *J. Appl. Meteorol. Climatol.* **2014**, *53*, 840–858.
31. Rendón, A.M.; Salazar, J.F.; Palacio, C.A.; Wirth, V. Temperature inversion breakup with impacts on air quality in urban valleys influenced by topographic shading. *J. Appl. Meteorol. Climatol.* **2015**, *54*, 302–321.
32. Ma, Y.J.; Wang, Y.F.; Liu, N.W.; Hong, Y. Analysis of inhalable particulate matter: PM_{10} and $PM_{2.5}$ pollution levels in grouped cities of central Liaoning province. *China Power Sci. Technol.* **2010**, *16*, 9–13.
33. Zhao, H.J.; Che, H.Z.; Ma, Y.J.; Xia, X.G.; Wang, Y.F.; Wang, P.; Wu, X.C. Temporal variability of the visibility, particulate matter mass concentration and aerosol optical properties over an urban site in Northeast China. *Atmos. Res.* **2015**, *166*, 204–212. [CrossRef]
34. Li, L.G.; Zhou, D.P.; Wang, Y.F.; Hong, Y.; Cui, J.; Jiang, P. An observational study of atmospheric ice nuclei number concentration during three fog-haze weather periods in Shenyang, northeastern China. *Atmos. Res.* **2017**, *188*, 11–19. [CrossRef]
35. Tao, Z.H.; Xie, S.Q.; He, W.N.; Yu, B.B.; Fang, C.; Ge, L.L.; Li, W.; Wang, Q.L.; Wang, X.Q. Pollution characteristics of $PM_{2.5}$ in Taizhou, Zhejiang Province. *J. Zhejiang Univ.* **2017**, *44*, 464–471.
36. Feng, W.F.; Zhou, J.L.; Zhang, J.S.; Shan, Y.X.; Li, Y.F.; Zhang, Y.H. PM_{10} and $PM_{2.5}$ pollution characteristics in different ecological function zones in Xinyang. *J. Northwest For. Univ.* **2018**, *33*, 269–275.

37. Li, L.G.; Wang, H.B.; Zhao, Z.Q.; Cai, F.; Zhao, X.L.; Xu, S.L. Characteristics of urban heat island (UHI) source and sink areas in urban region of Shenyang. In Proceedings of the 3rd International Workshop on Earth Observation and Remote Sensing Applications, EORSA, Changsha, China, 11–14 June 2014; pp. 62–66.
38. Zhao, Z.; Shen, L.; Li, L.; Wang, H.; He, B.J. Local climate zone classification scheme can also indicate local-scale urban ventilation performance: An evidence-based study. *Atmosphere* **2020**, *11*, 776. [CrossRef]
39. He, B.-J.; Zhao, Z.Q.; Shen, L.D.; Wang, H.B.; Li, L.G. An approach to examining performances of cool/hot sources in mitigating/enhancing land surface temperature under different temperature backgrounds based on landsat 8 image. *Sustain. Cities Soc.* **2019**, *44*, 416–427. [CrossRef]
40. Xie, X.J.; Li, J.; Wang, Z.F. Numerical simulati on of diurnal variations of the air pollutant s in winter of Lanzhou city. *Clim. Environ. Res.* **2010**, *15*, 695–703.
41. Zhang, X.M. Diurnal Variationand Correlation analysis of mass concentration of $PM_{2.5}$ and gaseous pollutants in a district in Guangzhou. *Guangzhou Environ. Sci.* **2017**, *32*, 8–11.
42. Shi, H.Q.; Zeng, S.L.; Li, H.N. Temporal and spatial distribution characteristics and influencing meteorological factors of air pollutants in Sichuan Basin. *Acta Sci. Circumstantiae* **2020**, *40*, 763–778.
43. Jia, J.; Yang, P.Q.; Jiang, H.M. Daily maximum height of the atmospheric boundary layer and its response to air quality over Urumqi city. *J. Meteorol. Environ.* **2019**, *35*, 55–60.
44. Wang, S.X.; Hao, J.X. Air quality management in China: Issues, challenges, and options. *J. Environ. Sci.* **2012**, *24*, 2–13. [CrossRef]
45. Chen, Z.; Cai, J.; Gao, B.; Xu, B.; Dai, S.; He, B.; Xie, X. Detecting the causality influence of individual meteorological factors on local $PM_{2.5}$ concentration in the Jing-Jin-Ji region. *Sci. Rep.* **2017**, *7*, 40735. [CrossRef] [PubMed]
46. Luan, T.; Guo, X.L.; Guo, L.J.; Zhang, T.H. Quantifying the relationship between $PM_{2.5}$ concentration, visibility and planetary boundary layer height for long-lasting haze and fog-haze mixed events in Beijing. *Atmos. Chem. Phys.* **2018**, *18*, 203–225. [CrossRef]
47. Zhang, B.; Jiao, L.M.; Xu, G.; Zhao, S.L.; Tang, X.; Zhou, Y.; Gong, C. Influences of wind and precipitation on different-sized particulate matter concentrations ($PM_{2.5}$, PM_{10}, $PM_{2.5-10}$). *Meteorol. Atmos. Phys.* **2018**, *130*, 383–392. [CrossRef]
48. Zhang, G.; He, B.J.; Dewancker, B.J. The maintenance of prefabricated green roofs for preserving cooling performance: A field measurement in the subtropical city of Hangzhou, China. *Sustain. Cities Soc.* **2020**, *61*, 102314. [CrossRef]
49. Wang, X.; Zhang, R.; Yu, W. The effects of $PM_{2.5}$ concentrations and relative humidity on atmospheric visibility in Beijing. *J. Geophys. Res.* **2019**, *124*, 2235–2259. [CrossRef]
50. Junk, J.; Helbig, A.; Luers, J. Urban climate and air quality in Trier Germany. *Int. J. Biometeorol.* **2003**, *47*, 230–238. [CrossRef]

atmosphere

MDPI

Article

Local Climate Zone Classification Scheme Can Also Indicate Local-Scale Urban Ventilation Performance: An Evidence-Based Study

Ziqi Zhao [1], Lidu Shen [2,*], Liguang Li [1], Hongbo Wang [1] and Bao-Jie He [3,*]

1 Institute of Atmospheric Environment, China Meteorological Administration, Shenyang 110166, China; ziqizhao@yahoo.com (Z.Z.); liliguangyjs@foxmail.com (L.L.); hb_wang@foxmail.com (H.W.)

2 Regional Climate Centre of Shenyang, Liaoning Meteorological Bureau, Shenyang 110166, China

3 Faculty of Built Environment, University of New South Wales, Sydney 2052, NSW, Australia

* Correspondence: shenlidu052@foxmail.com (L.S.); baojie.he@unsw.edu.au (B.-J.H.)

Received: 27 June 2020; Accepted: 21 July 2020; Published: 23 July 2020

Abstract: Studies on urban ventilation indicate that urban ventilation performance is highly dependent on urban morphology. Some studies have linked local-scale urban ventilation performance with the local climate zone (LCZ) that is proposed for surface temperature studies. However, there is a lack of evidence-based studies showing LCZ ventilation performance and affirming the reliability of using the LCZ classification scheme to demonstrate local-scale urban ventilation performance. Therefore, this study aims to analyse LCZ ventilation performances in order to understand the suitability of using the LCZ classification scheme to indicate local-scale urban ventilation performance. This study was conducted in Shenyang, China, with wind information at 16 weather stations in 2018. The results indicate that the Shenyang weather station had an annual mean wind speed of 2.07 m/s, while the mean wind speed of the overall 16 stations was much lower, only 1.44 m/s in value. The mean wind speed at Shenyang weather station and the 16 stations varied with seasons, day and night and precipitation conditions. The spring diurnal mean wind was strong with the speeds of 3.56 m/s and 2.21 m/s at Shenyang weather station and the 16 stations, respectively. The wind speed (2.21 m/s at Shenyang weather station) under precipitation conditions was higher than that (1.75 m/s at Shenyang weather station) under no precipitation conditions. Downtown ventilation performance was weaker than the approaching wind background, where the relative mean wind speed in the downtown area was only 0.53, much less than 1.0. The downtown ventilation performance also varied with seasons, day and night and precipitation conditions, where spring diurnal downtown ventilation performance was the weakest and the winter nocturnal downtown ventilation performance was the strongest. Moreover, the annual mean wind speed of the 16 zones decreased from the sparse, open low-rise zones to the compact midrise zones, indicating the suitability of using LCZ classification scheme to indicate local-scale urban ventilation performance. The high spatial correlation coefficients under different seasons, day and night and precipitation conditions, ranging between 0.68 and 0.99, further affirmed that LCZ classification scheme is also suitable to indicate local-scale urban ventilation performance, despite without the consideration of street structure like precinct ventilation zone scheme.

Keywords: urban ventilation performance; local climate zone; evidence-based analysis; downtown ventilation performance; relative mean wind speed

1. Introduction

Because of anthropogenic activities and global climate change [1], cities are experiencing a variety of environmental problems, such as urban warming [1,2], air and water pollution [3,4] and urban flooding [5,6]. Such critical problems have significant human health, social and economic consequences, such as increases in morbidity and mortality, reduction of economic productivity and increases in energy and water consumption [4,7]. Nevertheless, the urbanization remains a rapid upward trend [8], which means the intensity of anthropogenic influences on the urban environment will be intensified [1]. Meanwhile, the global climate change cannot be ceased in the current era and it can only be decelerated in the next several decades under a promising condition. Through this vision, environmental problems in cities will be inevitably aggravated.

Multi-dimensional efforts have been making to address environmental problems in cities. It is, for example, advocated to improve urban energy efficiency through utilising renewable energy and shortening the working distance for alleviating vehicle dependence [9]. Meanwhile, there is also a consensus in the improvement of green and blue infrastructures that play versatile roles in addressing local warming, urban flooding, air pollution and physiological, psychological and physical problems [2,6,10]. Moreover, urban texture and urban morphology have also been an important focus [11–16]. The urban texture, e.g., construction materials, can significantly affect urban temperature and urban flooding, through the interaction with solar radiation and heat transfer, as well as the interaction with soil moisture and evaporation [17,18]. Urban morphology which presents the typology of human settlement and the process of forming so has a close relationship with wind and solar radiation [14,16].

Understanding the relationship between urban morphology and wind environment is helpful in understanding the urban ventilation performance that indicates the capacity of wind flows out and into the urban canopy layer. Urban ventilation performance can be assessed by various indicators (e.g., wind velocity, wind velocity ratio, the age of air, pollutant concentration, purging flow rates) for demonstrating the capability of wind in dealing with different problems, such as urban warming mitigation, wind energy potential, pedestrian wind comfort air pollution alleviation and building energy efficiency [16,19–22]. Various studies have been carried out to investigate wind field around buildings [23,24], airflow mechanism in street canyons [25,26], wind environment in blocks [27–29], neighbourhoods and precincts [16,30,31] and urban ventilation corridors in cities [32,33]. In particular, analysing the relationship between local-scale (e.g., neighbourhoods, precincts) urban morphology and wind environment is increasingly important, as it can practically inform local governments, city authorities and developers to perform proper planning and design activities for resilient, healthy and sustainable communities [34].

However, existing studies are not sufficient to support the comprehensive understandings of the relationship between local-scale urban morphology and wind. For instance, Blocken et al. [35] conducted a study in the context of an education precinct in the Netherlands based on numerical simulation to understand its implications for wind comfort and wind safety. Antoniou et al. [36] investigated the wind environment of a compact precinct consisting of low-rise and midrise buildings in Nicosia, Cyprus. Priyadarsini et al. [37] investigated the wind environment in a CBD area of Singapore and pointed out the implications of wind for UHI mitigation that a 35% increase in wind speed can lead to a temperature reduction by 0.7 °C. In order to enhance the cooling potential of sea breeze, Rajagopalan et al. [38] analysed the variation in wind speed with the urban geometry in the Muar region of Malaysia.

Such studies are case-specific, while the local-scale urban morphology is quite complex because of the combination of morphological factors at micro and local scales [16,39]. To overcome this challenge, He et al. [16] proposed an innovative precinct (e.g., neighbourhood, community) morphology classification scheme (also called precinct ventilation zone) that is built upon the urban morphological characteristics of building height, compactness and street structure for characterizing precinct ventilation. Moreover, they applied the precinct morphology classification scheme into the Greater Sydney region of Australia for case studies [40–42]. They further compared the precinct ventilation performance in the compact high-rise [40], open midrise [41] and open low-rise precincts [42] with gridiron street structures. Another stream of studies on local-scale ventilation performance is based on the local climate zone (LCZ) scheme that is built upon building height and building surface fraction. For instance, Yang et al. [21] assessed the ventilation performance of distinct local climate zones in the megacity of Shanghai through analysing the frontal area index of different local climate zones. Zhou et al. [43] examined if LCZ scheme could be adopted for understanding the cooling potential of the sea breeze in Sendai, Japan. However, the LCZ scheme is originally developed for land surface temperature studies and it does not consider the potential impact of street structure on local-scale ventilation performance compared with the precinct ventilation zone [16]. Moreover, existing local-scale ventilation studies based on LCZ scheme have only shown ventilation potential or wind cooling potential, while the evidence for validating LCZ ventilation performance has not been given.

Therefore, in order to address this research gap, this study aims to analyse LCZ ventilation performance. This study was conducted in Shenyang, China, by installing weather stations in different local climate zones. This study is of importance to support urban planners and designers to make evidence-based decisions in performing wind-sensitive urban planning and design. The remainder of this paper is structured into several sections. Section 2 introduces the basic information of the case study area of Shenyang, China and the LCZ for analysis. Section 3 presents the data and methods to perform this study. Section 4 analyses and discusses the results and Section 5 concludes this paper.

2. Case Study Area

This study was conducted in Shenyang, the capital city of Liaoning Province, China. Shenyang, the only metropolitan city in Northeast China, has an area of 12,948 km^2 and a city area of about 3500 km^2. Shenyang experiences a temperate semi-humid continental monsoon climate with distinct seasons and large diurnal temperature differences. The spring and autumn seasons of Shenyang are short in the period during which time the temperatures shift rapidly towards the summer and winter seasons, respectively. The annual average temperature of Shenyang is 8.4 °C. It is really hot in summer. July is generally the hottest month with the average highest temperature of 24.5 °C. The extreme temperature in summer can reach 38.4 °C (2 August 2018). Meanwhile, it is also cold in winters and January is the coldest month with the average lowest temperature of −11.5 °C. In addition, statistics indicate Shenyang has an annual precipitation of 680.0 mm, which is mainly in summer, accounting for 63.2% of the annual total amount.

Shenyang has been experiencing rapid economic development and urbanization with the industrialisation. The urban agglomeration in central and southern Liaoning Province, with the core cities of Shenyang and Dalian, has gradually taken its shape. Under such circumstance, the city of Shenyang is facing the problem of high population density, with the population density reaching 10,952, 11,961 and 12,360 people/km^2 in Heping, Shenhe and Huanggu districts, respectively. The high population density has resulted in various issues, such as high building density, low vegetation proportion, air pollution and urban heat island. The wall effect related to the high building density is generally an important cause of urban wind speed reduction, thereby severe air pollution and urban temperature increase [17,44–46]. Accordingly, understanding the wind environment in Shenyang is practically meaningful for building resilient, healthy and comfortable cities.

2.1. Synoptic Wind Background

In the analysis of urban wind environment, He et al. pointed out that urban ventilation performance is a synthetic result of synoptic wind background and the urban morphological characteristics. The synoptic wind background is related to the geo-location of Shenyang city. As illustrated in Figure 1, Shenyang is located in the central part of Liaohe Plain, a large alluvial plain formed by the Bohai Bay in the southwest. In contrast, there are hills and mountains on the east and north borders of Shenyang. Under such conditions, the altitude of Shenyang shows large deviations, ranging between 5 and 441 m and exhibiting a gradually decreasing trend from the northeast to southwest. In summer, southerly and south-westerly winds can prevail, and in winter the wind blows from the north and northeast. In addition, there are hills and mountains in Shenyang so that the local air circulations can be compromised [47].

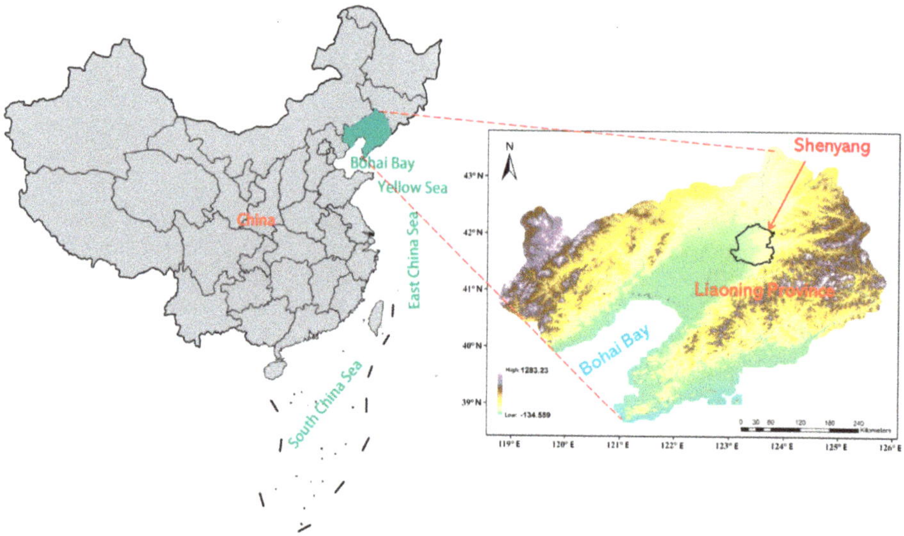

Figure 1. Location of Liaoning province, Shenyang city and the Bohai Bay.

2.2. *Urban Morphological Characteristics*

The urban morphological characteristics are represented by specific LCZ types in Shenyang. The local climate zones are identified [14], through artificial recognition and machine learning. The specific classification process is to first select appropriate training zone samples (google earth image on 2 August 2018, no cloud, 30 m resolution) through visual interpretation. Seventeen types of local climate zones were identified, and each type included 5–25 samples according to actual urban morphological characteristics of Shenyang. Moreover, the training area of each type of local climate zone was larger than 1 km^2, where the side length was longer than 200 m and the urban morphological characteristics were required to be homogeneous over 1 km^2, through these considerations some small areas with heterogeneous or irregular characteristics were excluded in training local climate zones. In addition, a buffer zone with a width of 100 m was reserved between different LCZ training zones in order to avoid uncertain recognition of boundaries. Finally, the LCZ distribution in Shenyang was generated, as illustrated in Figure 2.

Afterwards, sixteen zones, within each area a set of automatic anemometers was mounted and maintained by Liaoning Meteorological Bureau, were selected as the case study areas, as demonstrated in Figure 2. Locations of 16 local climate zones were different, which may show different external (local) wind background. A detailed description of the 16 local climate zones presented in Figure 2 is further provided in Table 1. The LCZ types of the 16 zones included LCZ-2, LCZ-2E, LCZ-2G, LCZ-4G, LCZ-5, LCZ-4,5 and LCZ-9. Such situations were caused by the limited LCZ types in the core area of Shenyang and the constraints of available sensors. Nevertheless, these LCZ types were sufficient to support the analysis of urban ventilation performance of different local climate zones under the synthetic influence of external wind background and urban morphological characteristics [16]. The 16 stations consist of one national weather station (Shenyang weather station) and 15 regional automatic weather stations. The Station 1 (its surrounding area is classified as LCZ-9) is Shenyang Meteorological Station of China Meteorological Administration. Shenyang weather station has been established for more than 30 years and was re-located for three times because of the change of surroundings. The latest re-location took place in 2010, and the new site lied within the suburban area with the sparsely built, low structural landscape. On the other hand, the other 15 stations were gradually deployed by local regional meteorological administration since 2016, and their locations were mainly within the downtown Shenyang (not re-located since then). These locations were selected not only for the maintenance, but also for not being close to the neighbourhood with large structural heterogeneity.

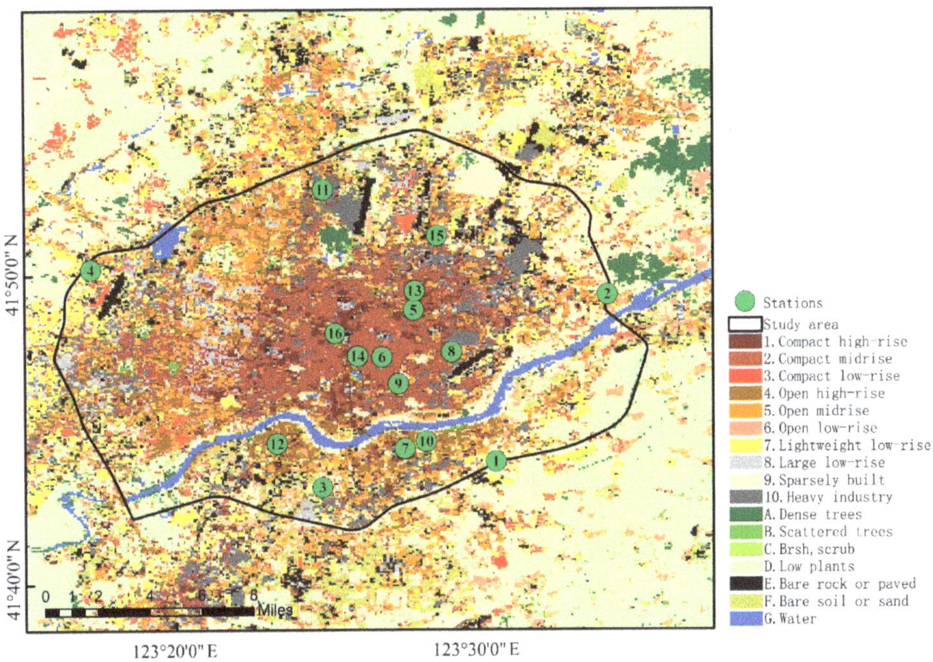

Figure 2. Local climate zone (LCZ) types in Shenyang and the location of 16 case study area.

Table 1. A description of the LCZ types and images of case study areas.

No.	LCZ (image)	Description	No.	LCZ (image)	Description
1	**LCZ-9** **Sparsely built** 	A few low-rise (1–3 stories) buildings; Mixed species of deciduous trees, shrubs and grasslands; Permeable but with a 30-m asphalt-paved surface.	2	**LCZ-5** **Open midrise** 	Large-scale stadium buildings and open midrise (3–9 stories) buildings; A few deciduous trees and shrubs along the street; Impervious surface.
3	**LCZ-6** **Open low-rise** 	Open low-rise buildings (1–3 stories); Deciduous trees; Permeable agricultural surface.	4	**LCZ-5** **Open midrise** 	Open midrise buildings (3–9 stories) and a large size stadium; Scattered grassland, low shrubs and deciduous trees; Impervious pavement.

Table 1. *Cont.*

No.	LCZ (image)	Description	No.	LCZ (image)	Description
5	**LCZ-4, G** **Open high-rise,** **Water** 	Open high-rise buildings (>9 stories); Scattered low shrubs and deciduous trees; A river; Impervious pavement.	6	**LCZ-2, E** **Compact midrise,** **Bare rock or paved** 	Dense midrise buildings (3–9 stories); Scattered low shrubs and deciduous trees; Impervious pavement.
7	**LCZ-5** **Open midrise** 	Large stadium and open midrise building (3–9 stories); A few deciduous trees, low shrubs and open grassland; Impervious pavement.	8	**LCZ-5** **Open midrise** 	Open midrise buildings (3–9 stories); Scattered low shrubs and deciduous trees; Impervious pavement.
9	**LCZ-2, G** **Compact midrise** **Water** 	Dense midrise buildings (3–9 stories); Scattered low shrubs and deciduous trees; Impervious pavement; A river with dense vegetation on the riverbank.	10	**LCZ-4** **Open high-rise** 	Open high-rise buildings (>9 stories); Scattered low shrubs and deciduous trees; Impervious pavement.
11	**LCZ-2** **Compact midrise** 	Dense midrise buildings (3–9 stories); Scattered low shrubs and deciduous trees; Impervious pavement.	12	**LCZ-4** **Open high-rise** 	Open high-rise buildings (>9 stories); Scattered low shrubs and deciduous trees; Impervious pavement.
13	**LCZ-2** **Compact midrise** 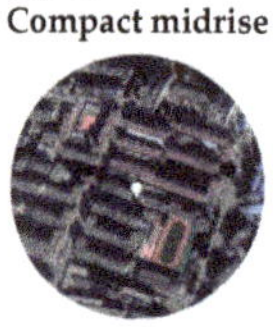	Dense midrise buildings (3–9 stories) and a high-rise building (>9 stories); Scattered low shrubs and deciduous trees; Impervious pavement.	14	**LCZ-2** **Compact midrise** 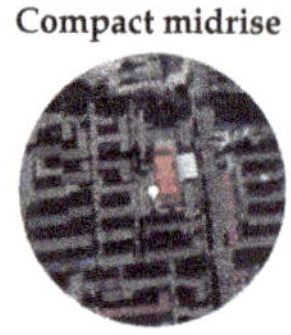	Dense midrise buildings (3–9 stories); Scattered low shrubs and deciduous trees; Impervious pavement.
15	**LCZ-1** **Compact high-rise** 	Dense high-rise buildings (>9 stories); Scattered low shrubs and deciduous trees; Impervious pavement.	16	**LCZ-1** **Compact high-rise** 	Dense high-rise buildings (>9 stories); Scattered low shrubs and deciduous trees; Impervious pavement.

3. Data and Methods

3.1. Data

This study adopts the mean values of the data recorded at a certain weather station to represent the microclimatic characteristics of its corresponding local climate zone. The regional automatic weather stations have been deployed in the downtown Shenyang one by one since 2016, and for each station, the measurement site and its environment were fully considered, especially for wind observation since wind is sensitive to urban morphological characteristics and thereby wind direction and velocity are easily affected by various structures (e.g., buildings, trees). The anemometers (EL15-1A model for wind speed sensor, and EL15-2D model for wind direction vane, both manufactured by Tianjin Meteorological Instrument Factory) were therefore set in relatively open areas (e.g., schools, stadiums), in order to avoid blockage and/or interference effects generated by surrounding buildings and trees and ensure the statistical stationarity and reliability of the measurement data. In addition, the weather stations also recorded the precipitation data (SL3 model manufactured by Shanghai Meteorological Instrument Factory), enabling us to analyse the wind variation under different precipitation conditions (Section 3.3). All anemometers were installed at the 10-m height above ground level. One-year (in 2018) hourly 10-m wind speed, wind direction and precipitation data at 16 weather stations (corresponding to different local climate zones given in Table 1 and Table S1) were obtained. Meanwhile, statistics indicate the data recorded had good completeness (>98%), which is reliable for subsequent analysis.

3.2. Assessment Indicators of Urban Ventilation Performance

Since urban ventilation is critical to several fields (e.g., airflow regime, wind comfort, urban thermal environment and thermal comfort, air pollution), various indicators have been developed to assess urban ventilation performance. In this study, the assessment indicators such as mean wind velocity and relative mean wind speed (RMWS, or relative wind velocity ratio) were adopted to assess the ventilation performance in each zone. The raw data used in this study is hourly collected data, which is averaged over 10 min right before every integral time according to the clock. The mean wind speed or RMWS in the paper is defined as the averaged wind speed (or ratio) over a certain longer period, such as one-year or one-season, using the hourly collected, 10-minute-averaged raw wind speeds recorded at 10 m above ground level.

The wind is the synthetic results of external wind background and LCZ types so that the wind environment in zones is changing with the fluctuating external wind background. The instantaneous wind speed, therefore, cannot accurately demonstrate the ventilation performance in a specific zone. Moreover, the instantaneous wind speed makes it difficult to compare the ventilation performances in different zones as external wind background of a specific zone also varies with different combinations of a larger-scale wind background and urban surfaces. Therefore, the mean wind speed was adopted to assess ventilation performances in different zones.

Relative mean wind speed was adopted to assess ventilation performances in different zones [48]. The RMWS can exclude the influence caused by the fluctuating approaching wind through averaging the ratio of wind velocities in a specific zone to the wind velocity at a reference station Equation (1) [48].

$$\text{RMWS} = \frac{\text{Mean wind speed in specific zones}}{\text{Mean wind speed at a reference station}} \tag{1}$$

The weather station 1 (LCZ-9) was set as the reference station, as it is the national meteorological station and has the sparsely built characteristics. The wind at this reference station is also less influenced by surrounding built environments and can well represent the approaching wind background. As illustrated in Figure 2, LCZ-1 and LCZ-2 accounts for a large proportion of urban morphology in the downtown of Shenyang. Therefore, the RMWS in LCZ-1 and LCZ-2 (No.13, No.14, No.15 and No.16) was calculated to present downtown ventilation performance. As LCZ-1 and LCZ-2 present the compact midrise built form, the highest urbanization level among the 16 locations, the RMWS in LCZ-1 and LCZ-2 can also indicate the largest ventilation performance differences between the city core and approaching wind.

According to the mean wind speed, the ventilation performances in different zones were compared, building upon which LCZ ventilation performances were further sorted. The LCZ ventilation performance is the combined result of external wind background and LCZ types. However, external wind backgrounds generally indicate seasonal and diurnal/nocturnal variations and urban morphology of different zones is heterogeneous. The combined results of external wind background and LCZ types may accordingly demonstrate temporal variations (e.g., season, day and night, weather condition). The spatial correlation coefficient (η) is therefore proposed to assess the short-term variations (e.g., season, day and night, weather condition) in LCZ ventilation performance compared with the annual LCZ ventilation performance. The spatial correlation coefficient (η) ranges between 0 and 1, where a larger η indicates a stronger consistency between short-term and annual ventilation performances and if η equals to 1.0, there is no difference in short-term and annual ventilation performances. In particular, the η equals to the R^2 when conducting the linear regression analysis. In this study, annual LCZ ventilation performance was set as the reference sequence.

3.3. LCZ Ventilation Performance Under Different Scenarios

The LCZ ventilation performance sequence (according to mean wind speed) was analysed at different temporal scales of year, season, day and night and precipitation weather conditions. Moreover, the RMWS in LCZ-1 and LCZ-2 was also analysed to understand the difference in downtown ventilation performance from the approaching wind background.

In the analysis, the diurnal and nocturnal LCZ ventilation performance, the diurnal and nocturnal times were different because of the seasonal variation. According to the sunrise and sunset times, the spring daytime is 6:00–17:00, summer daytime is 5:00–18:00, autumn daytime is 6:00–17:00 and winter daytime is 7:00–16:00.

Since the precipitation process is generally associated with wind, the LCZ ventilation performance was also analysed under different precipitation conditions. In particular, the hourly precipitation of 0.1 mm was adopted as a threshold to screen the conditions of no precipitation and precipitation. Furthermore, the precipitation condition was divided into light rain (≤2.5 mm), moderate rain (2.6–8.0 mm), heavy rain (8.1–15.0 mm) and torrential rain (≥16.0 mm) according to 1-h precipitation (without considering snowfall) [49,50]. Therefore, the LCZ ventilation performance was analysed under fair (no precipitation) and rainy conditions, and further under light, moderate, heavy and torrential rain conditions. Table 2 exhibits the precipitation condition in the 16 stations in Shenyang in 2018. The raining hour decreased gradually following the order of light, moderate, heavy and torrential rain. Moreover, there were limited hours in both heavy rain and torrential rain conditions.

Table 2. Hours of light rain, moderate rain, heavy rain and torrential rain in 2018 at each station.

	Local Climate Zone		Light Rain	Moderate Rain	Heavy Rain	Torrential Rain
	Class	Name				
No.1	LCZ-9	Sparsely built	220	54	4	2
No.2	LCZ-5	Open midrise	157	43	2	5
No.3	LCZ-6	Open low-rise	107	44	4	2
No.4	LCZ-5	Open midrise	192	37	6	2
No.5	LCZ-4, G	Open high-rise, Water	202	43	10	4
No.6	LCZ-2, E	Compact midrise, Bare rock or paved	118	32	2	1
No.7	LCZ-5	Open midrise	97	13	5	2
No.8	LCZ-5	Open midrise	158	31	8	5
No.9	LCZ-2, G	Compact midrise, Water	96	26	4	6
No.10	LCZ-4	Open high-rise	198	51	7	2
No.11	LCZ-2	Compact midrise	114	24	6	4
No.12	LCZ-4	Open midrise	190	30	4	1
No.13	LCZ-2	Compact midrise	203	40	8	5
No.14	LCZ-2	Compact midrise	188	47	7	8
No.15	LCZ-1	Compact high-rise	171	44	9	5
No.16	LCZ-1	Compact high-rise	179	42	11	5

4. Analysis and Discussion

4.1. Mean Wind Speed at Shenyang Weather Station and in Shenyang City

This subsection presents the variation in mean wind speed with several scenarios, such as year, season, day and night and weather conditions. The mean wind speed also exhibits the spatial variations by mean wind speed at Shenyang weather station (No.1, LCZ-9) and mean wind speed of the 16 stations for representing the Shenyang wind condition.

Table 3 presents annual and seasonal mean wind speed at the Shenyang weather station and 16 stations (also refer to Table S2). The annual mean wind speed at Shenyang weather station (No.1, LCZ-9) was 2.07 m/s. However, the annual mean value of the 16 stations was only 1.44 m/s, about 60% of the wind speed at the Shenyang weather stations. This result indicates the construction of urban structures has already lowered the urban ventilation performance. Moreover, the actual wind speed of Shenyang cannot be sufficiently represented by Shenyang weather station, although Shenyang weather station is a national meteorological station. The reduction of urban ventilation performance indicates that there is a need for refined and actual wind speed rather than wind information at only national meteorological stations when analysing the urban ventilation potential of cities.

Table 3. Annual and seasonal mean wind speed at Shenyang weather station and the 16 stations with standard deviations (2018).

Season	Mean Wind Speed (m/s)	
	Shenyang Weather Station	Shenyang City (16 Stations)
Spring	2.84 ± 1.78	1.85 ± 0.51
Summer	1.94 ± 1.26	1.36 ± 0.36
Autumn	1.63 ± 1.20	1.16 ± 0.25
Winter	1.87 ± 1.25	1.37 ± 0.26
Annual	2.07 ± 1.47	1.44 ± 0.33

The wind speed also exhibited seasonal variation, where the mean wind speed at both Shenyang weather station and the 16 stations in spring was the highest among four seasons, with the values of 2.84 m/s and 1.85 m/s. The strongest wind condition in spring is related to its geographic location as well as the macro wind background. Spring is a transitional season between winter and summer, during which period synoptic weather systems move fast. Shenyang is therefore susceptible to cold and warm air mass alternation and thereby high (continental high pressure from Siberia) and low air pressure change frequently. Such conditions result in a northeast vortex before the high pressure extends to the sea and leading to the strong geostrophic winds in Shenyang.

In comparison, the autumn experienced the weakest wind condition, where the wind speed at Shenyang weather station and the 16 stations were 1.63 m/s and 1.16 m/s, respectively. In the autumn, the Siberian air mass gradually strengthens southward and heating mass slowly retreats. However, the East China Sea and the South China Sea are mainly dominated by the low air pressure of retreating heating mass. There is no high and low air pressure alternation in Shenyang. Under such conditions, the air pressure gradient is low. The northeast cold vortex in autumn is, therefore, less than that in spring and the wind was weaker than that in spring. In addition, wind speed at the 16 stations was also lower than that at Shenyang weather station, indicating slower wind speeds in cities compared with the wind speed recorded at the meteorological stations.

Table 4 exhibits the diurnal and nocturnal mean wind speed at Shenyang weather station and the 16 stations (Shenyang city) in four seasons. The wind speed presented by the 16 stations was lower than the wind speed recorded at Shenyang weather station. Both nocturnal and diurnal wind speed reached their largest values in spring, while they experienced the lowest values in autumn. For instance, the diurnal and nocturnal wind speeds at Shenyang weather station were 3.56 m/s and 2.14 m/s, respectively. However, the ones in the autumn were only 2.18 m/s and 1.07 m/s, respectively.

Table 4. Diurnal and nocturnal mean wind speed at Shenyang weather station and the 16 stations (2018).

Time	Seasons	Mean Wind Speed (m/s)	
		Shenyang Weather Station	**Shenyang City (16 Stations)**
Diurnal	Spring	3.56 ± 1.75	2.21 ± 0.69
	Summer	2.35 ± 1.30	1.56 ± 0.46
	Autumn	2.18 ± 1.22	1.41 ± 0.37
	Winter	2.36 ± 1.52	1.66 ± 0.37
Nocturnal	Spring	2.14 ± 1.38	1.48 ± 0.34
	Summer	1.37 ± 0.87	1.07 ± 0.24
	Autumn	1.07 ± 0.86	0.90 ± 0.17
	Winter	1.52 ± 1.14	1.15 ± 0.19

More importantly, the diurnal mean wind speed in four seasons was consistently higher than the nocturnal mean wind speed. For instance, the diurnal mean wind speed at Shenyang weather station in spring reached 3.56 m/s, about 1.5 m/s higher than the nocturnal mean wind speed of 2.14 m/s. The weaker wind condition in the night may be related to the stable atmospheric boundary layer stratification and weaker thermal turbulence. As such, the momentum downward transmission is mainly transported by mechanical turbulence. In comparison, in the daytime, thermal turbulence develops vigorously and then the combined effects of thermal and mechanical turbulences are conducive to the upper momentum downward transmission, resulting in stronger winds near the surface.

Table 5 presents the variation in mean wind speed at Shenyang weather station and the 16 stations with precipitation conditions. Under all scenarios, both no precipitation and different precipitation conditions, mean wind speed at the 16 stations (Shenyang city) was consistently lower than that at the Shenyang weather stations. Meanwhile, the wind speed under precipitation conditions was 2.21 m/s, 0.46 m/s higher than the wind speed (1.75 m/s) under the no precipitation conditions. Under no precipitation conditions, namely under sunny weather conditions that are generally under the control of mesoscale high pressure, background wind speed is even, local strong winds rarely occur and mean wind speed is generally weak. In comparison, under precipitation conditions that generally occur in summer, precipitation, especially moderate, heavy and torrential rain conditions, can be accompanied with local convective weather, resulting in stronger winds near the surface. The wind speed reached 2.33 m/s under torrential rain conditions.

Table 5. Mean wind speed at Shenyang weather station and the 16 stations with standard deviations under different precipitation conditions (2018).

Precipitation Conditions	Mean Wind Speed (m/s)	
	Shenyang Weather Station	**Shenyang City (16 Stations)**
No precipitation	1.75 ± 1.47	1.34 ± 0.23
Precipitation	2.21 ± 1.15	1.48 ± 0.39
Light	1.92 ± 1.18	1.36 ± 0.29
Moderate	2.13 ± 1.03	1.49 ± 0.38
Heavy	1.98 ± 0.87	1.42 ± 0.31
Torrential	2.33 ± 0.99	1.52 ± 0.44

4.2. Downtown Ventilation Performance

The analysis in Section 4.1 indicates that the ventilation performance in Shenyang city was worse than that at the Shenyang weather station (No.1, LCZ-9). Therefore, further analysis was conducted to examine the downtown ventilation performance through the indicator of RMWS. In general, the greater the RMWS is, the better the downtown ventilation performance is, and vice versa. The variation in downtown ventilation performance was analysed in yearly, seasonally, diurnal and nocturnal and different precipitation scenarios.

Table 6 presents the annual and seasonal RMWS. The annual RMWS value of 0.53 (Table 6) shows that the annual downtown ventilation performance was not good, only about 50% of the wind condition at the Shenyang weather station. Among four seasons, the downtown Shenyang experienced the best ventilation performance in winter with the RMWS of 0.60 and experienced the worst ventilation performance in spring with the RMWS of 0.47. In combination with the mean wind speed in different seasons, the RMWS had the practical implications for better urban environments. In spring, the mean wind speed was high and frequent, a low RMWS indicates low wind speed in the downtown area. As such, weak ventilation performance can help reduce the discomfort caused by strong winds. In comparison, in winter, a high RMWS indicates that more cold air could penetrate the downtown areas, increasing the discomfort caused by the cold wind. In summer and autumn, the downtown ventilation performance was less than 0.60, where the potential of wind for addressing urban environmental problems (e.g., UHI) and improving urban liveability was limited. Therefore, to address such problems, it is essential to improve ventilation efficiency in Shenyang.

Table 6. Variation in downtown ventilation performance (RMWS: relative mean wind speed) in Shenyang in 2018.

Annual/Seasonal	RMWS	Day and Night	RMWS	Precipitation	RMWS
Year	0.53	Day (Spring)	0.41	No precipitation	0.65
Spring	0.47	Day (Summer)	0.47	Precipitation	0.49
Summer	0.51	Day (Autumn)	0.47	Light rain	0.57
Autumn	0.57	Day (Winter)	0.56	Moderate rain	0.52
Winter	0.60	Night (Spring)	0.53	Heavy rain	0.55
		Night (Summer)	0.61	Torrential rain	0.46
		Night (Autumn)	0.65		
		Night (Winter)	0.65		

Table 6 further presents the diurnal and nocturnal downtown ventilation performance in four seasons. The diurnal and nocturnal downtown ventilation performance also indicates the evident seasonal variations, where winter experienced the best ventilation performance and spring underwent the worst ventilation performance, according to the RMWS. Table 6 also indicates that nocturnal ventilation performance was better than the diurnal ventilation performance, where the diurnal RMWS was 0.41, 0.47, 0.49 and 0.56 in spring, summer, autumn and winter, respectively, and the nocturnal RMWS reached to 0.53, 0.61, 0.65 and 0.65, respectively. The weaker downtown ventilation performance may be related to the insignificant thermal difference (weaker UHI phenomenon). Therefore, the ventilation performance in the downtown area is determined by the surface roughness. In comparison, nocturnal UHI can be stronger compared with the diurnal UHI, at which time the heat stored in cities can enhance the thermal turbulence and thereby enhance the downtown ventilation performance.

Table 6 also compares downtown ventilation performance under no precipitation and various precipitation conditions. The downtown ventilation performance (0.65) under no precipitation conditions was better than that (0.49) under precipitation conditions. Downtown ventilation performance weakened with an increase in precipitation level. Such variations in downtown ventilation performance with precipitation conditions may be due to the combined effects of localised precipitation characteristics (Table 2), the obstruction of surface roughness in downtown and thermal uplift led by the downtown surface layer. In particular, the worse downtown ventilation performance may reduce the risks of heavy rain and torrential rain.

4.3. LCZ Ventilation Performance Sequence

4.3.1. Annual LCZ Ventilation Performance and Sequence

Mean wind speed at the 16 stations is compared in this subsection in order to analyse the LCZ ventilation performance. Figure 3 demonstrates the annual mean wind speed at the 16 stations and the LCZ ventilation performance sequence (original data given in Table S2). The wind speed at stations of No.1, No.2, No.3 and No.4 was high (above 1.9 m/s), higher than the wind speed at the remaining 12 stations. The four stations (e.g., No.1, No.2, No.3, No.4) were located near the city boundary (the black line given in Figure 2). The peripheral locations with low construction levels indicate that the approaching wind was less affected by the urban structures compared with the remaining 12 stations approaching the city centre more. The four stations had LCZ types of LCZ-9, LCZ-5, LCZ-6 and LCZ-5 (Table 1). The sparse and open morphologies were another cause of the higher wind speed at the four stations, compared with the compact morphologies the remaining 12 stations had. LCZ ventilation performance was ranked according to annual mean wind speed at the 16 stations (Figure 3). The results indicate that the wind speed decreased with the weather stations from No.1 to No.16. This indicates that the LCZ ventilation performance decreased with the LCZ types following the order of sparse, open low-rise, open midrise and compact midrise. Such consistent relationships between wind speed and LCZ types confirm the reliability of using the LCZ classification scheme to indicate the local-scale urban ventilation performance.

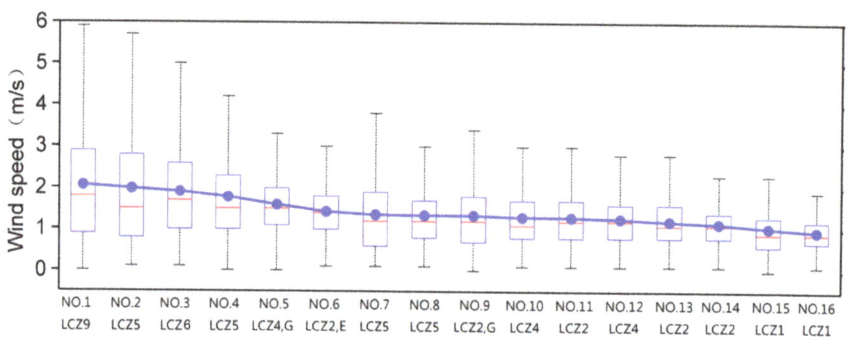

Figure 3. Boxplot and mean wind speed (blue point-fold line) of the 16 stations.

Although there is a clear relationship between local-scale urban ventilation performance and LCZ types, we keep the argument that LCZ ventilation performance can also be affected by other factors, such as the location of weather stations in a city, the approaching wind condition and the altitude of different stations. It should be noted that there were mixed LCZ types in the 16 stations, such as No.5 (LCZ-4,G), No.6 (LCZ-2,E) and No.9 (LCZ-2,G). The ventilation performance at such four stations was affected by compactness and building height, while the types of G (water) and E (bare soil or paved) were open in morphological characteristics, enabling the increase in mean wind speed in the certain direction from wind blows over the underlying types G or E.

4.3.2. Seasonal LCZ ventilation Performance and Sequence

Figure 4 exhibits seasonal mean wind speed and LCZ ventilation performance sequence through the spatial correlation coefficient (η, or R^2) (also refer to Table S3). First, the fitted line of spring wind speed (Figure 4) was above the reference line (1:1, annual wind speed), indicating the spring wind speed was higher than the annual scenario. The autumn wind speed was much lower, and summer and winter wind speeds were similar to the annual wind speed. This result indicates the seasonal variation in LCZ ventilation performance.

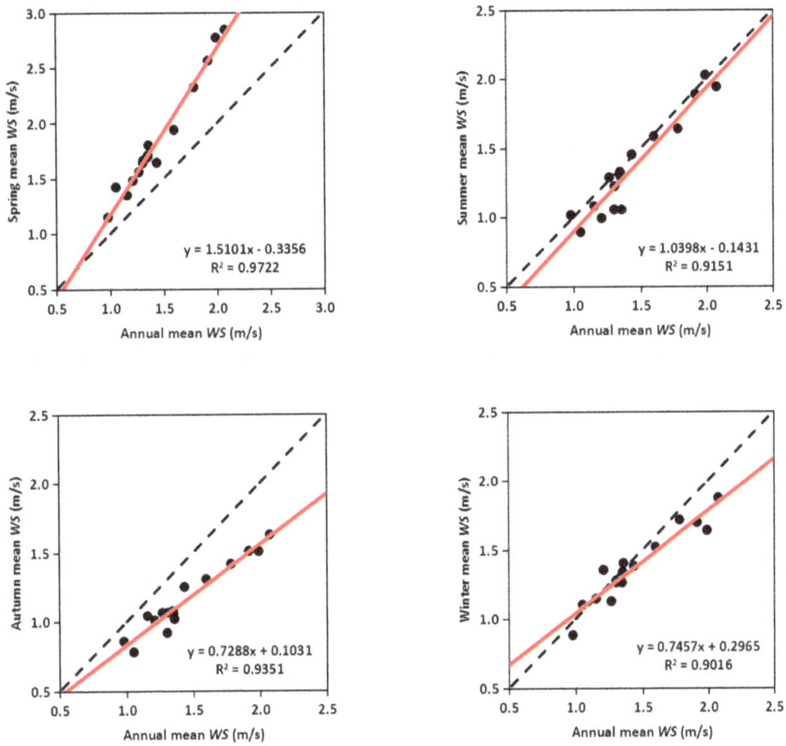

Figure 4. Seasonal mean wind speed at the 16 stations and local climate zone ventilation performance sequences.

Nevertheless, LCZ ventilation performance sequences in four seasons were generally consistent with annual LCZ ventilation performance sequence, with the R^2 of 0.97, 0.92, 0.94 and 0.90. However, the value of the spatial correlation coefficient did not equal to 1.0, indicating the change of LCZ ventilation performance sequence at specific LCZ types. In particular, the inconsistency between open spaces around specific zones (e.g., No.6, LCZ-2,E) and seasonal approaching wind direction is a critical cause, as the weather station may be in the frontal or backward areas according to wind direction alternation. Moreover, vegetation canopy varies with seasonal alternation, especially for zones with dense trees (No.7, LCZ-5), resulting in the seasonal change in wind speed.

4.3.3. Diurnal and Nocturnal LCZ Ventilation Performance and Sequence

Figures 5 and 6 present the diurnal and nocturnal mean wind speed and LCZ ventilation performance sequence in four seasons, respectively (also refer to Tables S4 and S5). The fitted line of spring diurnal wind speed was above the reference line (1:1, annual wind speed), indicating the stronger spring wind environment in Shenyang. The winter diurnal wind was also stronger than annual scenario, the summer diurnal wind was similar to the annual scenario and the autumn diurnal wind was weaker than the annual scenario. This result indicates the seasonal variation in diurnal LCZ ventilation performance.

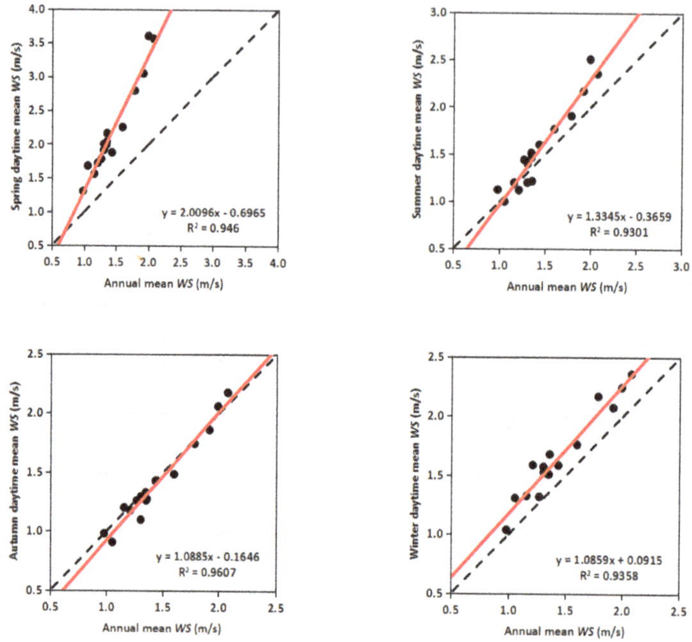

Figure 5. Diurnal mean wind speed at the 16 stations and the local climate zone ventilation performance sequences in four seasons.

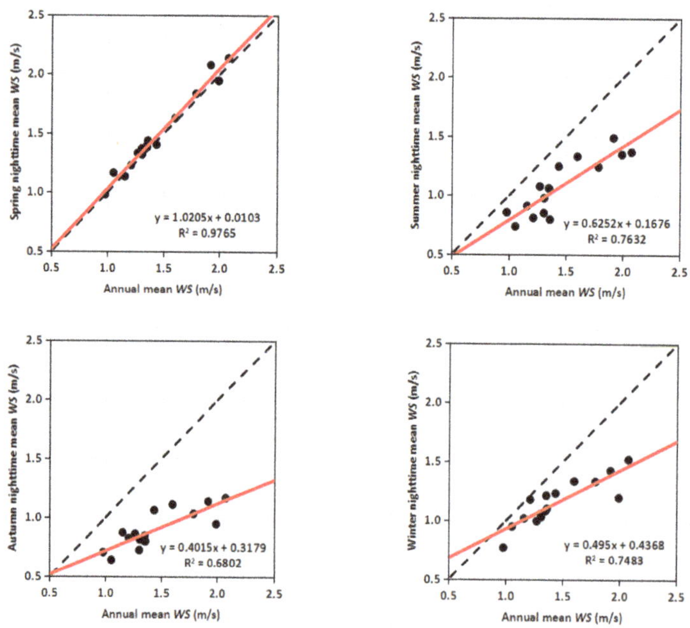

Figure 6. Nocturnal mean wind speed at the 16 stations and the local climate zone ventilation performance sequences in four seasons.

However, the results also indicate that diurnal LCZ ventilation performance sequences in four seasons (Figure 5) were generally consistent with the annual LCZ ventilation performance sequence, with the R^2 of 0.95, 0.93, 0.96 and 0.94. This result further affirms the reliability of using the LCZ classification scheme to indicate diurnal local-scale urban ventilation performance. Nevertheless, there were some slight fluctuations as the spatial correlation coefficients did not equal to 1.0.

The fitted line of spring nocturnal wind speed roughly coincided with the reference line (1:1, annual wind speed), indicating the spring nocturnal wind was similar to the annual wind condition (Figure 6). However, the nocturnal wind speed in summer, autumn and winter was weaker than the annual wind condition in Shenyang. Moreover, the LCZ classification scheme could also indicate local-scale urban ventilation performance, especially in spring with the spatial correlation coefficient of 0.98. The capacity of LCZ classification scheme of indicating local-scale urban ventilation performance weakens, with the spatial correlation coefficient of 0.76, 0.68 and 0.75 in summer, autumn and winter, respectively.

4.3.4. LCZ Ventilation Performance and Sequence under Various Precipitation Conditions

Figure 7 demonstrates mean wind speed and LCZ ventilation performance sequence under different precipitation conditions. First, the fitted line of wind speed under no precipitation condition was below the reference line (1:1, annual wind speed), indicating the wind under no precipitation condition was weaker. The wind under precipitation condition was stronger than the annual scenario. Moreover, LCZ classification scheme could also indicate local-scale urban ventilation performance, with the R^2 of 0.99 and 0.80 under precipitation and no precipitation conditions, respectively. Moreover, under light, moderate, heavy and torrential rain conditions, mean wind speeds were generally similar to the annual wind speed, as fitted lines roughly coincided with the reference line (1:1, annual wind speed). Under these rain conditions, the LCZ classification scheme was also robust to indicate local-scale urban ventilation performance with the R^2 of 0.83, 0.86, 0.95 and 0.96, respectively (Figure 7).

Overall, this section compared the ventilation performance of different local climate zones under the annual, seasonal, diurnal and nocturnal and different precipitation conditions. The annual mean wind speed consequence indicated that LCZ ventilation performance decreased following the order of open and sparse zone, open midrise, compact midrise, compact high-rise zones. The spatial correlation coefficients were calculated, where the coefficients generally did not equal to 1.0 because of the heterogeneity of real local climate zones. This result indicates that the LCZ ventilation performance varied with season, day and night and precipitation conditions. Nevertheless, the value of spatial correlation coefficients approached to 1.0, indicating the LCZ ventilation performance patterns under various scenarios were generally consistent with the annual LCZ ventilation performance pattern. Namely, LCZ ventilation performance could also indicate the decreasing trend from open and spare zones to compact midrise zones and then compact high-rise zones. Such results indicate the suitability of using the LCZ classification scheme to indicate local-scale urban ventilation performance.

Even though, the pattern of LCZ ventilation performance was not critically consistent with the precinct ventilation performance reported in [40–42], where the open midrise gridiron precinct ventilation zone had the best ventilation performance, followed by the open low-rise gridiron precinct ventilation zone and then the compact high-rise gridiron precinct ventilation zone. This might be due to the influence of the street structure or the influence of local fluctuations in the open midrise gridiron contexts [40]. Such results further indicate the necessity of analysing the impact of street structure on the local-scale ventilation performance, apart from the LCZ ventilation performance studies.

Moreover, in precinct ventilation performance studies [40–42], the ventilation performance was obtained through averaging wind speeds at 52 points, which provided a high-resolution wind distribution by catching all possible phenomena (e.g., corner effect, interference effect, blockage effect, wake effect) at different points. In comparison, in current LCZ ventilation performance study, one-point wind condition might be not representative enough to exhibit local-scale ventilation performance, as one-point method ignored all possible phenomena at other points. Therefore, it is essential to set more wind sensors to generate high-resolution wind distribution to further validate the suitability of LCZ classification scheme for local-scale ventilation performance indication.

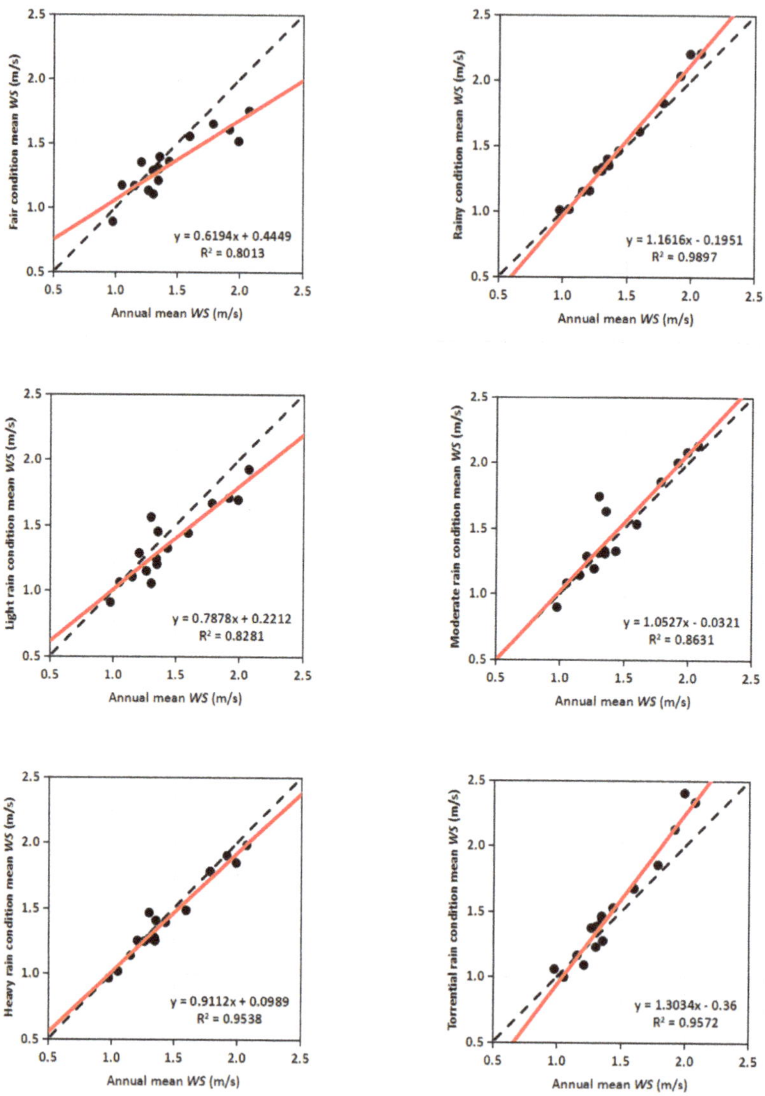

Figure 7. Mean wind speed at the 16 stations and local climate zone ventilation performance sequences under different precipitation conditions.

5. Summary and Conclusions

This study analysed the local-scale urban ventilation performance in Shenyang, China through analysing the wind speed (e.g., mean wind speed, relative mean wind speed, spatial correlation coefficient) in 2018 at the 16 weather stations that were located in different local climate zones. Through the analysis, several key conclusions were obtained.

Shenyang weather station had an annual mean wind speed of 2.07 m/s, while the mean wind speed of Shenyang city was only 60% of the wind speed Shenyang weather station, about 1.44 m/s. Moreover, both mean wind speeds at Shenyang weather station and the 16 stations varied with seasons, day and night and precipitation conditions. The spring was the windiest season, where the mean wind speeds at Shenyang weather station and the 16 stations were 2.84 m/s and 1.85 m/s, respectively. However, autumn experienced the weakest wind condition, with the mean wind speeds at Shenyang weather station and the 16 stations of 1.63 m/s and 1.16 m/s, respectively. Moreover, the diurnal wind was much stronger than the nocturnal wind, across four seasons. The spring diurnal wind was the strongest, with the wind speeds of 3.56 m/s and 2.21 m/s at Shenyang weather station and the 16 stations, respectively. The autumn nocturnal wind was the weakest, with the wind speeds of 1.07 m/s and 0.90 m/s at Shenyang weather station and the 16 stations, respectively. In addition, the wind was under precipitation condition generally stronger than that under no precipitation condition. In particular, the wind under torrential rain condition was the strongest, with the mean wind speeds at Shenyang weather station and the 16 stations were 2.33 m/s and 1.52 m/s, respectively.

Wind speed at Shenyang weather station was consistently higher than that of Shenyang city (16 stations) across all scenarios this study presented (e.g., year, season, day and night, no precipitation and precipitation). This implies that the necessity of concerning refined and actual wind condition in cities and it is improper to only adopt wind recorded at the national meteorological station when conducting studies related to local-scale wind. An analysis of downtown ventilation performance based on RMWS of LCZ-1 and LCZ-2 stations indicated that the downtown ventilation performance was weaker than approaching wind background, with the annual RMWS value of 0.53. Downtown ventilation performance varied with seasons, day and night and precipitation. The downtown ventilation performance, according to the RMWS value, was much weaker, especially in the spring season and under diurnal scenarios. The downtown ventilation performance, in comparison, was much stronger in the winter season and under nocturnal scenarios. The spring diurnal RMWS was only 0.41, but the winter nocturnal RWMS reached 0.65. The downtown ventilation performance (0.65) under no precipitation conditions was better than that (0.49) under precipitation conditions. In addition, with an increase in rain level, downtown ventilation performance decreased.

LCZ ventilation performance was ranked according to mean wind speed. The results indicate that annual mean wind speed of each zone decreased gradually from the sparse and open low-rise zones to compact midrise zones. This finding affirms that the LCZ classification scheme can be adopted to indicate local-scale urban ventilation performance, although it was originally proposed to indicate surface temperature [14] and it did not consider the impact of street structure on the local-scale urban ventilation performance like precinct ventilation zone scheme [16]. Moreover, despite some changes in the order of the ventilation performance of specific zones, the spatial correlation coefficients under several scenarios (e.g., seasons, day and night, no precipitation and precipitation) ranged between 0.68 and 0.99, affirming the LCZ classification scheme was robust to indicate the local-scale urban ventilation performance.

Overall, this study offered an evidence-based analysis of local-scale urban ventilation performance in Shenyang, China. The study directly affirms that the LCZ classification scheme was effective in indicating local-scale urban ventilation performance, although it did not consider the street structures. This study also indicates actual ventilation performance in a city should be much weaker than the approaching wind information the national meteorological stations provide. Only using the wind information given by national meteorological stations to analyse ventilation performance in cities may overestimate the results.

Supplementary Materials: The following are available online at http://www.mdpi.com/2073-4433/11/8/776/s1, Table S1: The variables of different local climate zones in Shenyang, China; Table S2. Annual and Seasonal mean wind speed and standard deviations of 16 stations (m/s); Table S3. *P*-values of Mann-Whitney U-test of wind speeds between Station No.1 and other stations for a year and four seasons; Table S4. *P*-values of Mann-Whitney U-test of wind speeds between Station No.1 and other stations for daytime of four seasons. Table S5. *P*-values of Mann-Whitney U-test of wind speeds between Station No.1 and other stations for night-time of four seasons.

Author Contributions: Conceptualisation, B.-J.H. and L.S.; methodology, Z.Z, B.-J.H. and L.S.; software, Z.Z. and L.S.; validation, Z.Z. and L.S.; formal analysis, Z.Z., B.-J.H. and L.S.; investigation, Z.Z., L.S. and B.-J.H.; resources, Z.Z., L.S., H.W. and L.L; data curation, L.S.; writing—original draft preparation, Z.Z., B.-J.H. and L.S.; writing—review and editing, Z.Z., B.-J.H., L.S., H.W. and L.L.; project administration, H.W. and L.L.; supplement materials, Z.Z, L.S., L.L. and H.W. All authors have read and agreed to the published version of the manuscript.

Funding: This research was funded by Key Program of Science Foundation of Liaoning Meteorological Office (grant number: 202010, 201904), Basic Scientific & Research Business Expenses for Scientific and Scientific Institutions for Central Public Welfare (Grant number: 2018SYIAEMS2). Program of Natural Science Foundation of Liaoning Province (2020-MS-350). Scientific research project of Shenyang Institute of Atmospheric Environment, China Meteorological Administration (Grant No. 2020SYIAEJY12).

Conflicts of Interest: The authors declare no conflict of interest.

References

1. IPCC. Summary for Policymakers of IPCC Special Report on Global Warming of 1.5 °C Approved by Governments. Intergovernmental Panel on Climate Change. Available online: https://www.ipcc.ch/2018/10/08/summary-for-policymakers-of-ipcc-special-report-on-global-warming-of-1-5c-approved-by-governments/ (accessed on 8 October 2018).
2. Zhang, G.; He, B.-J.; Dewancker, B.J. The maintenance of prefabricated green roofs for preserving cooling performance: A field measurement in the subtropical city of Hangzhou, China. *Sustain. Cities Soc.* **2020**, *61*, 102314. [CrossRef]
3. Kumar, P.; Saroj, D.P. Water–energy–pollution nexus for growing cities. *Urban Clim.* **2014**, *10*, 846–853. [CrossRef]
4. Rosenzweig, C.; Solecki, W.D.; Romero-Lankao, P.; Mehrotra, S.; Dhakal, S.; Ibrahim, S.A. *Climate Change and Cities: Second Assessment Report of the Urban Climate Change Research Network*; Cambridge University Press: Cambridge, UK, 2018; pp. 27–60.
5. Hammond, M.J.; Chen, A.S.; Djordjević, S.; Butler, D.; Mark, O. Urban flood impact assessment: A state-of-the-art review. *Urban Water J.* **2015**, *12*, 14–29. [CrossRef]
6. He, B.-J.; Zhu, J.; Zhao, D.-X.; Gou, Z.-H.; Qi, J.-D.; Wang, J. Co-benefits approach: Opportunities for implementing sponge city and urban heat island mitigation. *Land Use Policy* **2019**, *86*, 147–157. [CrossRef]
7. Anderson, G.B.; Bell, M.L. Heat waves in the United States: Mortality risk during heat waves and effect modification by heat wave characteristics in 43 US communities. *Environ. Health Perspect.* **2011**, *119*, 210–218. [CrossRef]
8. United Nations Department. *World Urbanization Prospects: The 2014 Revision*; United Nations Department of Economics and Social Affairs, Population Division: New York, NY, USA, 2015.
9. Raven, J.; Stone, B.; Mills, G.; Towers, J.; Katzschner, L.; Leone, M.; Gaborit, P.; Georgescu, M.; Hariri, M. Urban planning and urban design. In *Climate Change and Cities: Second Assessment Report of the Urban Climate Change Research Network*; Rosenzweig, C., Solecki, W., Romero-Lankao, P., Mehrotra, S., Dhakal, S., Ali Ibrahim, S., Eds.; Cambridge University Press: New York, NY, USA, 2018; pp. 139–172.
10. Alves, A.; Gersonius, B.; Kapelan, Z.; Vojinovic, Z.; Sanchez, A. Assessing the Co-Benefits of green-blue-grey infrastructure for sustainable urban flood risk management. *J. Environ. Manag.* **2019**, *239*, 244–254. [CrossRef]
11. Xu, Y.; Ren, C.; Ma, P.; Ho, J.; Wang, W.; Lau, K.K.L.; Lin, H.; Ng, E. Urban morphology detection and computation for urban climate research. *Landsc. Urban Plan.* **2017**, *167*, 212–224. [CrossRef]
12. Yang, J.; Wang, Y.; Xiao, X.; Jin, C.; Xia, J.C.; Li, X. Spatial differentiation of urban wind and thermal environment in different grid sizes. *Urban Clim.* **2019**, *28*, 100458. [CrossRef]
13. Oke, T.R.; Mills, G.; Christen, A.; Voogt, J.A. *Urban Climates*; Cambridge University Press: Cambridge, UK, 2017.
14. Stewart, I.D.; Oke, T.R. Local climate zones for urban temperature studies. *Bull. Am. Meteorol. Soc.* **2012**, *93*, 1879–1900. [CrossRef]
15. Adolphe, L. A simplified model of urban morphology: Application to an analysis of the environmental performance of cities. *Environ. Plan. B Plan. Des.* **2001**, *28*, 183–200. [CrossRef]
16. He, B.-J.; Ding, L.; Prasad, D. Enhancing urban ventilation performance through the development of precinct ventilation zones: A case study based on the Greater Sydney. *Aust. Sustain. Cities Soc.* **2019**, *47*, 101472. [CrossRef]

17. He, B.J.; Zhao, Z.Q.; Shen, L.D.; Wang, H.B.; Li, L.G. An approach to examining performances of cool/hot sources in mitigating/enhancing land surface temperature under different temperature backgrounds based on landsat 8 image. *Sustain. Cities Soc.* **2019**, *44*, 416–427. [CrossRef]
18. Azhdari, A.; Soltani, A.; Alidadi, M. Urban morphology and landscape structure effect on land surface temperature: Evidence from Shiraz, a semi-arid city. *Sustain. Cities Soc.* **2018**, *41*, 853–864. [CrossRef]
19. Ng, E.; Yuan, C.; Chen, L.; Ren, C.; Fung, J.C. Improving the wind environment in high-density cities by understanding urban morphology and surface roughness: A study in Hong Kong. *Landsc. Urban Plan.* **2011**, *101*, 59–74. [CrossRef]
20. Peng, Y.; Gao, Z.; Buccolieri, R.; Ding, W. An Investigation of the Quantitative Correlation between Urban Morphology Parameters and Outdoor Ventilation Efficiency Indices. *Atmosphere* **2019**, *10*. [CrossRef]
21. Yang, J.; Jin, S.; Xiao, X.; Jin, C.; Xia, J.C.; Li, X.; Wang, S. Local climate zone ventilation and urban land surface temperatures: Towards a performance-based and wind-sensitive planning proposal in megacities. *Sustain. Cities Soc.* **2019**, *47*, 101487. [CrossRef]
22. He, B.-J. Potentials of meteorological characteristics and synoptic conditions to mitigate urban heat island effects. *Urban Clim.* **2018**, *24*, 26–33. [CrossRef]
23. Zhao, D.-X.; He, B.-J. Effects of architectural shapes on surface wind pressure distribution: Case studies of oval-shaped tall buildings. *J. Build. Eng.* **2017**, *12*, 219–228. [CrossRef]
24. Meng, F.-Q.; He, B.-J.; Zhu, J.; Zhao, D.-X.; Darko, A.; Zhao, Z.-Q. Sensitivity analysis of wind pressure coefficients on CAARC standard tall buildings in CFD simulations. *J. Build. Eng.* **2018**, *16*, 146–158. [CrossRef]
25. Oke, T.R. Street design and urban canopy layer climate. *Energy Build.* **1988**, *11*, 103–113. [CrossRef]
26. Ng, E. Designing for urban ventilation. In *Designing High Density Cities-For Social and Environmental Sustainability*; Earthscan Publications Ltd.: London, UK, 2010.
27. Abd Razak, A.; Hagishima, A.; Ikegaya, N.; Tanimoto, J. Analysis of airflow over building arrays for assessment of urban wind environment. *Build. Environ.* **2013**, *59*, 56–65. [CrossRef]
28. Xu, X.; Luo, F.; Wang, W.; Hong, T.; Fu, X. Performance-Based Evaluation of Courtyard Design in China's Cold-Winter Hot-Summer Climate Regions. *Sustainability* **2018**, *10*. [CrossRef]
29. Ying, X.; Wang, Y.; Li, W.; Liu, Z.; Ding, G. Group Layout Pattern and Outdoor Wind Environment of Enclosed Office Buildings in Hangzhou. *Energies* **2020**, *13*. [CrossRef]
30. Kubota, T.; Miura, M.; Tominaga, Y.; Mochida, A. Wind tunnel tests on the relationship between building density and pedestrian-level wind velocity: Development of guidelines for realizing acceptable wind environment in residential neighborhoods. *Build. Environ.* **2008**, *43*, 1699–1708. [CrossRef]
31. DeKay, M.; Brown, G. *Sun, Wind, and Light: Architectural Design Strategies*; John Wiley & Sons: Hoboken, NJ, USA, 2013.
32. Wong, M.S.; Nichol, J.E.; To, P.H.; Wang, J. A simple method for designation of urban ventilation corridors and its application to urban heat island analysis. *Build. Environ.* **2010**, *45*, 1880–1889. [CrossRef]
33. Qiao, Z.; Xu, X.; Wu, F.; Luo, W.; Wang, F.; Liu, L.; Sun, Z. Urban ventilation network model: A case study of the core zone of capital function in Beijing metropolitan area. *J. Clean. Prod.* **2017**, *168*, 526–535. [CrossRef]
34. Coutts, A.; Harris, R. Urban Heat Island Report: 'A Multi-Scale Assessment of Urban Heating in Melbourne during an Extreme Heat Event and Policy Approaches for Adaptation. Published by VCCCAR, Melbourne, Australia, 2013. Available online: https://watersensitivecities.org.au/content/vcccar-report-multi-scale-assessment-urban-heating-melbourne-extreme-heat-event-policy-approaches-adaptation/ (accessed on 1 November 2012).
35. Blocken, B.; Janssen, W.D.; van Hooff, T. CFD simulation for pedestrian wind comfort and wind safety in urban areas: General decision framework and case study for the Eindhoven University campus. *Environ. Model. Softw.* **2012**, *30*, 15–34. [CrossRef]
36. Antoniou, N.; Montazeri, H.; Wigo, H.; Neophytou, M.K.A.; Blocken, B.; Sandberg, M. CFD and wind-tunnel analysis of outdoor ventilation in a real compact heterogeneous urban area: Evaluation using "air delay". *Build. Environ.* **2017**, *126*, 355–372. [CrossRef]
37. Priyadarsini, R.; Hien, W.N.; David, C.K.W. Microclimatic modeling of the urban thermal environment of Singapore to mitigate urban heat island. *Sol. Energy* **2008**, *82*, 727–745. [CrossRef]
38. Rajagopalan, P.; Lim, K.C.; Jamei, E. Urban heat island and wind flow characteristics of a tropical city. *Sol. Energy* **2014**, *107*, 159–170. [CrossRef]

39. Edussuriya, P.; Chan, A.; Ye, A. Urban morphology and air quality in dense residential environments in Hong Kong. Part I: District-level analysis. *Atmos. Environ.* **2011**, *45*, 4789–4803. [CrossRef]
40. He, B.-J.; Ding, L.; Prasad, D. Relationships among local-scale urban morphology, urban ventilation, urban heat island and outdoor thermal comfort under sea breeze influence. *Sustain. Cities Soc.* **2020**, *60*, 102289. [CrossRef]
41. He, B.-J.; Ding, L.; Prasad, D. Wind-sensitive urban planning and design: Precinct ventilation performance and its potential for local warming mitigation in an open midrise gridiron precinct. *J. Build. Eng.* **2020**, *29*, 101145. [CrossRef]
42. He, B.-J.; Ding, L.; Prasad, D. Urban ventilation and its potential for local warming mitigation: A field experiment in an open low-rise gridiron precinct. *Sustain. Cities Soc.* **2020**, *55*, 102028. [CrossRef]
43. Zhou, X.; Okaze, T.; Ren, C.; Cai, M.; Ishida, Y.; Watanabe, H.; Mochida, A. Evaluation of urban heat islands using local climate zones and the influence of sea-land breeze. *Sustain. Cities Soc.* **2020**, *55*, 102060. [CrossRef]
44. Zhao, Z.-Q.; He, B.-J.; Li, L.-G.; Wang, H.-B.; Darko, A. Profile and concentric zonal analysis of relationships between land use/land cover and land surface temperature: Case study of Shenyang, China. *Energy Build.* **2017**, *155*, 282–295. [CrossRef]
45. Xu, Z.; Yu, D.; Jing, L.; Xu, X. Air pollution and daily mortality in Shenyang, China. *Arch. Environ. Health Int. J.* **2000**, *55*, 115–120. [CrossRef]
46. Ma, Y.; Chen, R.; Pan, G.; Xu, X.; Song, W.; Chen, B.; Kan, H. Fine particulate air pollution and daily mortality in Shenyang, China. *Sci. Total. Environ.* **2011**, *409*, 2473–2477. [CrossRef]
47. Bryant, E.A. Local climate processes in the Illawarra. *Wollongong Stud. Geogr.* **1982**, *11*, 1–4.
48. He, B.-J.; Ding, L.; Prasad, D. Outdoor thermal environment of an open space under sea breeze: A mobile experience in a coastal city of Sydney, Australia. *Urban Clim.* **2020**, *31*, 100567. [CrossRef]
49. Yu, M.; Miao, S.; Li, Q. Synoptic analysis and urban signatures of a heavy rainfall on 7 August 2015 in Beijing. *J. Geophys. Res. Atmos.* **2017**, *122*, 65–78. [CrossRef]
50. Yu, M.; Miao, S.; Zhang, H. Uncertainties in the impact of urbanization on heavy rainfall: Case study of a rainfall event in Beijing on 7 August 2015. *J. Geophys. Res. Atmos.* **2018**, *123*, 6005–6021. [CrossRef]

atmosphere

Article

Study on the Accelerated Aging Test Method in the Development of a Self-Cleaning Topcoat for Cool Roofs

Taizo Aoyama [1,*], Takeshi Sonoda [1] and Hideki Takebayashi [2]

1 Kaneka Corporation, 1–8, Miyamae-cho, Takasago-cho, Takasago Hyogo 676-8688, Japan; Takeshi.Sonoda@kaneka.co.jp
2 Department of Architecture, Graduate School of Engineering, Kobe University, Kobe Hyogo 657-8501, Japan; thideki@kobe-u.ac.jp
* Correspondence: Taizo.Aoyama@kaneka.co.jp; Tel.: +81-50-3133-6683

Received: 6 May 2020; Accepted: 4 June 2020; Published: 8 June 2020

Abstract: The heat island phenomenon in urban areas has become a problem in the recent years. One measure to overcome this, which has been attracting attention, is painting the cool roof with a paint that has high solar reflectance. However, the solar reflectance has been reported to decrease over time owing to staining from the paint. Our investigations reveal that a self-cleaning paint, based on an acrylic silicone polymer, is effective in maintaining a high solar reflectance and preventing dirt from adhering. These evaluations have been conducted on the cool roof of an actual building by conducting a pollution test under outdoor exposure conditions. In this study, we investigated an accelerated aging test and reported the results. ASTM D7897-15 (ASTM method) is an accelerated pollution test method for cool roofs, which has been established in the United States. In Japan, a similar accelerated pollution test method exists for antifouling civil engineering materials introduced by the Public Works Research Institute in Japan (PWRI method). The ASTM method and PWRI method were compared by conducting a stain acceleration test using a self-cleaning- and conventional-type cool-roof coating. The substitutability of the exposure test, correlation with the exposure test, performance evaluation of the paint, and the effect of the difference in the pretreatment conditions were verified. The results of these tests were reported.

Keywords: heat island; cool roof; solar reflectance; self-cleaning; accelerated aging test

1. Introduction

Cool-roof coatings have received considerable attention worldwide as a solution for the "heat island" effect in urban areas. The effects of cool-roof coatings and the reduction in the air-conditioning load have been studied [1–6]. Recently, many important studies on the evaluation of building roofs with conventional and reflective coating, and the energy gain of building roofs with many color types of waterproofing materials have been reported [7–9].

However, it has been pointed out that the solar reflectance of cool-roof coatings decreases shortly after installation [10–15]. Supposedly, the decrease in the solar reflectance is due to the soiling of the coating, chemical–physical modifications, and biological growth. Various studies on evaluation methods for cool-roof coatings have been carried out [16,17]. The reduction in reflectance has been studied in exposure tests using various materials. Changes in solar reflectance are measured every three months and are affected by the location of exposure, and it has been reported that the reflectance decreases by approximately 10% to 20% within 3 years [18,19]. Regarding the high-reflectance paint on the wall, the contamination status in the exposure test and the cleaning effect by precipitation have been reported [20]. In Japan, the main component of soil is carbon; it has been reported that, after

16 weeks of exposure of the white test piece, the reflectance in the visible light region is reduced by about 5–10% [21].

Self-cleaning paint is used for the wall coatings of buildings, and its ability to minimize soiling, especially dirt pick-up, has been reported [22].

There are two types of self-cleaning systems: a photocatalyst system with titanium oxide [17] and an alkyl silicate system [23–25]. A photocatalyst system has high hydrophilicity but cannot be used on organic-based paint because decomposition occurs. In addition to the usual paint layer, two more layers, i.e., a protective layer and photocatalyst layer, are required. Therefore, the photocatalyst system is somewhat expensive. On the other hand, the alkyl silicate system works as well on ordinary paint, and the self-cleaning layer with base-coating binder resin is applied via a one-coat process. An acrylic silicon polymer is a base resin that is useful for formulating self-cleaning paint with alkyl silicate. Therefore, acrylic silicon polymer and the alkyl silicate system is a useful self-cleaning paint that has a simple coating process and high cost-performance. We report the application of this system for cool-roof coatings.

We have investigated the effect of self-cleaning technology with an acrylic silicon polymer-based alkyl silicate system on the solar reflectance of a cool-roof coating. From data obtained through real outdoor exposure, we calculated the energy-saving effects of self-cleaning technology using the thermal load calculation model (Energy Plus) for sites in Japan (Osaka), Malaysia (Kuala Lumpur), and Thailand (Bangkok) [24]. In addition, we have found the influence of the annual cycle of the solar altitude in the change of reflectance, and it was suggested that there is a possibility of increase in reflectance due to coating deterioration with the whitening of chalking [26]. The deterioration of high-reflectance paint has been reported in connection with a loss of gloss and chalking of the coating surface [27].

Studies have been conducted to investigate the decrease in the solar reflectance due to soiling of the coating and the relationship between outdoor exposure and accelerated aging tests in particular [19]. The method examined here was termed as ASTM D7897-15 (ASTM method) [28]. A QUV (QUV Accelerated Weathering Tester) device was used to promote surface degradation, and the decrease in reflectance due to the spray coating of contaminants was evaluated. The pollutant was based on the average of the analysis values of the exposure results at three locations in the United States, and a mixture of dust, salts, humic acid, and carbon black was used at a low concentration. The target material was a material with a wide reflectance range. In addition, various accelerated aging methods were reviewed, including ASTM D7897-15, and the results of the trials of accelerated tests were reported with proper citations [21]. There is a method introduced by the Public Works Research Institute in Japan (PWRI method), which evaluates the dirt on antifouling materials in civil engineering structures [29]. The JIS proposals method and Kithutaka method [30] also aim to evaluate the stains on building walls. In the PWRI method, only high-concentration carbon black is used as a pollutant based on the Japanese exposure test environment. In addition, the antifouling material is the target, which is treated at high humidity in the pretreatment, and the contaminants are washed off during the post-treatment. In addition, Osaka HITEC (Osaka Heat Island Countermeasure Technology Consortium) has recently proposed a method in which the pretreatment and post-treatment are performed in accordance with the Japanese equipment environment, using a pollutant similar to that used in the ASTM method [31]. The Japan Federation of Construction Contractors has proposed a pretreatment method that involves immersion in water, followed by usage of a contaminant similar to that used in the PWIR method, and post-treatment, which involves washing it with an ultrasonic cleaner and running water through it [32,33].

It should be noted that the results of the evaluation of self-cleaning-type high-reflectance paints by these stain acceleration test methods have hardly been reported. This study investigates the PWRI method, which is an accelerated aging test method for antifouling materials that is compatible with the Japanese environment. It also evaluates the ASTM method, which is set for a wide range of materials, in accordance with the environment in the United States. Substitutes for the exposure test in

Japan and their correlation with the exposure test were investigated using self-cleaning-type paint and conventional-type paint as evaluation target materials. In addition, the performance of the paints was evaluated and the effects of variance in pretreatment conditions were verified. These results have been reported.

In the United States, an accelerated aging test method has been proposed and applied to various cool-roof materials. However, it is necessary to study its applicability in areas having different weather and atmospheric conditions. In addition, we must study the application to typical cool-roof materials in areas having different surface-coating characteristics. Therefore, in this study, we examined the applicability and problems of the test method for paint products commonly used in Japan, under the typical weather and atmospheric conditions in Japan.

2. Experiments

2.1. Outline of Materials

The coating materials used in the test were listed in Table 1. Most of the materials contained an acrylic silicone polymer coating. Three solvent systems, including two self-cleaning types, and four water systems, including one self-cleaning type, were used.

Table 1. Cool-roof coating materials used in the test.

Mark	Self-Cleaning	Polymer Component	Solvent System	Comparison
(1) AS-S	Conventional	Acrylic silicon	Solvent borne	conventional acrylic silicon polymer coating
(2) SC-AS-S	Self-cleaning	Acrylic silicon	Solvent borne	comparison with (1) AS-S for Self-Cleaning
(3) SC-F-S	Self-cleaning	Fluorine-based	Solvent borne	comparison with (2) SC-AS-S for using fluorine as the polymer species
(4) AS-W1	Conventional	Acrylic silicon	Water borne	water borne, conventional acrylic silicon polymer coating
(5) AS-W2	Conventional	Acrylic silicon	Water borne	comparison with (4) AS-W1 for higher hardness and lower contamination
(6) A-W	Conventional	Acrylic	Water borne	comparison with (4) AS-W1 for using acrylic as the polymer species
(7) SC-AS-W	Self-cleaning	Acrylic silicon	Water borne	comparison with (4)AS-W1 for Self-Cleaning

The composition and cleaning mechanism of the self-cleaning type of paint were detailed in the previous report [24,26]. It is composed of two liquids containing a base material made of an acrylic resin containing an alkoxysilyl group and a self-cleaning imparting agent containing an alkyl silicate, a hydrolysis, and a condensation catalyst. When the paint is used, the two liquids mix, and the paint is crosslinked and hardened by hydrolysis. A hydrophilic layer of an alkyl silicate is simultaneously formed on the paint surface, thereby exhibiting a low staining property. Figure 1 illustrates the details of the water-based self-cleaning two-component acrylic silicone paint.

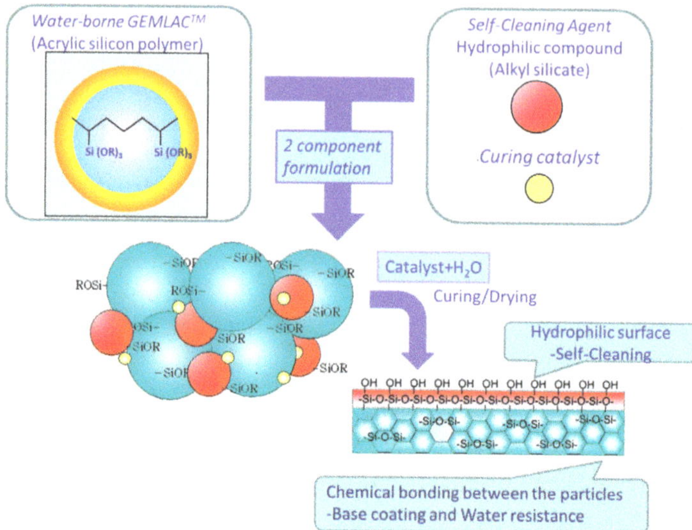

Figure 1. Waterborne two-component acrylic silicon polymer system and hydrophilic surface formation.

The mechanism of the self-cleaning property is explained. The hydrophilic layer formed on the surface of the paint makes it difficult for hydrophobic contaminants to adhere, and the hydrophobic stains adhered on the surface are simultaneously washed with rainwater. The details are illustrated in Figure 2.

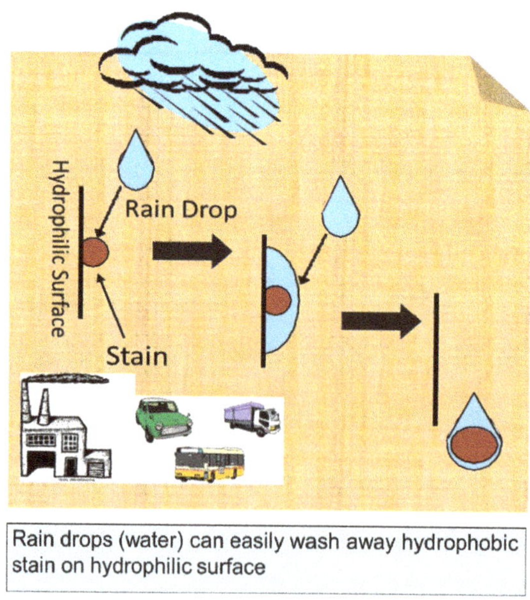

Figure 2. Self-cleaning mechanism of a hydrophilic surface.

As detailed in Table 1, the self-cleaning ability of the material used in the solvent system was investigated by comparing SC-AS-S and AS-S. In addition, SC-F-S was used to investigate the influence of fluorine-based polymer in comparison with SC-AS-S. In the water system, AS-W2 possessed a higher degree of cross-linking after curing the coating material than AS-W1. Therefore, it was a material having a high hardness and stain resistance. A-W was compared with AS-W1 in order to investigate the effect of using a general-purpose acrylic polymer as the polymer species. SC-AS-W was compared with AS-W1 in order to investigate the effect of a self-cleaning property. In the acrylic silicone-based paint, the two-component curing type is typically used in the solvent system, and the two-component curing type is generally used in the self-cleaning type in the water system. Assigning and abbreviation of coating materials are given in Appendix B.

2.2. Test Specimen

An anodic oxidized aluminum plate (70 mm × 150 mm × 0.8 mm) was painted with an epoxy-based sealer and a commercially available white cool-roof paint to make the base plate. The test paints were applied to the base plate and cured for no less than 14 days at room temperature to create the test specimens. Influence of the sample size and the substrate on reflectance are given in Appendix C.

2.3. Exposure Testing Method

In order to enable comparison, each specimen was exposed to outdoor conditions at 45° in Settsu City (Figure 3), Osaka. The exposure period was about 3 years, starting from 26 April 2012 to 27 April 2015. During this period, substantial exposure was carried out for 30 months (36 months including the loss time for measurement). The weather information (temperature, humidity, precipitation, solar radiation) and pollutant information (total hydrocarbon concentration, nitrogen oxide concentration) during this period are shown in the information of Higashinari Ward, Osaka City, which is close to Settsu City, Osaka Prefecture in Figures 4–6 (http:/taiki.kankyo.pref.osaka.jp/taikikanshi/). Information related to outdoor exposure is given in Appendix B. It tends to be hot and humid in summer but cold and dry in winter. Precipitation tends to increase as the rainy season (June) progresses to autumn. The total amount of hydrocarbons and nitrogen oxides tends to increase in the winter, because the ground surface cools in the winter and the vertical mixing becomes low, and pollutants are likely to accumulate. The total hydrocarbon content is approximately two orders of magnitude higher than the nitrogen oxide content. Osaka is located in western Japan, and is hotter and more humid than eastern Japan.

Figure 3. Outdoor exposure test at Osaka.

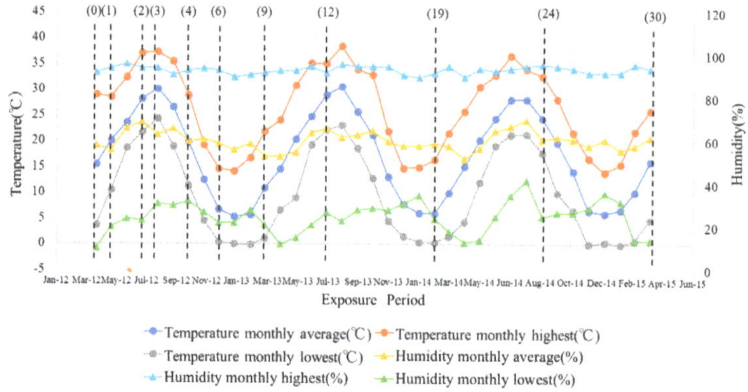

Figure 4. Weather information during the exposure period in Osaka (1).

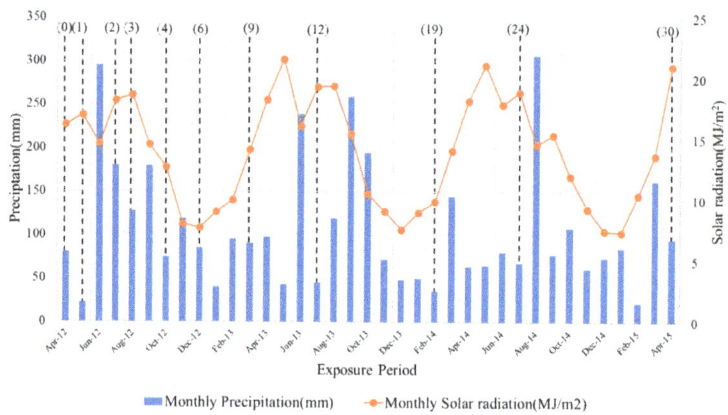

Figure 5. Weather information during the exposure period in Osaka (2).

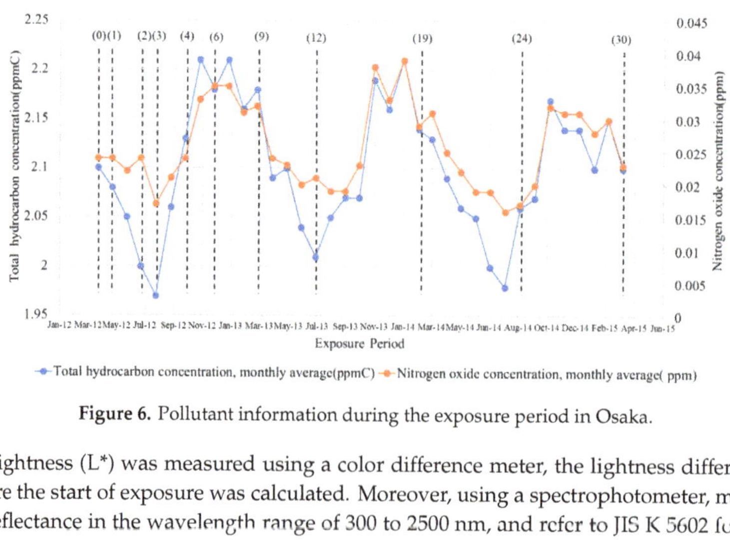

Figure 6. Pollutant information during the exposure period in Osaka.

The lightness (L*) was measured using a color difference meter, the lightness difference (ΔL*) from before the start of exposure was calculated. Moreover, using a spectrophotometer, measure the spectral reflectance in the wavelength range of 300 to 2500 nm, and refer to JIS K 5602 for the solar reflectance of all wavelengths, calculation was performed.

2.4. Accelerated Testing Method

The accelerated test consists of three steps: pretreatment, pollution treatment, and post-treatment. All three steps use the ASTM method [28], in order to investigate the effect of pretreatment of the ASTM method, after pretreatment by the ASTM method, contamination treatment and post-treatment by the PWRI method (combined method), and all three processes are PWRI methods (PWRI method) [29] were performed. These test methods were compared in Table 2. The composition of the soiling solution was outlined in Table 3. The lightness (L*) and spectral reflectance were measured in the same manner as in the exposure test, and the lightness difference (ΔL*) and solar reflectance retention were calculated. Further details of the accelerated aging test are given in Appendix A.

Table 2. Accelerated aging test methods.

Method	ASTM Method [28]	Combined Method	PWRI Method [29]
Pretreatment	Conditioning by QUV: ASTM G154 Cycle * 2 times, 24 h		Conditioning: Wet conditions, 50 ± 1 °C, humidity of 95% or more, 24 h
Soiling treatment	Mixing soiling solution (see Table 3) spray coating Uniformly applying an 8 ± 1 mg/cm^2 2–10 min dried using an infrared lamp (250 W) **	Carbon black aqueous dispersion (see Table 3) spray coating Uniformly on the surface After 1 h drying at 60 °C, allowed to cool to 23 °C	
Post-treatment	Conditioning by QUV: ASTM G154 Cycle * 2 times, 24 h	Wash under running water until the dirt substance is no longer fall with a gauze (BEMCOT M-3 "manufactured by Asahi Kasei Co., Ltd.")	

* Cycle: Irradiation: 8 h (UVA-340, Irradiation strength: 0.89 W/m^2/nm, 60 °C), Condensation: 4 h (50 °C) ** That the surface temperature does not exceed 80 °C.

Table 3. Soiling composition.

	Soiling Materials	ASTM Method (%)	PWRI Method (%)
Soot	Aquablack 001 (TOKAI Carbon, Inc.)	0.0065	
	Color Black FW200 (Orion Engineered Carbons)		5.0000
Dust	Iron oxide(Fe_2O_3)	0.0075	
	Montmorionite	0.0250	
	Bentonite	0.0250	
Salts	Sodium chloride	0.0075	
	Sodium nitrate	0.0075	
	Calcium sulfate dehydrate	0.0100	
POM	Humic Acid	0.0350	
Solvent	Deionized water	99.8760	95.0000
	Total	100.0000	100.0000

3. Results

3.1. Result of Exposure Test

Changes in lightness (L*) and difference in lightness (ΔL*) (30 months) are shown in Figures 7 and 8. The initial value of the lightness (L*) varies depending on the paint type, but the SC-AS-S, SC-F-S, and SC-AS-W self-cleaning-type paint exhibit a smaller fluctuation in lightness than conventional paint. The self-cleaning-type paint showed good results even in the difference in lightness. It is considered that the hydrophilic layer on the surface of the self-cleaning-type paint causes less adherence of pollutants; the adhered pollutants are washed by rainwater. In addition, as shown in Figure 8, SC-F-S, which is a solvent-based self-cleaning-type fluoropolymer, exhibits smaller difference in lightness than SC-AS-S, which is a self-cleaning acrylic silicon-based polymer. It is considered that the adhesion of pollutants due to the oil-repellent fluoropolymer was suppressed and the effect of cleaning by the self-cleaning type. In addition, in the water system, the conventional acrylic silicon-based AS-W2, which has a high hardness, has a smaller difference in lightness than the conventional acrylic silicon-based AS-W1. It is conceivable that, due to the higher hardness, the adherence of contaminants was reduced. In addition, there is a large difference in lightness between the conventional and general-purpose acrylic A-W, and the difference in lightness is the largest at 30 months, and the acrylic polymer is inferior in weather resistance to acrylic silicon. It is considered that the effect is caused by the decrease in gloss and the adhesion of dirt.

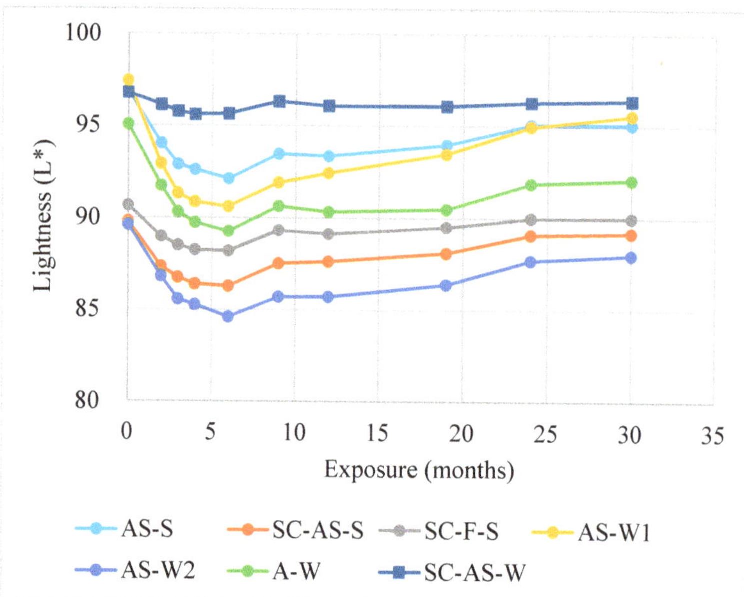

Figure 7. Lightness of exposure test, AS-S (conventional, acrylic silicon, solvent borne), SC-AS-S (self-cleaning, acrylic silicon, solvent borne), SC-F-S (self-cleaning, fluorine based, solvent borne), AS-W1 (conventional, acrylic silicon, water borne), AS-W2 (conventional, acrylic silicon, water borne, higher hardness), A-W (conventional, acrylic, water borne), SC-AS-W (self-cleaning, acrylic silicon, water borne).

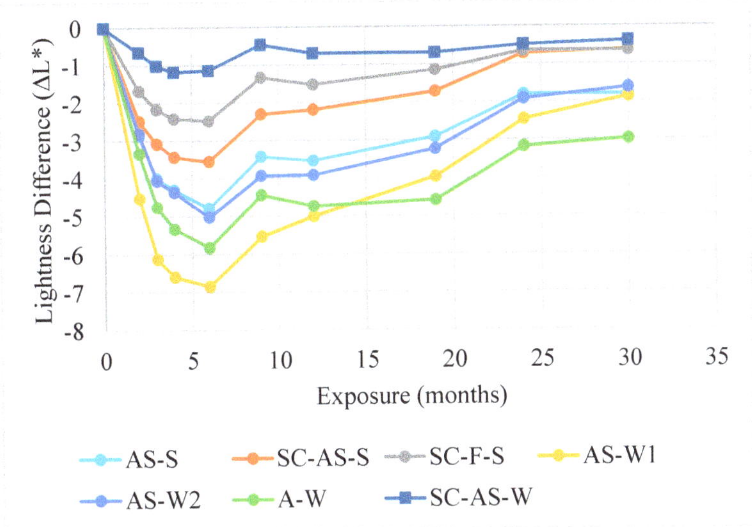

Figure 8. Difference of exposure test, AS-S (conventional, acrylic silicon, solvent borne), SC-AS-S (self-cleaning, acrylic silicon, solvent borne), SC-F-S (self-cleaning, fluorine based, solvent borne), AS-W1 (conventional, acrylic silicon, water borne), AS-W2 (conventional, acrylic silicon, water borne, higher hardness), A-W (conventional, acrylic, water borne), SC-AS-W (self-cleaning, acrylic silicon, water borne).

The solar reflectance and solar reflectance retention of the exposure test (30 months) was illustrated in Figures 9 and 10. There was a small decrease in the reflectance of the self-cleaning-type coatings of SC-AS-S, SC-F-S, and SC-AS-W. The retention of the reflectance was also high for the self-cleaning-type coating. The reflectance retention was reduced by approximately 10–15% for conventional coatings and 5–10% for self-cleaning coatings. Furthermore, as shown in Figure 10, in the same manner as the lightness difference in the reflectance retention, the SC-F-S of the fluoropolymer is higher than the SC-AS-S of the acrylic silicone type. AS-W2 has higher hardness than AS-W1 and has higher reflectance retention than AS-W1. A-W is a general-purpose acrylic, which has a lower reflectance retention for a long period.

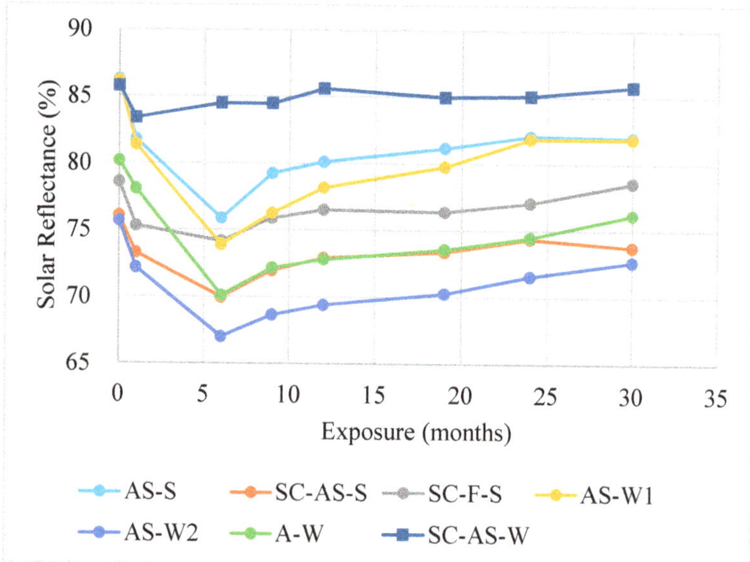

Figure 9. Reflectance of exposure test, AS-S (conventional, acrylic silicon, solvent borne), SC-AS-S (self-cleaning, acrylic silicon, solvent borne), SC-F-S (self-cleaning, fluorine based, solvent borne), AS-W1 (conventional, acrylic silicon, water borne), AS-W2 (conventional, acrylic silicon, water borne, higher hardness), A-W (conventional, acrylic, water borne), SC-AS-W (self-cleaning, acrylic silicon, water borne).

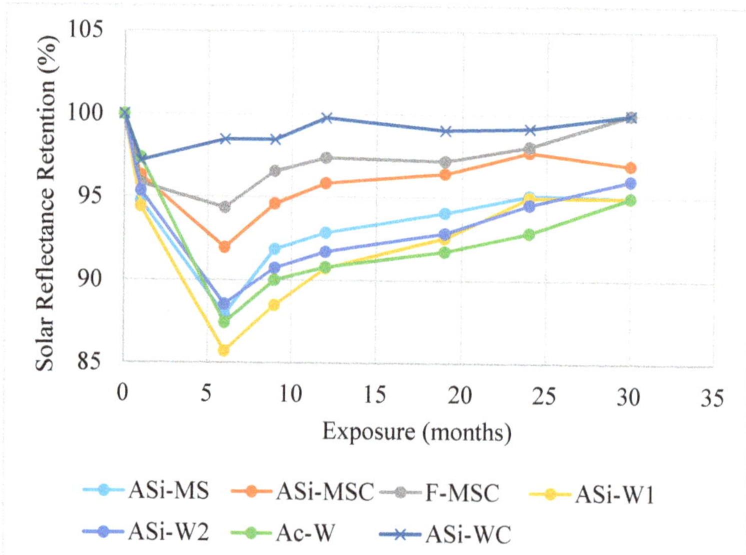

Figure 10. Reflectance retention of exposure test, AS-S (conventional, acrylic silicon, solvent borne), SC-AS-S (self-cleaning, acrylic silicon, solvent borne), SC-F-S (self-cleaning, fluorine based, solvent borne), AS-W1 (conventional, acrylic silicon, water borne), AS-W2 (conventional, acrylic silicon, water borne, higher hardness), A-W (conventional, acrylic, water borne), SC-AS-W (self-cleaning, acrylic silicon, water borne).

The reflection spectrum, which is the basic data of Figure 9, is shown in Figures 11 and 12 as a representative sample. The self-cleaning-type SC-AS-W hardly changed, even after 6 months of exposure, whereas the conventional-type AS-W1 showed a decrease in reflectance in the visible to the near-infrared wavelength region.

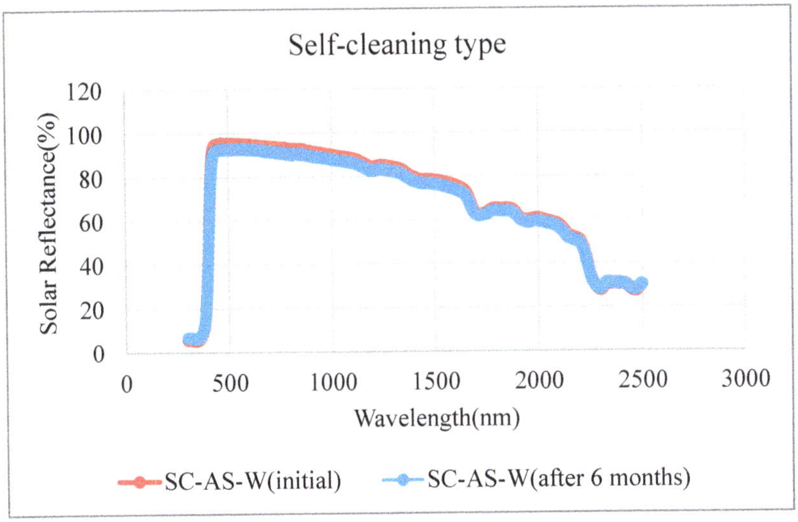

Figure 11. Spectral reflectance of SC-AS-W (self-cleaning, acrylic silicon, water borne), initial and after 6 months exposure.

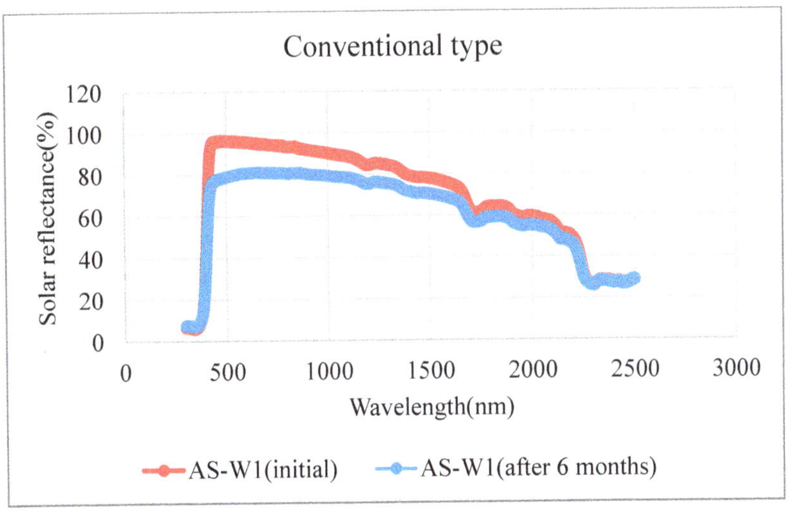

Figure 12. Spectral reflectance of AS-W1 (conventional, acrylic silicon, water borne), initial and after 6 months exposure.

3.2. Result of Accelerated Aging Test

3.2.1. The State of Each Test Piece after the Exposure and Accelerated Test

Photographs of the initial test pieces and accelerated test pieces from the PWRI method, combined method, and ASTM method, and test pieces after 30-month outdoor exposure are shown in

Tables 4 and 5. In addition, photographs of the processing status in each process of the PWRI method and ASTM method are presented in Tables 6 and 7.

Table 4. Result of the accelerated test and exposure test for the solvent-borne coating.

	Initial	PWRI Method	Combined Method	ASTM Method	Outdoor Exposure 30 Months in Osaka
(1) AS-S				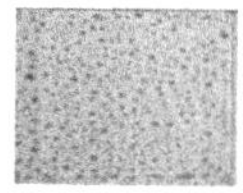	
(2) SC-AS-S				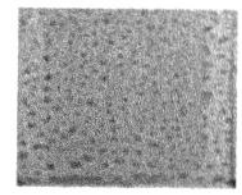	
(3) SC-F-S			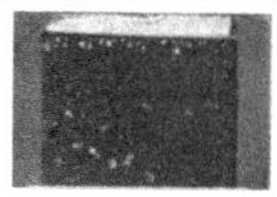		

(1) AS-S (Conventional, acrylic silicon, solvent borne), (2) SC-AS-S (Self-cleaning, acrylic silicon, solvent borne), (3) SC-F-S (Self-cleaning, fluorine-based, solvent borne).

Table 5. Result of the accelerated test and exposure test for the water-borne coating.

	Initial	PWRI Method	Combined Method	ASTM Method	Outdoor Exposure 30 Months in Osaka
(4) AS-W1				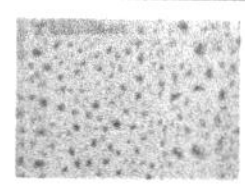	
(5) AS-W2				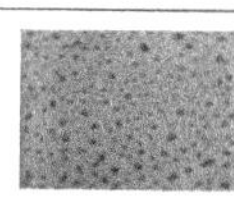	
(6) A-W		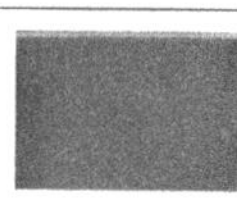	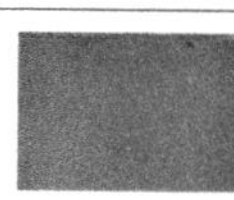	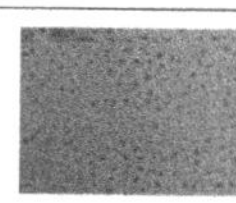	
(7) SC-AS-W		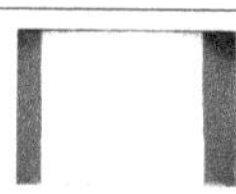		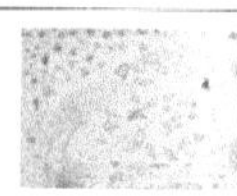	

(4) AS-W1 (conventional, acrylic silicon, water borne), (5) AS-W2 (conventional, acrylic silicon, water borne, higher hardness), (6) A-W (conventional, acrylic, water borne), (7) SC-AS-W (self-cleaning, acrylic silicon, water borne).

Table 6. Status of each treatment under the Public Works Research Institute (PWRI) method.

After Pretreatment	After Sprayinga Pollutant	After Dryinga Pollutant	AfterPost-Treatment
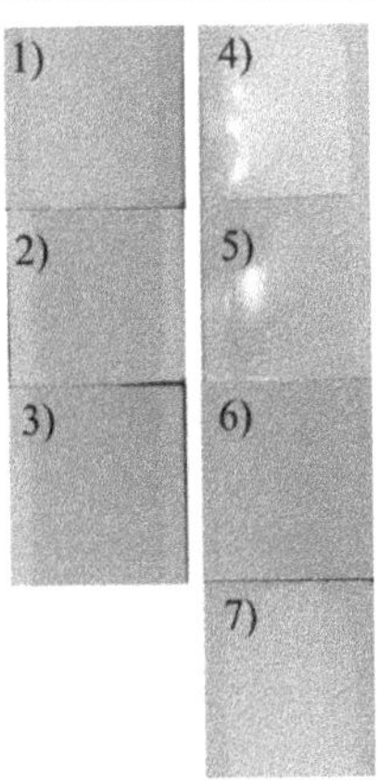		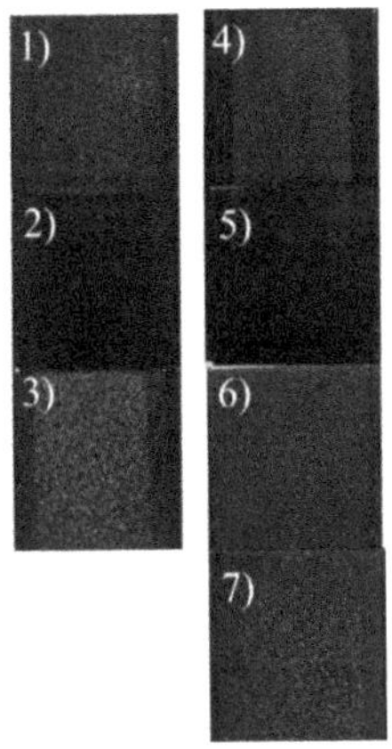	

(1) AS-S (conventional, acrylic silicon, solvent borne), (2) SC-AS-S (self-cleaning, acrylic silicon, solvent borne), (3) SC-F-S (self-cleaning, fluorine based, solvent borne), (4) AS-W1 (conventional, acrylic silicon, water borne), (5) AS-W2 (conventional, acrylic silicon, water borne, higher hardness), (6) A-W (conventional, acrylic, water borne), (7) SC-AS-W (self-cleaning, acrylic silicon, water borne).

Table 7. Status of each treatment under ASTM D7897-15 (ASTM method).

After Pretreatment	After Sprayinga Pollutant	After Dryinga Pollutant	AfterPost-Treatment
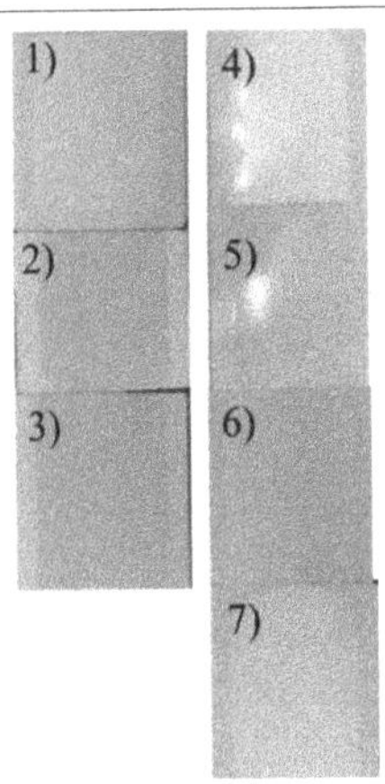	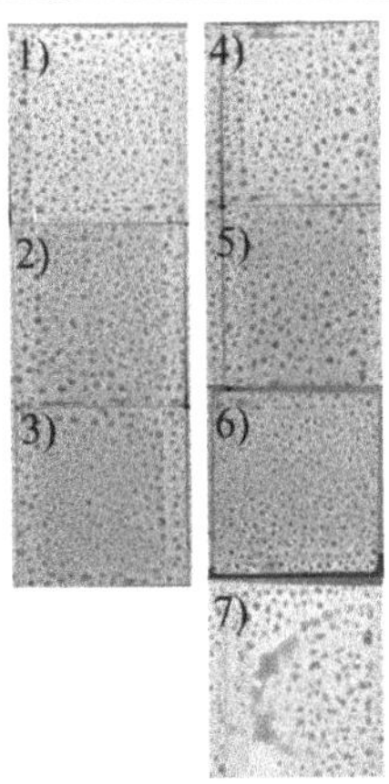	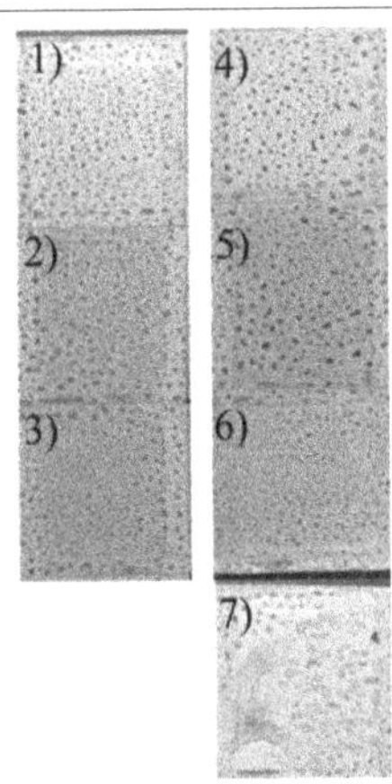	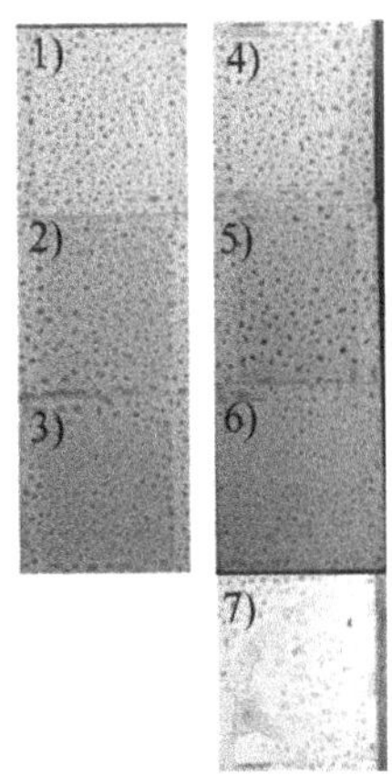

(1) AS-S (conventional, acrylic silicon, solvent borne), (2) SC-AS-S (self-cleaning, acrylic silicon, solvent borne), (3) SC-F-S (self-cleaning, fluorine based, solvent borne), (4) AS-W1 (conventional, acrylic silicon, water borne), (5) AS-W2 (conventional, acrylic silicon, water borne, higher hardness), (6) A-W (conventional, acrylic, water borne), (7) SC-AS-W (self-cleaning, acrylic silicon, water borne).

The PWRI method illustrates that the self-cleaning-type paint has less contamination than conventional paint, and it is possible to detect the characteristics of the self-cleaning sample. However, the degree of contamination of conventional paints was found to be significantly higher than that of outdoor exposure. As shown in Table 6, in the PWRI method, the high-concentration pollutant consisting of carbon black is considered to be uniformly sprayed.

As depicted in Tables 4 and 5, the combined method exhibited the identical tendency as the PWRI method, but the contamination was found to increase significantly in the fluorine-based self-cleaning-type paint. In the pretreatment process, UV irradiation and heating under the same conditions as in the ASTM method are considered to have affected the development of self-cleaning property due to the hydrophilicity of the coating surface.

In addition, although the contamination level in the ASTM method is similar to that in the exposure test, the difference between the self-cleaning type and the conventional type is small. Therefore, it is difficult to detect the difference between the samples, as shown in Table 7. It was found that a uniform coating is difficult to attain on a highly glossy surface, because contaminants repel it after spraying.

3.2.2. Relationship between Outdoor Exposure Test and Accelerated Aging Test

Table 8 illustrates the initial reflectance of each test piece, reflectance after the accelerated aging test, and reflectance after the outdoor exposure test for 6 months and 30 months. Furthermore, Table 9 shows the reflectance retention after the outdoor exposure test for 6 months and 30 months and after the accelerated pollution test. Figures 13 and 14 indicate the correlation between the reflectance retention rate in the outdoor exposure test and the reflectance retention rate in the accelerated contamination test.

Table 8. Solar reflectance of initial, outdoor exposure test and accelerating test.

Mark	Initial Solar Reflectance	Solar Reflectance (%)				
		PWRI Method	Combined Method	ASTM Method	Outdoor Exposure Test (6 months)	Outdoor Exposure Test (30 months)
(1) AS-S	86	49	51	77	76	82
(2) SC-AS-S	76	44	41	67	70	74
(3) SC-F-S	79	76	25	73	76	79
(4) AS-W1	86	27	27	79	74	82
(5) AS-W2	76	27	28	66	68	73
(6) A-W	80	32	31	74	71	76
(7) SC-AS-W	86	82	79	80	85	86

(1) AS-S (conventional, acrylic silicon, solvent borne), (2) SC-AS-S (self-cleaning, acrylic silicon, solvent borne), (3) SC-F-S (self-cleaning, fluorine based, solvent borne), (4) AS-W1 (conventional, acrylic silicon, water borne), (5) AS-W2 (conventional, acrylic silicon, water borne, higher hardness), (6) A-W (conventional, acrylic, water borne), (7) SC-AS-W (self-cleaning, acrylic silicon, water borne).

Table 9. Solar reflectance retention obtained by outdoor exposure test and accelerating test.

Mark	Solar Reflectance Retention (%)				
	PWRI Method	Combined Method	ASTM Method	Outdoor Exposure Test (6 months)	Outdoor Exposure Test (30 months)
(1) AS-S	57	59	89	88	95
(2) SC-AS-S	58	54	89	92	97
(3) SC-F-S	96	31	92	96	100
(4) AS-W1	31	31	91	86	95
(5) AS-W2	36	37	87	89	96
(6) A-W	41	39	92	89	95
(7) SC-AS-W	95	92	93	99	100

(1) AS-S (conventional, acrylic silicon, solvent borne), (2) SC-AS-S (self-cleaning, acrylic silicon, solvent borne), (3) SC-F-S (self-cleaning, fluorine based, solvent borne), (4) AS-W1 (conventional, acrylic silicon, water borne), (5) AS-W2 (conventional, acrylic silicon, water borne, higher hardness), (6) A-W (conventional, acrylic, water borne), (7) SC-AS-W (self-cleaning, acrylic silicon, water borne).

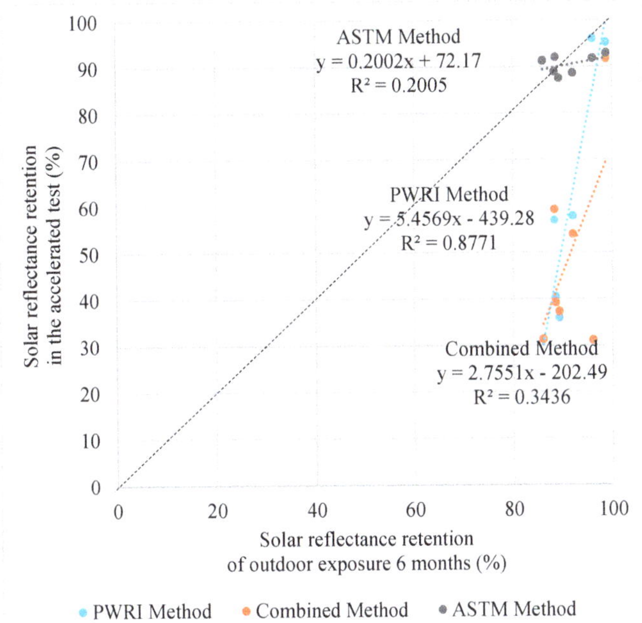

Figure 13. Correlation between the accelerated test and outdoor exposure test (6 months; solar reflectance retention).

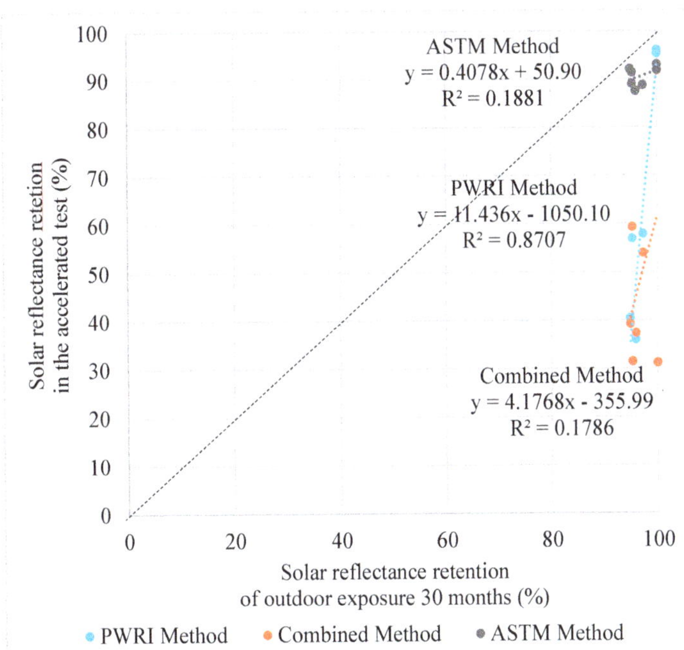

Figure 14. Correlation between the accelerated test and outdoor exposure test (30 months; solar reflectance retention).

1. Replaceability of the outdoor exposure test by the accelerated aging test method

The rate of solar reflectance decreases by approximately 0–15% in the ASTM method and the exposure test. However, there is an extremely large change in the PWRI method and the combined method. Therefore, the results of the PWRI method and combined method cannot be used as a substitute for the exposure test. In addition, the solar reflectance in the exposure test tends to recover from 6 months to 30 months, but the result of ASTM is close to that before recovery (6 months). The recovery of solar reflectance was also depicted in the previous report [26], but it is thought that the chalking (whitening) is due to the deterioration of the paint surface.

In the ASTM method, the reflectance retention in the accelerated pollution test was found to be close to the reflectance retention in the outdoor exposure test, and the result was closer to the approximate straight line of 1:1 in Figures 13 and 14. On the other hand, in the PWIR method and the combined method, the decrease in the solar reflectance retention in the conventional type is remarkably large, and it greatly deviates from the approximate straight line of 1:1 in outdoor exposure and accelerated test in Figures 13 and 14. It may be necessary to dilute the concentration of pollutants to match the reflectance retention of outdoor exposure.

2. Correlation between outdoor exposure test results and accelerated aging test results

A correlation was confirmed between the outdoor exposure test results and the PWRI method results. However, there was no correlation obtained between the ASTM method and combined method. Therefore, it is concluded that the results of the ASTM method and the combined method cannot be used to evaluate the performance of paints.

In the ASTM method, the difference in reflectance retention between the self-cleaning type and the conventional type was small, and it was difficult to detect the difference between the properties of samples, such as the self-cleaning property. In Figures 13 and 14, the coefficient of determination (R^2) of the approximate straight line in the outdoor exposure test and the accelerated pollution test was small. This is because, in the ASTM method, (1) there is little treatment at a high temperature and high humidity in order to make the surface of self-cleaning-type paint hydrophilic by pretreatment, and (2) pollutants are based on the analysis results in the United States and the mixture of pollutants was used at a low concentration, and (3) there was almost no wash-out removal of pollutants in the post-treatment, as shown in Table 7. Furthermore, the concentration of the pollutant is low. This makes uniform coating difficult because the contaminant is repelled on the surface of the high-gloss paint.

Furthermore, as described above, the PWRI method exhibited a tendency that the reflectance retention in the self-cleaning-type paint was high, and the difference between the samples could be detected. Figures 13 and 14 indicate that the coefficient of determination (R^2) of the approximate straight line was higher than that of the combined method and ASTM method in the relationship between the retention rate in the outdoor exposure test and the accelerated aging test. This is because (1) the hydrophilization of the paint surface in self-cleaning-type paint by pretreatment occurs at high humidity according to the Japanese meteorological environment, (2) in accordance with information on pollutants in the exposed environment of Japan, it is considered that carbon black is used at a high concentration as a pollutant, and (3) cleaning and removal with water was performed after spraying the pollutant, according to the cleaning mechanism of the self-cleaning-type paint.

In the combined method, the fluorine self-cleaning-type paint had a low reflectance retention value in the accelerated test method, and, due to this effect, the correlation was low as in the ASTM method in Figures 13 and 14. As a pretreatment condition for the accelerated contamination evaluation of self-cleaning-type paint, it is considered that the hydrophilicity of the paint surface was affected by light irradiation at a high temperature by the ASTM method QUV. As a pretreatment condition for the self-cleaning-type paint, it is considered that the high temperature and humidity conditions of the PWRI method are more suitable.

3. Relationship with weather information and pollution information

From the weather information and pollution information described in the experimental section, it is evident that the period from early summer (June) to autumn (October) in Japan has high levels of temperature, humidity, and precipitation. A high temperature causes the occurrence of upstream and downstream on the ground surface. It also decreases the total amount of hydrocarbons and the concentration of nitrogen oxide pollutants. Furthermore, high temperature and high humidity cause the surface of the self-cleaning paint to become hydrophilic, and the self-cleaning paint is easily washed by rainwater. High precipitation gives rise to the possibility that the pollution will be cleaned by rainwater and will thereby be suppressed to a low level. On the other hand, the winter season (November to March) is a period characterized by low temperature and a lack of moisture. Low temperatures affect the upstream and downstream of the ground surface decrease and increase the concentration of pollutants. As for the effect of drying, it is considered that as the amount of precipitation decreases, the cleaning of pollution decreases. In the self-cleaning-type paint used in this research, the hydrolysis, cross-linking, and curing of the paint is promoted by the high temperature and high humidity, and a hydrophilic layer is formed on the paint surface to suppress the adhesion of hydrophobic contaminants. It is considered that low pollution is achieved because the pollutants are cleaned by precipitation.

For the self-cleaning-type paint, the difference from the conventional type increases during the period of high temperature, high humidity, and high precipitation from summer to autumn, and the initial 6 months in the exposure test correspond to this period. The difference in reflectance tends to widen between the self-cleaning type and the conventional type.

4. Improvement of accelerated aging test method

The PWRI method will be used as the base method to improve the accelerated aging test method in Japan. It can detect the difference between self-cleaning-type paint and conventional-type paint and has a high correlation with the exposure test. By reducing the pollutant concentration of the PWRI method, we believe that it is possible to maintain the characteristics of correlation with the exposure test. It can also be made more coincident with the reflectance in the exposure test, and the substitutability of the exposure test can be enhanced.

The high correlation with the exposure test in the PWRI method is attributed to the hydrophilicity of the paint surface in the pretreatment, the use of carbon black corresponding to Japan as a pollutant, and the cleaning removal of the pollutant in the post-treatment. As for the reduction of pollutant concentration, the reflectance reduction rate of the conventional paint is about 40 to 70% by the PWRI method, while the reduction rate in the exposure test is approximately 5%–10%. It is approximately 1/8–1/7 of the acceleration test, and the carbon black concentration of the current PWRI method can be modified from the current 5% to 1/8–1/7, which is 0.6–0.7%. Since there is concern that the viscosity of the pollutant may be reduced and the coating property may be changed due to the reduction in the pollutant concentration, it is necessary to confirm these values before setting the proper pollutant concentration.

4. Conclusions

As previously reported, self-cleaning-type paints, based on acrylic silicone polymers, are effective in preventing dirt from adhering and maintaining high solar reflectance. In this study, the self-cleaning-type paint and conventional-type paint were used in the accelerated aging test method. This involved the ASTM method formulated in the United States and the PWRI method formulated for antifouling materials, in accordance with the contaminated environment in Japan. Both these accelerated aging test methods were compared.

The accelerated aging test method that was proposed in the United States was applied to various cool-roof materials. However, it is imperative to examine its application to other areas having different weather and atmospheric conditions and variation in the surface-coating characteristics of typical cool-roof materials. Therefore, in this study, we examined the applicability of the test method in Japan. We investigated the problems arising from using paint products commonly used in Japan and the typical weather and atmospheric conditions in Japan.

The decrease in solar reflectance in the outdoor exposure test and the ASTM method was approximately 0–15%, but there was a significant decrease in the PWRI method and the combined method. Therefore, the results of the PWRI method and the combined method cannot be used as a substitute for the exposure test. In addition, there is a correlation between the exposure test results and PWRI results, but not between the ASTM method and the combined method. Therefore, the results of the ASTM method and the combined method cannot be used to evaluate the performance of paints.

The PWRI method was established as an evaluation method for antifouling materials. It reflects the Japanese environment that has a high temperature, high humidity, and high concentration of pollutants mainly composed of carbon components. This affects the cleaning and removal of pollution due to precipitation, etc. It was possible to detect the difference between self-cleaning paint and conventional paint. It is difficult to detect differences between paints in the ASTM method because it uses a low-concentration mixture based on the US pollutant analysis results and has negligible cleaning and removal of contaminants. It was found that in the combined method (in which the pretreatment of the ASTM method is applied to the PWRI method), some paints are affected by the self-cleaning property by the pretreatment, and the correlation with the exposure test results is low.

According to the weather information and pollution information of Japan, the difference between the self-cleaning type and the conventional type is amplified due to the cleaning effect of pollutants caused by the high temperature, humidity, and precipitation in early summer to autumn. The early results of the exposure test showed that tendency.

The accelerated aging test method in Japan can be improved based on the PWRI method, which has a high correlation with the exposure test. The high correlation is due to the hydrophilicity of the paint surface during pretreatment, use of carbon black as a contaminant in Japan, and cleaning and removal of the contaminant during post-treatment. By reducing the current pollutant concentration from 5% to 0.6–0.7%, the reflectance of the accelerated test becomes coincident to that of the exposure test and the substitution of the exposure test is enhanced.

Author Contributions: Conceptualization, T.A., T.S. and H.T.; methodology, T.A. and T.S.; data curation, T.A. and T.S., writing—original draft preparation, T.A.; writing—review and editing, T.A. and H.T. All authors have read and agreed to the published version of the manuscript.

Funding: This research received no external funding.

Conflicts of Interest: The authors declare no conflict of interest.

Appendix A

Details and an abbreviation list of the accelerated aging test are shown below. The ASTM method involves pretreatment and post-treatment with QVA. It uses a mixture of dust, salts, humic acid, and carbon black at a low concentration based on the average value of the analysis results of the pollutants from the US exposure test. The ASTM method aims to coincide with the exposure test result based on the data of a wide range of reflectance of asphalt, tile, and metal. On the other hand, the PWRI method is an accelerated test method for antifouling materials that is suitable for the Japanese environment. Due to the high temperature and humidity in Japan, the surface of paints, such as the self-cleaning type, is made hydrophilic, and contaminants are washed away due to the high precipitation from the rainy season in June to autumn. In addition, the above-mentioned meteorological information indicates that pollutants in Japan contain a large amount of hydrocarbons; the previous report [21] also showed that the majority of the pollutants contain carbon components. Therefore, in the PWRI method, pretreatment is performed at a high temperature and humidity. A high-concentration contaminant of only carbon black is applied, and post-treatment is performed to wash the contaminants with water. In addition to information on pollutants in Japan, the PWIR method assumes that the exposure location will be along a main road, and that high-concentration carbon as a contaminant is used in order to make it easy for differences in antifouling material and is washed off with water after application. The pollutant information in Figure 6 indicates that the amount of total hydrocarbons is approximately 1.9–2.2 ppmC/month, and the amount of nitrogen oxides is 0.015–0.04 ppm/month. In the PWRI method, only carbon black is used in a highly concentrated 5% aqueous solution.

In this study, we evaluated the PWRI method, which is an accelerated pollution test method based on the environment in Japan. The ASTM method is set according to the environment in the United States. Self-cleaning-type and conventional-type paints were used as evaluation materials. This study was conducted for verifying the substitutability of the exposure test for the accelerated contamination test method in Japan, correlation between the accelerated test method and the exposure test, and the evaluation of the performance of paints.

Table A1. Abbreviation list of accelerated aging test.

	Pretreatment	Soiling	Post-Treatment
ASTM Method	Irradiation: 8 h QUV, 60 °C condensation: 4 h, 50 °C 2 times, 24 h	mixture of dust, humic acid, carbon black, low concentration	Irradiation: 8 h QUV, 60 °C condensation: 4 h, 50 °C 2 times, 24 h
Combined Method	Irradiation: 8 h QUV, 60 °C condensation: 4 h, 50 °C 2 times, 24 h	only carbon black, high concentration	Wash under running water
PWRI Method	Wet conditions, 50 ± 1 °C humidity of 95% or more 24 h	only carbon black, high concentration	Wash under running water

Appendix B

The assigning of abbreviation for coating materials, abbreviation list of coating materials and information related to outdoor exposure are shown in Tables A2 and A3.

(1) to (7) were the target materials in this study, that is solvent borne and water borne, self-cleaning type and conventional type, mainly acrylic silicone polymers. (3) and (6) were used as representative samples for polymer-type examination, and other fluorine-based and acrylic-based samples were not used. Acrylic solvent-borne and water-borne self-cleaning types have not been commercialized. In addition, SC-F-S was used to investigate the influence of fluorine-based polymer in comparison with SC-AS-S. A-W was compared with AS-W1 in order to investigate the effect of using a general-purpose acrylic polymer as the polymer species.

Table A2. Assigning of abbreviation for coating materials.

Solvent System	Cleaning Ability	Acrylic Silicon	Fluorine	Acrylic
Solvent borne	Conventional	(1) AS-S	F-S	A-S
	Self-Cleaning	(2) SC-AS-S	(3) SC-F-S	SC-A-S *
Water borne	Conventional	(4,5) AS-W1,2	F-W	(6) A-W
	Self-Cleaning	(7) SC-AS-W	SC-F-W	SC-A-W *

Self-Cleaning: conventional (-) or self-cleaning (SC) polymer component: acrylic silicon (AS), fluorine (F), acrylic (A) solvent system: solvent (S) or water (W) borne * SC-A-S, SC-A-W: not commercialized.

Table A3. Abbreviation list of coating materials and information related to outdoor exposure.

Abbreviation, Item	Content
AS-S	Solvent-borne acrylic silicon polymer coating, conventional
SC-AS-S	Solvent-borne acrylic silicon polymer coating, self-cleaning
SC-F-S	Solvent-borne fluorine-based polymer coating, self-cleaning
AS-W1	Water-borne acrylic silicon polymer coating, conventional
AS-W2	Water-borne acrylic silicon polymer coating, higher hardness, conventional
A-W	Water-borne acrylic polymer coating, conventional
SC-AS-W	Water-borne acrylic silicon polymer coating, self-cleaning
Outdoor exposure	Outdoor conditions at 45° in Settsu City, Osaka From 26 April 2012 to 27 April 2015, about 3 years
Weather, Pollutant information	Temperature, humidity (Figure 4), precipitation, solar radiation (Figure 5) Total hydrocarbon concentration, nitrogen oxide concentration (Figure 6) Higashinari ward, Osaka City, which is close to Settsu City (http://taiki.kankyo.pref.osaka.jp/taikikanshi/)

Appendix C

It was confirmed that there was sufficient evaluation accuracy and reproducibility even with the small size of the test plate (70 mm × 150 mm). Figure A1 shows the results of two exposure tests carried out on the small test plates (Test plate-1, 2). Sufficient accuracy and reproducible results of solar reflectance were obtained with the small test plates for the self-cleaning type and the conventional type.

The results of examining the influence of sample size on reflectance are shown in Figure A2. It showed almost the same reflectance behavior in the case of a small test plate (70 mm × 150 mm) and in a large area of an actual building (folded plate roof, the size of one flat part was 200 mm × 4000 mm) [34]. The same tendency was shown for both the self-cleaning type and the conventional type.

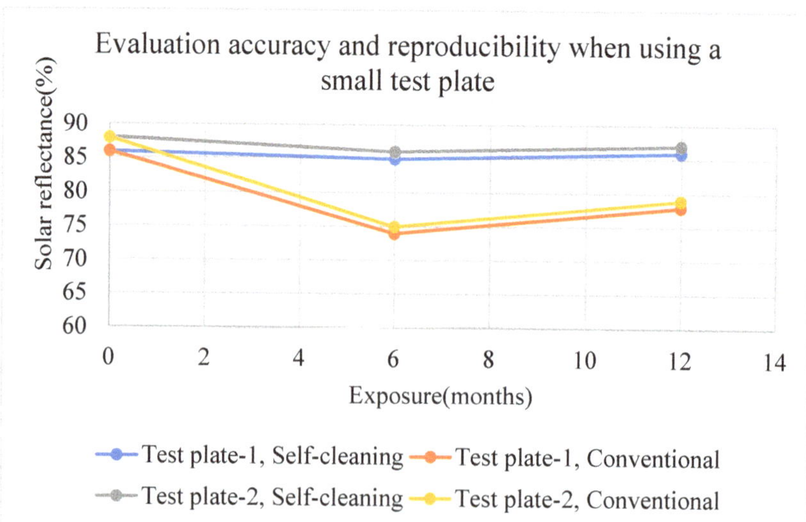

Figure A1. Evaluation accuracy and reproducibility when using a small test plate, Test plate-1, 2: 70 × 150 mm.

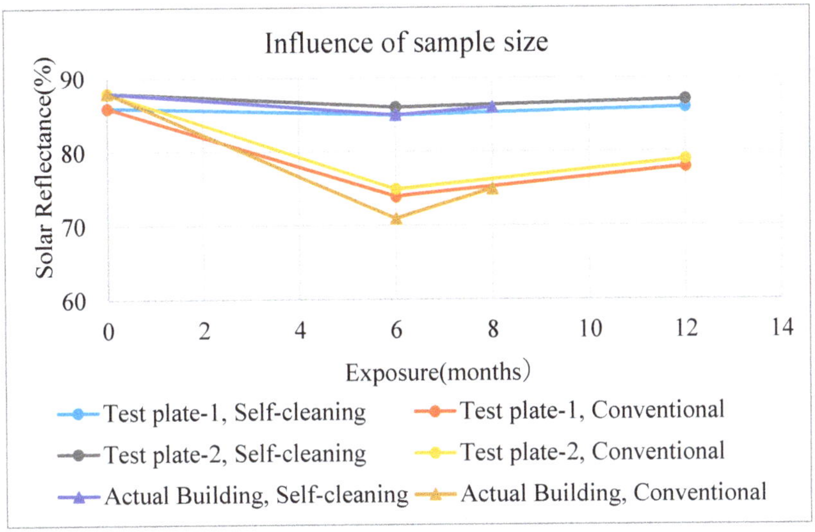

Figure A2. Influence of sample size on reflectance, test plate-1, 2: 70 × 150 mm, actual building: 200 × 4000 mm.

The results of examining the influence of substrate on reflectance are shown in Figure A3. The same behavior was observed between the slate plate and the aluminum plate with the self-cleaning type and the conventional type.

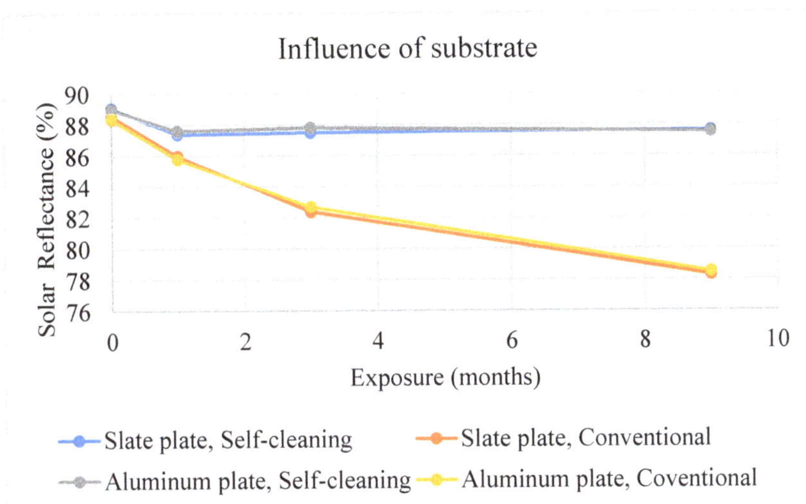

Figure A3. Influence of substrate on reflectance, slate and aluminum plate: 70 × 150 mm.

References

1. Akbari, H.; Konopacki, S.; Pomerantz, M. Cooling energy saving potential of reflective roofs for residential and commercial building in the United States. *Energy* **1999**, *24*, 391–407. [CrossRef]
2. Akbari, H.; Konopacki, S. Calculating energy-saving potentials of heat-island reduction strategies. *Energy Policy* **2005**, *33*, 721–756. [CrossRef]

3. Akbari, H.; Menon, S.; Rosenfeld, A. Global cooling: Increasing world-wide urban albedos to offset CO_2. *Climatic Chang.* **2009**, *94*, 275–286. [CrossRef]
4. Romeo, C.; Zinzi, M. Impact of a cool roof application on the energy and comfort performance in an existing non-residential building. A Sicilian case study. *Energy Build.* **2013**, *67*, 647–657. [CrossRef]
5. Synnefa, A.; Saliari, M.; Santamouris, M. Experimental and numerical assessment of the impact of increased reflectance on a school building in Athens. *Energy Build.* **2012**, *55*, 7–15. [CrossRef]
6. Takebayashi, H.; Yamada, C. Field observation of cooling energy saving due to high-reflectance paints. *Buildings* **2015**, *5*, 310–317. [CrossRef]
7. Hernandez-Perez, I.; Alvarez, G.; Xaman, J.; Zavala-Guillen, I.; Arce, J.; Sima, E. Thermal Performance of Reflective Materials Applied to Exterior Buildings Components—A Review. *Energy Build.* **2014**, *80*, 81–105. [CrossRef]
8. Hernandez-Perez, I.; Xaman, J.; Macias-Melo, E.V.; Aguilar-Castro, K.M.; Zavala-Guillen, I.; Hernandez-Lopez, I.; Sima, E. Experimental thermal evaluation of building roofs with conventional and reflective coatings. *Energy Build.* **2018**, *158*, 569–579. [CrossRef]
9. Hernandez-Perez, I.; Zavala-Guillen, I.; Xaman, J.; Belman-Flores, J.M.; Macias-Melo, E.V.; Aguilar-Castro, K.M. Test box experiment to assess the impact of waterproofing materials on the energy gain of building roofs in Mexico. *Energy* **2019**, *186*, 115847. [CrossRef]
10. Akbari, H.; Berhe, A.A.; Levinson, R. *Aging and Weathering of Cool Roofing Membranes;* Lawrence Berkeley National Laboratory: Berkeley, CA, USA, 2005.
11. Berdahl, P.A.; Akbari, H.; Levinson, R.; Miller, A. Weathering of roofing material—An overview. *Constr. Build. Mater.* **2008**, *22*, 423–433. [CrossRef]
12. Bretz, S.E.; Akbari, H. Long-term performance of high albedo roof coating. *Energy Build.* **1997**, *25*, 159–167. [CrossRef]
13. Mastrapostoli, E.; Santamouris, M.; Kolokotsa, D.; Vassilis, P.; Venieri, D.; Gompakis, K. A numerical and experimental analysis of the aging of the cool roofs for buildings in Greece. In Proceedings of the Summaries of Technical Papers of Third International Conference on Countermeasures to UHI, Venice, Italy, 13–15 October 2014; pp. 71–80.
14. Paolini, R.; Sleiman, M.; Terraneo, G.; Poli, T.; Zinzi, M.; Levinson, R.; Destaillats, H. Solar spectral reflectance of building envelope materials after natural exposure in Rome and Milano, and after accelerated aging. In Proceedings of the Summaries of Technical Papers of Third International Conference on Countermeasures to UHI, Venice, Italy, 13–15 October 2014; pp. 498–509.
15. Itoh, D.; Takeda, H.; Fujimoto, T.; Kondo, Y.; Ashie, Y. Study on change of performance by outdoor exposure of reflective paints. *AIJ J. Technol. Des.* **2011**, *17*, 217–220. [CrossRef]
16. Ferran, C.; Touchaei, A.G.; Sleiman, M.; Libbra, A.; Muscio, A.; Siligardi, C.; Akbari, H. Effect of ageing processes on solar reflectivity of clay roof tiles. In Proceedings of the Summaries of Technical Papers of 34th AIVC Conference, Athens, Greece, 25–26 September 2013; pp. 77–79.
17. Synnefa, A.; Pantazaras, A.; Santamouris, M.; Bozonner, E.; Doya, M.; Zinzi, M.; Muscio, A.; Libbra, A.; Ferrari, C.; Coccia, V.; et al. Interlaboratory comparison of cool roofing material measurement methods. In Proceedings of the Summaries of Technical Papers of 34th AIVC Conference, Athens, Greece, 25–26 September 2013; pp. 52–54.
18. Sleiman, M.; Weiss, G.B.; Gilbert, H.E.; François, D.; Berdahl, P.; Kirchstetter, T.W.; Destaillats, H.; Levinson, R. Soiling of building envelope surfaces and its effect on solar reflectance-Part I: Analysis of roofing product databases. *Sol. Energy Mater. Sol. Cells* **2011**, *95*, 3385–3399. [CrossRef]
19. Sleiman, M.; Kirchstetter, T.W.; Berdahl, P.; Gilbert, H.E.; Quelen, S.; Marlot, L.; Preble, C.V.; Chen, S.; Montalbano, A.; Rosseler, O.; et al. Soiling of building envelope surfaces and its effect on solar reflectance—Part II: Development of an accelerated aging method for roofing materials. *Solar Energy Mater. Solar Cells* **2014**, *122*, 271–281. [CrossRef]
20. Paolini, R.; Zani, A.; Poli, T.; Antretter, F.; Zinzi, M. Natural aging of cool walls: Impact on solar reflectance, sensitivity to thermal shocks and building energy needs. *Energy Build.* **2017**, *153*, 287–296. [CrossRef]
21. Takebayashi, H.; Miki, K.; Sakai, K.; Murata, Y.; Matsumoto, T.; Wada, S.; Aoyama, T. Examination on aging experiment and accelerated aging test method of solar reflectance of the high reflectance paint in Japan. In Proceedings of the Summaries of Technical Papers of Third International Conference on Countermeasures to UHI, Venice, Italy, 13–15 October 2014; pp. 630–641.

22. Akbari, H. Advance in developing standards for accelerated aging of cool roofing materials. In Proceedings of the Summaries of Technical Papers of Roof Coating Manufacturers Association International Roof Coatings Conference, Baltimore, MD, USA, 14–17 July 2014.
23. Kakuta, M. *Technology Trends of Anti-Soiling and Anti-Bacterium*; CMC Publishing: Tokyo, Japan, 2004; pp. 99–116.
24. Aoyama, T.; Sonoda, T.; Takebayashi, H. Study on aging of solar reflectance of the self-cleaning high reflectance coating. *Energy Build.* **2017**, *157*, 92–100. [CrossRef]
25. Fukaumi, H.; Hamamura, T.; Sonoda, T. Coating composition and coating film obtained from coating composition. *Eur. Pat. Appl.* **2015**, *506*, EP2821450A1.
26. Aoyama, T.; Sonoda, T.; Takebayashi, H. Influence of dirt and coating deterioration on the aging of solar reflectance of high reflectance paint. *AIMS Mater. Sci.* **2019**, *6*, 997–1009. [CrossRef]
27. Tamura, M.; Motohashi, K.; Shimizu, R.; Miura, M.; Hayashi, A. Study on performance of high reflectance paint for building Part 7, Solar reflection after weathering test. In Proceedings of the Japan Society for Finishing Technology, Summaries of Technical Papers of Annual Meeting, Tokyo, Japan, 18–19 October 2012; pp. 16–19.
28. American Society for Testing and Materials. *Standard Practice for Laboratory Soiling and Weathering of Roofing Materials to Simulate Effects of Natural Exposure on Solar Reflectance and Thermal Emittance*; ASTM D7897-15: Philadelphia, PA, USA, 2015.
29. Nishizaki, I.; Moriya, S. *Survey Report on Antifouling Materials for Civil Engineering, Appendix-1, Anti-Fouling Materials Evaluation Accelerated Test Method I*; Public Works Research Institute of the Ministry of Construction of Japan: Thukuba, Ibaraki, Japan, 2011. Available online: https://www.pwri.go.jp/team/imarrc/research/tech-info/tech4180.pdf (accessed on 31 May 2020).
30. Kitsutaka, Y. Accelerating test method for soiling on finishing materials of external walls; study on an evaluating method for the soiling on finishing materials of external building walls (Part 3). *J. Struct. Constr. Eng. AIJ* **1989**, *404*, 15–24.
31. *Heat Island Countermeasure Technology Certification System, Related Documents 4, Acceleration Aging Test Method (Summary Version)*; Osaka Heat Island Countermeasure Technology Consortium (Osaka HITEC): Osaka, Japan, 2018. Available online: http://osakahitec.com/cert/ninsyou_mousikomi.pdf (accessed on 31 May 2020).
32. Itatani, T. Study on evaluation method for stain on exterior coating materials, part 1–4. In Proceedings of the AIJ Summaries of Technical Papers of Annual Meeting, Kobe, Hyogo, Japan, 12–14 September 2014; pp. 845–852.
33. Ozano, T. Study on evaluation method for stain on exterior coating materials, part 5–8. In Proceedings of the AIJ Summaries of Technical Papers of Annual Meeting, Hirathuka, Kanagawa, Japan, 4–6 September 2015; pp. 875–887.
34. Tanabe, J.; Takebayashi, H.; Sonoda, T.; Aoyama, T. Study on aging of solar reflectance on high reflectance paint by exposure experiment. In Proceedings of the AIJ Summaries of Technical Papers of Annual Meeting, Hirathuka, Kanagawa, Japan, 4–6 September 2015; pp. 723–724.

MDPI
St. Alban-Anlage 66
4052 Basel
Switzerland
Tel. +41 61 683 77 34
Fax +41 61 302 89 18
www.mdpi.com

Atmosphere Editorial Office
E-mail: atmosphere@mdpi.com
www.mdpi.com/journal/atmosphere

www.ingramcontent.com/pod-product-compliance
Lightning Source LLC
Chambersburg PA
CBHW040312270926
51850CB00007BC/1147
9783036563411